Ticks and Tick-borne Diseases

Geographical Distribution and Control Strategies in the Euro-Asia Region

MIX
Paper from responsible sources
FSC® C018575
FSC
www.fsc.org

Ticks and Tick-borne Diseases

Geographical Distribution and Control Strategies in the Euro-Asia Region

Edited by

Mo Salman and Jordi Tarrés-Call

Authors:

Agustín Estrada-Peña, Róbert Farkas, Thomas G.T. Jaenson, Frank Koenen, Maxime Madder, Ilaria Pascucci, Mo Salman, Rita de Sousa and Alan R. Walker

www.cabi.org

CABI is a trading name of CAB International

CABI
Nosworthy Way
Wallingford
Oxfordshire OX10 8DE
UK

Tel: +44 (0)1491 832111
Fax: +44 (0)1491 833508
E-mail: info@cabi.org
Website: www.cabi.org

CABI
38 Chauncey Street
Suite 1002
Boston, MA 02111
USA

Tel: +1 800 552 3083 (toll free)
Tel: +1 (0)617 395 4051
E-mail: cabi-nao@cabi.org

A catalogue record for this book is available from the British Library, London, UK.

Library of Congress Cataloging-in-Publication Data

Salman, Mowafak Dauod.
Ticks and tick-borne diseases : geographical distribution and control strategies in the Euro-Asia region / edited by Mo Salman and Jordi Tarrés-Call; authors, Agustín Estrada-Peña ... [et al.].
p. ; cm.
Includes bibliographical references and index.
ISBN 978-1-84593-853-6 (alk. paper)
I. Tarrés-Call, Jordi. II. Estrada-Peña, Agustín. III. Title.
[DNLM: 1. Tick-Borne Diseases--epidemiology--Asia. 2. Tick-Borne Diseases--epidemiology--Europe. 3. Tick-Borne Diseases--prevention & control--Asia. 4. Tick-Borne Diseases--prevention & control--Europe. 5. Ticks--pathogenicity--Asia. 6. Ticks--pathogenicity--Europe. 7. Topography, Medical--Asia. 8. Topography, Medical--Europe. WC 600]

614.5'263094--dc23

2012024123

ISBN-13: 978 1 84593 853 6

Commissioning editor: Sarah Hulbert
Editorial assistant: Alexandra Lainsbury
Production editor: Shankari Wilford

Typeset by SPi, Pondicherry, India
Printed and bound in the UK by the MPG Books Group

Contents

The colour plate section can be found following p. 8

Contributors

Agustín Estrada-Peña, Professor of Parasitology, Department of Animal Pathology (Parasitology), Faculty of Veterinary Medicine, University of Zaragoza, Miguel Servet 177, 50013-Zaragoza, Spain. E-mail: AEstrada@unizar.es

Róbert Farkas, Professor in Parasitology, Department of Parasitology and Zoology, Faculty of Veterinary Science, Szent István University, H-1078 Budapest VII, István u. 2, Hungary. E-mail: Farkas.Robert@aotk.szie.hu

Thomas G.T. Jaenson, Professor in Medical Entomology, Medical Entomology Unit, Department of Systematic Biology, Evolutionary Biology Centre, Uppsala University, Norbyvägen 18d, SE-752 36 Uppsala, Sweden. E-mail: Thomas.Jaenson@ebc.uu.se

Frank Koenen, Operational Director, Interactions and Surveillance, Veterinary and Agrochemical Research Centre (Centrum voor Onderzoek in Diergeneeskunde en Agrochemie-Centre d'Etude et de Recherches Vétérinaires et Agrochimiques) (CODA-CERVA), Groeselenberg 99, B-1180 Brussels, Belgium. E-mail: Frkoe@var.fgov.be

Maxime Madder, Senior Researcher, Biologist, PhD, Veterinary Entomology Group, Department of Biomedical Sciences, Institute of Tropical Medicine, Nationalestraat 155B-2000 Antwerp, Belgium, and Extra-ordinary Lecturer, Department of Veterinary Tropical Diseases, Faculty of Veterinary Science, University of Pretoria, Onderstepoort 0110, South Africa. E-mail: Mmadder@itg.be

Ilaria Pascucci, DMV, PhD, Istituto Zooprofilattico Sperimentale dell'Abruzzo e del Molise "G. Caporale", Campo Boario, I-64100 Teramo, Italy. E-mail: I.Pascucci@izs.it

Mo Salman, Professor of Veterinary Epidemiology, Campus Stop 1644, Animal Population Health Institute, College of Veterinary Medicine and Biomedical Sciences, Colorado State University, Fort Collins, CO 80523-1644, USA. E-mail: M.D.Salman@ColoState.edu

Rita de Sousa, Centro de Estudos de Vectores e Doenças Infecciosas Doutor Francisco Cambournac (Center for the Study of Vectors and Infectious Disease Research Doutor Francisco Cambournac) (CEVDI), Instituto Nacional de Saúde Doutor Ricardo Jorge, IP (National Institute of Health Doutor Ricardo Jorge, IP) (INSA), Av. da Liberdade 5, 2965-575 Águas de Moura, Portugal. E-mail: Rita.Sousa@insa.min-saude.pt

Jordi Tarrés-Call, DMV, PhD, Scientific Officer of the FEED Unit (formerly Scientific Officer in the Animal Health and Animal Welfare Unit), European Food Safety Agency (EFSA), Via Carlo Magno 1/A, I-43126 Parma, Italy. E-mail: Jordi.Tarres-Call@efsa.europa.eu[1]
Alan R. Walker, Royal (Dick) School of Veterinary Studies, University of Edinburgh, Edinburgh, EH25 9RG, Scotland, UK. E-mail: Alan.R.Walker@ed.ac.uk

[1] The editor Jordi Tarrés-Call is employed by the European Food Safety Authority (EFSA). The positions and opinions presented in this monograph are those of the authors and editors alone and are not intended to represent the views or scientific works of EFSA.

Foreword

Jeremy Gray
Emeritus Professor of Animal Parasitology, University College Dublin, Republic of Ireland

Ticks are the most important ectoparasites worldwide in veterinary science, both as parasites in their own right and also as vectors of many different pathogens that cause diseases in livestock and companion animals – such as anaplasmosis, babesiosis, cowdriosis, ehrlichiosis and theileriosis. Although few tick species are adapted to parasitize mankind, they are also of great significance in human medicine as vectors of zoonotic pathogens, notably in the temperate northern hemisphere, where Lyme borreliosis, caused by the spirochaete species complex, *Borrelia burgdorferi* sensu lato, is now acknowledged as the most prevalent vector-borne disease. Despite long-standing knowledge of the subject, relatively few books have been devoted to ticks and tick-borne diseases, with the two-volume 1992/3 publication *Biology of Ticks* by D.E. Sonenshine constituting the most important recent contribution, and a new multi-author version currently in preparation. Several other books such as A. Bowman and P. Nuttall's *Ticks: Biology, Disease and Control* and *Tick-borne Diseases of Humans* by J.L. Goodman *et al.* address current 'hot topics', and there are several published proceedings of conferences and, reflecting growing public interest, a plethora of popular books. However, none of these publications are concerned with the very important subject of the geographical distribution of tick-borne pathogens and their vectors. This topic has acquired special significance in light of the predicted effects of climate and environmental change brought about by global warming and other anthropogenic factors, especially in the context of remote sensing and risk mapping. For such information pertaining to Europe and the Mediterranean, tick biologists have had to rely in the past on texts such as *Ticks of North-west Europe* by P.D. Hillyard and *Ticks of the Domestic Animals in the Mediterranean Region* by A. Estrada-Peña *et al.*, though both are quite short publications and are primarily concerned with taxonomy and identification rather than distribution. Not since the outdated and relatively limited monographs compiled by the great Harry Hoogstraal and co-workers in the 1950s has such a comprehensive record of tick distribution been published as is presented in this monograph, and to my knowledge no such attempt to record the distribution of tick-borne diseases in these particular geographical regions has been attempted before.

This monograph comes in two parts: first is the main text with descriptions and accompanying maps, applying to both the ticks and to the diseases they transmit, and second is an appendix detailing the sources of the data and other reference material. The main text starts with an introduction that focuses on the significance of tick-borne diseases, the geographic scope of the work, the methods used for reviewing the literature and a brief account of tick identification methods. Information on the literature was extracted from PubMed and the ISI Web of Knowledge using several databases including those of CABI, *Current Contents, Food Science and Technology Abstracts, Journal Citation Reports*, MedLine and Web of Science. This resulted in the citation of more than 1400 papers concerning the distribution of ticks and tick-borne diseases, with many others addressing other topics, and the vast majority published in English.

The first chapter consists of an account of the factors behind the emergence and re-emergence of tick-borne diseases, highlighting the theme of environmental and climatic change and also the renewed interest in ticks and the diseases they transmit, which has been stimulated by an increased awareness of tick-borne zoonoses. Chapter 2 describes the basic biology of a total of 25 important tick species endemic to part or all of the geographical region under consideration, and also includes short accounts of their life cycles, geographical distributions, and significance as vectors. It contains original photographs of examples of the vast majority of the tick species considered in the publication, some of which have evidently been collected from host animals. In addition there are comments on some non-endemic species such as *Amblyomma* (*Am.*) *variegatum* and *Am. hebraeum*, which are of immense significance south of the Sahara and whose distribution may change as a result of accidental introduction, as in the Caribbean. In Chapter 3 the factors responsible for the spread and distribution of ticks are considered, which include climate, land use, animal movement – both wild and domestic, and the importation of exotic vertebrates. Tick-borne infections are reviewed in Chapter 4, and within each section of this chapter topics as varied as molecular taxonomy, pathology and epidemiology are covered. Approximately 29 bacterial species, 12 protozoan and three viral species are considered at varying levels of detail depending on their significance, and these descriptions provide useful thumbnail sketches of the species biology to put alongside their distributions. The geographical distribution of tick-borne pathogens is the focus of Chapter 5, in the form of maps with accompanying qualifying and illustrative comments; Chapter 6 addresses the distributions of the vector ticks in the same way. In both cases, the longitude and latitude coordinates were utilized; if these were not available, the nomenclature of territorial units for statistics (NUTS) was used for EU countries, while for non-EU countries the name of the administrative area given in the publication was used. Chapter 7 addresses the surveillance and control of ticks and tick-borne diseases. It includes a brief description of tick sampling methods, an introduction to the principles of surveillance and monitoring, and control options for both ixodids and argasids.

The text of the book is supported by an extremely extensive appendix. Nearly all of this (as Appendix 1) consists of tables of geographic data for individual tick-borne diseases and tick species that have been derived from the very numerous bibliographic records (Appendices 2 and 3; available at http://www.cabi.org/vetmedresource/) from which the distribution data were extracted. Finally, Appendix 4 presents illustrated identification keys for the genera and species of ticks in the Mediterranean basin. Most of the focus is on identification to genus level but there are notes on selected species accompanying the illustrations. All the important tick species that occur in northern Europe and that are mentioned in the main text are also included here, with the exception of *Ixodes persulcatus*, which is primarily a tick of temperate Asia.

There is no question that this publication will prove to be extremely useful for all those working with ticks and tick-borne diseases in Europe and around the shores of the Mediterranean. It fills an obvious gap in the tick literature and will complement the more specialized content of other recent publications, including the forthcoming new edition of *Biology of Ticks*. It will be a useful addition to the library of every European university with a biology department, and it should also find a place in all institutions in this geographical region that are concerned with the surveillance and monitoring of vector-borne diseases, which are likely to become increasingly relevant to those concerned with the determination of public health policies and the implementation of control measures.

Acknowledgements

The initial mandate for this work was conducted under the auspices of the Animal Health and Animal Welfare Panel of European Food Safety Authority (EFSA). The editors and the contributors of this monograph would like to express their appreciation for the full support of EFSA for producing this monograph.

The authors acknowledge the valuable assistance of Drs Milen Georgiev and Sofie Dhollander (Scientific Officers of the Animal Health and Animal Welfare Unit of EFSA); and of Dr Nely Sánchez (Veterinary Parasitologist of the Department of Animal Pathology – Parasitology – of the University of Zaragoza) during the systematic literature review.

Thomas Jaenson's contributions were related to his research on ticks and tick-borne infections which is funded by Carl Trygger's Stiftelse (Norwegian Foundation) and Magnus Bergvall's Stiftelse.

The data set produced for this monograph was a joint effort made possible in part by the International Consortium of Ticks and Tick-Borne Diseases, funded by the European Union and led by Frans Jongejan (University of Utrecht, The Netherlands), and EFSA.

Parts of the data set were compiled with records curated in the collection of the Institute of Parasitology of the Czech Republic, with the help of Františ ek Dusbábek and the support of the European Concerted Action on Lyme Borreliosis.

We express our thanks to the people in these concerted actions who contributed with expert assistance to the compilation: Danielle de Meneghi, Peter Zemman and Milan Daniel. We also express our thanks to Ali Bouattour, Laurence Vial, Miguel A. Habela, Ramón Juste, Ana García-Pérez, Joaquim Castellà, Margarida Santos Silva and Sofia Núncio for providing advice and lists of records in the collections of ticks under their charge. Byron Papadopoulos contributed with records in Macedonia. Lise Gern contributed with much help on data on tick distribution in Switzerland. John Scharlemann provided a list with records of *Ixodes ricinus* in the UK. Zati Vatansever compiled unreported reports of ticks in Turkey and produced the list of species for the country as included in the final data set. Jean-Louis Camicas and Ali Bouattour examined earlier maps of the species reported in the Mediterranean region. We want to extend our gratitude to collectors and systematists who made the creation of this data set possible.

Abbreviations

A.: *Anaplasma*
AF: Astrakhan fever
Am.: *Amblyomma*
Ar.: *Argas*
ASF(V): African swine fever (virus)
B.: *Borrelia* or *Babesia*
Bo.: *Boophilus*
CCHF(V): Crimean-Congo haemorrhagic fever (virus)
D.: *Dermacentor*
DEBONEL: *Dermacentor*-borne necrosis erythema lymphadenopathy
DEET: diethyl toluamide
E.: *Ehrlichia*
EPNs: Entomopathogenic nematodes
H.: *Hepatozoon*
Ha.: *Haemaphysalis*
Hy.: *Hyalomma*
I.: *Ixodes*
ISF: Israeli spotted fever
ICTTD: Integrated Consortium on Ticks and Tick-borne Diseases (European project)
LAR: Lymphangitis-associated rickettsiosis
LI(V): Louping ill (virus)
MSF: Mediterranean spotted fever
NUTS: Nomenclature for territorial units for statistics
O.: *Ornithodoros*
R.: *Rickettsia*
RF: Russian Federation
Rh.: *Rhipicephalus*
T.: *Theileria*
TBD: Tick-borne disease
TBE(V): Tick-borne encephalitis (virus)
TCC: Transcaucasian Countries
TIBOLA: Tick-borne lymphadenopathy

Introduction

Mo Salman[1] and Jordi Tarrés-Call[2]
[1]*Colorado State University, USA; and* [2]*European Food Safety Agency (EFSA), Parma, Italy*

Significance of Tick-borne Diseases

The aim of this monograph is to provide a general overview of the geographical distribution of various tick species which have proven their involvement in the transmission of the pathogens causing animal diseases and zoonoses in Europe, the Middle East and the Mediterranean basin. As blood-feeding parasites, ticks are able to transmit to their hosts a wide variety of pathogens which may cause tick-borne infections and tick-borne diseases (TBDs) that affect wild and domestic animals, including companion animals. The transmission of pathogens among ticks may occur transovarially, i.e. the pathogen is transmitted via the eggs from females to their offspring, trans-stadially from larva to nymph and/or from nymph to adult, and venereally during copulation from male to female tick. Vectorial competence is the overall ability of a vector tick species to transmit a pathogen to a range of receptive vertebrate hosts in a given location at a specific time. The TBDs usually are geographically distributed within the range of their vectors. Some of these infections/diseases can give clinical signs which can be severe (in the acute phase) but can also present as subclinical forms (mainly in endemic areas) in animals or humans. Furthermore, co-infection with different pathogens can occur in the same vertebrate animal when the same tick species transmits more than one pathogen (e.g. *Hepatozoon canis*, *Ehrlichia canis* and *Anaplasma platys* transmitted by *Rhipicephalus sanguineus*), or when two or more tick species infest an animal or human at the same time (Belongia, 2002; Stańczak *et al.*, 2002; Bremer *et al.*, 2005; Halos *et al.*, 2005; Swanson *et al.*, 2006). TBD co-infections by ticks are frequent in companion animals living in endemic areas and this may often impair an appropriate aetiological diagnosis (EFSA, 2007).

Several recent reports have indicated the wider spread of specific tick species as vectors in geographic areas that were not considered previously infested with these tick species (e.g. Jääskelainen *et al.*, 2006, for *Ixodes persulcatus*; Nijhof *et al.*, 2007, for *Dermacentor reticulatus*; Gray *et al.*, 2009, Jaenson *et al.*, 2012, for *I. ricinus* and *D. reticulatus*). Furthermore, other reports have shown that the host range of some tick species was wider than previously known. There are limited data on vector competence for many tick species and a lack of information on the effect of environmental factors, including climatic factors, on the transmission of pathogens.

The taxonomy of tick-borne pathogens is also evolving and subject to change. Based on recent molecular studies, the taxonomy of certain tick species has been adapted accordingly. This monograph uses the taxonomy proposed by Uilenberg (2006).

Geographic Scope

The list of countries that are covered in this monograph includes the European countries (Albania, Andorra, Austria, Belgium, Bosnia and Herzegovina, Bulgaria, Croatia, Cyprus, the Czech Republic, Denmark, Estonia, Faroe Islands, Finland (including Åland Islands), France, Germany, Greece, Hungary, Ireland, Italy, Kosovo, Latvia, Liechtenstein, Lithuania, Luxembourg, Macedonia, Malta, Montenegro, The Netherlands, Norway, Poland, Portugal, Romania, San Marino, Serbia, Slovakia, Slovenia, Spain, Sweden, Switzerland, the UK). In addition, the following countries are also included: Algeria, Egypt, Israel, Jordan, Lebanon, Libya, Morocco, Syria, Tunisia and Turkey, and the geographic areas covering Palestine and the Western Sahara. These neighbouring countries were included owing to their proximity to Europe and the potential for the spread of ticks from them. The eastern border of Europe (Armenia, Belarus, Georgia, Moldova and Ukraine, and a buffer zone of some 600 km in European Russia) is also included to account for the important tick species recorded in that area which are of potential interest for Europe under future scenarios.

Methods Used for the Systematic Review of the Literature

A systematic literature review was carried out based on the general principles of the systematic review methodology (CRD, 2009; Higgins and Green, 2009) with the aim of gathering information about the geographical distribution of tick species and tick-borne pathogens. The literature review was based on scientific papers published from January 2000 to March 2010 (some of the most recent data published or generated after this monograph was initiated are not present in the maps).These papers were retrieved from the citation indexing service of ISI Web of Knowledge℠ (Thomson Reuters) and from PubMed (US National Library of Medicine).

The active databases during the search in the ISI Web of Knowledge were:

- Web of Science®
- Current Contents Connect®
- CAB Direct
- Food Science and Technology Abstracts™
- MEDLINE®
- Journal Citation Reports®

Additional publications within which the topics were considered relevant to the review, were submitted by the other contributors to this monograph. Two different sets of search strings were used – one for ticks and another for tick-borne pathogens. These strings were applied to the title and the abstract (Table I.1) of the publications. No language restrictions were set. The geographical limitation was applied as per the rustication to countries and regions listed above. Published validated data from the European Union Project Integrated Consortium on Ticks and Tick-borne Diseases (ICTTD-3), collected by one of the contributors to this monograph, were also considered. All data derived from these sources were collated in a relational database (using FileMaker software, see http://www.filemaker.com), from which the maps were created.

Duplicate references were deleted automatically by means of a reference management system[i]. References were then checked manually and duplicates removed. The search in ISI Web of Knowledge and in PubMed produced a list of 2197 references.

The title and abstract of each reference were screened following the criteria listed in Table I.2.

After the first screening, 1222 references were considered relevant, 309 doubtful, and 666 non-relevant. The doubtful references (title and abstract) were further revised by two contributors to this monograph who still found ten relevant references. By checking the full text,

Table I.1. The search strings for publications on ticks and tick-borne pathogens.

Search terms related either to ticks or tick-borne pathogens		AND	AND (search terms related to the geographic region)
Tick	*Argas* OR *Ornithodoros* OR *Dermacentor* OR *Haemaphysalis* OR *Hyalomma* OR *Ixodes* OR *Rhipicephalus* OR *Boophilus*		Aland OR Albania OR Andorra OR Austria OR Belgium OR Bosnia and Herzegovina OR Bulgaria OR Croatia OR Cyprus OR Czech Republic OR Denmark OR Germany OR Spain OR Estonia OR Faroe Islands OR Finland OR France OR Greece OR Hungary OR Ireland OR Italy OR Kosovo OR Latvia OR Liechtenstein OR Lithuania OR Luxembourg OR Macedonia OR Malta OR Montenegro OR The Netherlands OR Norway OR Poland OR Portugal OR Romania OR San Marino OR Serbia OR Slovakia OR Slovenia OR Switzerland OR Sweden OR United Kingdom OR Algeria OR Egypt OR Israel OR Jordan OR Lebanon OR Libya OR Morocco OR Syria OR Tunisia OR Turkey OR Palestine OR Western Sahara OR Armenia OR Belarus OR Georgia OR Moldova OR Ukraine OR Russia OR USSR
Tick-borne pathogen	African Swine Fever virus OR ASF virus OR ASFV OR *Anaplasma* OR *A phagocytophilum* OR *Ehrlichia phagocytophila* OR *Babesia* OR Crimean Congo Haemorrhagic Fever virus OR CCHF virus OR CCHFV OR Hepatozoon OR Lyme disease agent OR *Borrelia* OR *B burgdorferi* OR *B garinii* OR *B spielmanii* OR *B lusitaniae* OR *Rickettsia* OR *R conorii* OR *Ehrlichia canis* OR *Borrelia hispanica* OR *B hispanica* OR *Theileria* OR *T parva* OR *T hirci* OR *T ovis* OR *T lestoquardi* OR Tick borne encephalitis virus OR Louping ill virus OR TBE virus OR TBEV OR tick borne flavivirus OR TBEF OR TBEFV OR TBE group OR *Francisella* OR *F tularensis* OR *Bartonella* OR Q fever OR *Coxiella* OR African horse sickness virus OR AHSV OR tick borne orbivirus	Distribution OR presence OR occurrence OR reported	

Table I.2. Criteria used in the first screening for relevance of the selected references.

Criterion	Included	Excluded
Concerns occurrence of a tick species or a tick-borne pathogen in the area considered	Yes	No
Concerns a tick species with proven involvement in transmitting pathogens of animal diseases or zoonoses	Yes	No
Contains geographic information on the distribution of the tick species or the tick-borne pathogen	Yes	No

43 other initially doubtful references were considered as relevant. This produced a total of 1275 relevant references and 822 non-relevant references. The full articles of the relevant references were then retrieved, but some could not be found. There were 125 scientific articles (six in English language and 119 in other languages) that were missed.

The second screening of articles was performed in parallel with the data extraction by two experienced veterinary parasitologists of the Veterinary School of the University of Zaragoza (Spain), one of whom is a contributor to this monograph. All contributors were consulted when doubts arose. Apart from the three criteria considered in the first screening, the contributors were asked to also consider the secondary criteria listed in Table I.3.

This second screening resulted in a total of 637 scientific papers considered appropriate to be included in the review. These are listed in Appendix 2, sorted by the identification number of the reference.

Data were extracted to an Excel spreadsheet containing the following fields:

- Tick genus and species, recorded as in the original paper.
- Location of the tick species: based on the nomenclature of territorial units for statistics (NUTS) for the EU (European Union) countries. For non-EU countries, locations were recorded by the name provided in the original report, at the equivalent level of geographical precision. The coordinates (latitude/longitude) were recorded if given in the article.
- Whether molecular techniques were used to identify the tick species (yes/no).
- The source of the tick specimen: free living (questing), livestock, pet, human, wildlife (taxonomic order of the host).
- The corresponding bibliographic reference.
- The tick-borne pathogen genus and species (as it appeared in the original work).
- Location of the pathogen: NUTS for European countries. For the countries that do not have the NUTS, the name of the administrative region at the level of precision provided in the scientific paper. Geographical coordinates if provided.
- Diagnostic/identification method of the pathogen: isolation, molecular, serology.

Table I.3. Criteria used in the second screening for relevance.

Criterion	Included	Excluded
Language publication	English abstract and text available in English, French, German, Italian, Spanish, Hungarian, Bulgarian or Dutch	Abstract not available or text not in English, French, German, Italian, Spanish, Hungarian, Bulgarian or Dutch
Original work (not a review document)	Yes	No
Contains geographic information on the distribution of the tick species or the tick-borne pathogen	Yes	No
Does not specify a concrete geographic location. Rather refers it to the entire country	No	Yes
Concerns a prompt importation of a tick species that does not reach sufficient epidemiological threshold for its establishment in the specified location	No	Yes
Case reports of human infections that, in order to protect the privacy of personal data, provide the address of a reference hospital instead of the residence of the infected patient	No	Yes
Tick identification is unequivocal and appropriate	Yes	No
The diagnostic method for the tick-borne pathogen is appropriate (serology, isolation, biological methods such as PCR)	Yes	No

- Source of the sample (pathogen): livestock, pet, human, wildlife (taxonomic order of the host), tick.
- The corresponding bibliographic reference.

A more detailed description of the systematic literature review process is given in EFSA (2010).

Identification Methods for Tick Species, Including the Most Recent Advanced Techniques and their Limitations

Identification of ticks can be based on different criteria. Morphology can be the only criterion in some occasions but, in a more holistic approach, identification may be based on host species; predilection sites, geographical occurrence, seasonality, and other criteria. Morphological identification is based on: presence/absence of eyes; the presence of anal plates; ornamentation on the legs and scutum; shape, size and the patterns of the scutum; and other unique characteristics. Figure I.1 describes these morphological terms, and Tables I.4 and I.5 show how these morphological characteristics are present in hard and soft ticks, respectively.

The taxonomy of ticks is not exempt from complications. There are problems of morphological identification and sympatric tick speciation. Identification of tick species requires extensive experience with the subject over many species. Molecular biology has reshuffled many names at both species and genus level, although it must be stressed that the 'conventional' identification was in many cases able to identify the different tick entities. For instance, in the case of the changes of *Boophilus microplus* to *Rhipicephalus* (*Rh.*) *microplus*, and of *Hyalomma* (*Hy.*) *marginatum marginatum* to *Hy. marginatum*, it was only the name that changed, not the species identification. In other situations, the identification changed owing to tick physical appearance and not because of molecular characteristics, such as the change of *Haemaphysalis* (*Ha.*) *leachi* to *Ha. elliptica* in South Africa.

The colour plate section contains photographs of some hard and soft tick species involved in the transmission of animal diseases and zoonoses, and Appendix 4 contains identification keys for the morphological identification of hard and soft ticks in the Mediterranean basin.

Limitations

This monograph addresses data on the distribution of ticks and TBDs in Europe and the Mediterranean basin that were retrieved from existing literature and technical reports. The report's findings, however, are limited as a result of the following issues:

- The presence of pathogens was based on antigen detection assays without consideration of serological evidence.
- Cases were only reported from the literature without consideration of other sources that may be available through other means of searching.
- Literature was retrieved mainly in the English language, with only limited inclusion of other languages.
- Although intended to be comprehensive in the literature search strings, the search may have missed some publications.
- The maps have not been updated since March 2010, so some of the most recent data published or generated after the completion of this monograph are not included.

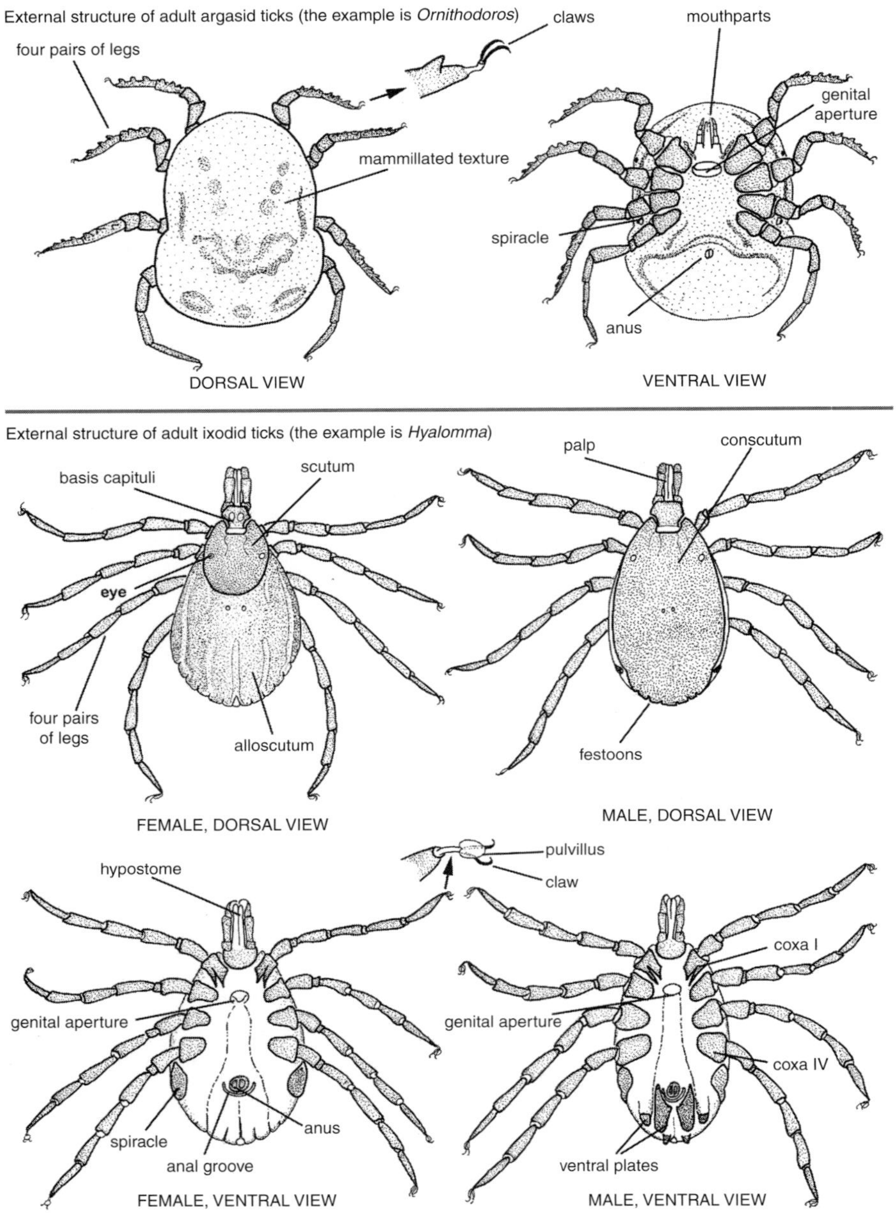

Fig. I.1. Main anatomical characteristics and terms used to describe the morphology of adult hard ticks, from Walker *et al.*, 2003.

Note

[i] EndNote X1.0.1 [Bld 2682], Thomson. http://www.endnote.com

Table I.4. Generic distinguishing morphological features of different genera of hard ticks.

Feature	*Hyalomma*	*Rhipicephalus*	*Rhipicephalus (Boophilus)*	*Ixodes*	*Dermacentor*	*Haemaphysalis*	*Amblyomma*	*Rhipicentor**	*Aponomma** *(Amblyomma)*	*Margaropus**
Size (unfed adults, total length)	Large (0.5–0.8 cm)	Medium (0.3–0.6 cm)	Small (0.1–0.3 cm)	Small (0.2–0.3 cm)	Medium-large (0.4–0.9 cm)	Small (0.1–0.3 cm)	Very large (0.5–1.0 cm)	Medium (0.3–0.5 cm)	Large (0.5–0.6 cm)	Small (0.1–0.3 cm)
Mouthparts	Long	Short to medium	Very short	Long	Medium	Short	Very long	Medium	Long	Very short
Basis capituli		Hexagonal	Hexagonal	With ventral auriculae	Rectangular	Rectangular		Hexagonal		
Ornate	No	No, except for 4 species	No	No	Yes	No	Yes	No	Some species	No
Eyes	Yes	Yes	Yes	No	Yes	No	Yes	Yes	No	Yes
Festoons	Yes	Yes	No	No	Yes	Yes	Yes	Yes	Yes	No
Adanal plates	Yes	Yes	Yes	No	No	No	None, or very small	No	No	Yes
Sub-anal plates	Yes	No	No	No	No	No	No	No	No	No
Anal groove	Posterior, chalice-shaped	Posterior, chalice-shaped	Posterior, vertical line	Anterior, semicircular or U-shaped	Posterior, chalice-shaped	Posterior, chalice-shaped	Posterior, semicircular	Posterior, chalice-shaped	Posterior, chalice-shaped	Absent
Caudal process	No	Present in some species	Yes, on some	No	No	No	No	No	No	Yes, pilose posterior margin
Coxae I	Bifid	Bifid			Large, bifid			Large, bifid		
Coxae IV					Very large			Very large, with spurs		
Legs	Banded			Grouped anteriorly	Banded		Banded			Banded, enlarged segments in males

(*) These genera are currently not present in European member states, but they may be imported in the future.

Table I.5. Generic distinguishing morphological features of different genera of soft ticks.

Feature	*Argas*	*Ornithodoros*
Body	Flattened	Thick
Presence of suture (differentiated tegumental tissue between dorsal and ventral surface)	No	Yes
Presence of cuticle extensions rounded hook-like cones shaped	No	Yes
Eyes	No	Yes
Tegument	Granular	Squamous
Discs of tegument	Usually in radial rows	Not in radial rows

References

Belongia E.A. (2002) Epidemiology and impact of coinfections acquired from *Ixodes* ticks. *Vector-Borne and Zoonotic Diseases* 2, 265–273.

Bremer, W.G., Schaefer, J.J., Wagner, E.R., Ewing, S.A., Rikihisa, Y., Needham, G.R., Jittapalapong, S., Moore, D.L. and Stich R.W. (2005) Transstadial and intrastadial experimental transmission of *Ehrlichia canis* by male *Rhipicephalus sanguineus*. *Veterinary Parasitology* 131, 95–105.

CRD (2009) *Systematic Reviews, CRD's Guidance for Undertaking Reviews in Health Care*. January 2009, Centre for Reviews and Dissemination, University of York, UK.

EFSA (European Food Safety Authority) (2007) Opinion of the Scientific Panel on AHAW on a request from the European Commission on the risk of tick introduction into the UK, Ireland, and Malta as a consequence of abandoning the national rule. *EFSA Journal* 469, 1–102.

EFSA Panel on Animal Health and Welfare (AHAW) (2010) Scientific opinion on the role of tick vectors in the epidemiology of Crimean Congo hemorrhagic fever and African swine fever in Eurasia. *EFSA Journal* 8(8): 1703.

Gray, J.S., Dautel, H., Estrada-Peña, A., Kahl, A. and Lindgren, E. (2009) Effects of climate change on ticks and tick-borne diseases in Europe. *Interdisciplinary Perspectives on Infectious Diseases* 2009, 1–12, Article ID 593232.

Halos, L., Jamal, T., Maillard, R., Beugnet, F., Menach, A.L., Boulouis, H.J. and Vayssier-Taussat, M. (2005) Evidence of *Bartonella* sp. in questing adult and nymphal *Ixodes ricinus* ticks from France and co-infection with *Borrelia burgdorferi sensu lato* and *Babesia* sp. *Veterinary Research* 36, 79–87.

Higgins, J.P.T. and Green, S. (eds) (2009) *Cochrane Handbook for Systematic Reviews of Interventions Version 5.0.2* (updated September 2009). The Cochrane Collaboration, Oxford, UK.

Jääskeläinen, A.E., Tikkakoski, T., Uzcátegui, N.Y., Alekseev, A.N., Vaheri, A. and Vapalahti, O. (2006) Siberian subtype tickborne encephalitis virus, Finland. *Emerging Infectious Diseases* 12, 10. Available at: http://dx.doi.org/10.3201/eid1210.060320 (accessed 22 June 2012).

Jaenson, T.G.T., Jaenson, D.G.E., Eisen, L., Petersson, E. and Lindgren, E. (2012) Changes in the geographical distribution and abundance of the tick *Ixodes ricinus* during the past 30 years in Sweden. *Parasites and Vectors* 5, 8, doi:10.1186/1756-3305-5-8. Available at: http://www.parasitesandvectors.com/content/5/1/8 (accessed 22 June 2012).

Nijhof, A.M., Bodaan, C., Postigo, M., Nieuwenhuijs, H., Opsteegh, M., Franssen, L., Jebbink, F. and Jongejan, F. (2007) Ticks and associated pathogens collected from domestic animals in the Netherlands. *Vector-borne and Zoonotic Diseases* 7, 585–595.

Stańczak, J., Racewicz, M., Kruminis-Lozowska, W. and Kubica-Biernat, B. (2002) Coinfection of *Ixodes ricinus* (Acari: Ixodidae) in northern Poland with the agents of Lyme borreliosis (LB) and human granulocytic ehrlichiosis (HGE). *International Journal of Medical Microbiology* 291(Supplement 33), 198–201.

Swanson S.J., Neitzel, D., Reed K.D. and Belongia, E.A. (2006) Coinfections acquired from *Ixodes* ticks. *Clinical Microbiological Reviews* 19, 708–727.

Uilenberg, G. (2006). *Babesia* – a historical overview. *Veterinary Parasitology* 138, 3–10.

Walker, A.R., Bouattour, A., Camicas, J.-L., Estrada Peña, A., Horak, I.G., Latif, A.A., Pegram, R.G. and Preston, P.M. (2003) *Ticks of Domestic Animals in Africa: A Guide to Identification of Species*. Bioscience Reports, The University of Edinburgh, Edinburgh, UK.

1 Emerging (Re-emerging) Tick-borne Infections and the Dissemination of Ticks

Mo Salman[1] and Agustín Estrada-Peña[2]
[1]*Colorado State University, USA; and* [2]*University of Zaragoza, Spain*

Ticks are arthropods and therefore they are susceptible to changes in climate. All stages of the life cycle of ticks are associated with vegetation, so small changes in the prevailing environment may have important consequences for the overall impact of ticks and tick-borne infections in surrounding areas. Nevertheless, this issue is not as simple as it appears. On completion of their life cycles, ticks are also sensitive to the availability of hosts and to several other factors related to the surrounding landscape. For instance, the fragmentation of a forest into smaller patches of vegetation surrounded by a matrix of non-habitat landscape may deeply affect the behaviour of the hosts, and this might have profound consequences for the life cycle of the ticks.

The different life stages of the tick are sensitive to temperature, i.e. they occur faster if a temperature near the optimum for that stage is available in the very few centimetres above the ground. While a higher temperature will result in a faster completion of a given stage, it will probably also generate a higher level of evaporation. This effect is hazardous for tick survival, as ticks are sensitive to desiccation. A higher temperature is, therefore, not always better for the overall performance of the tick population. The effect of temperature and desiccation are dependent upon the preferences of the species. For example, limited variations in temperature may have serious consequences for populations of ticks that colonize cold and wet environments in the forests of northern Europe, but are probably not so disastrous for species living in Mediterranean conditions, for which a high seasonal variability of climatic characteristics is the norm. Climatic and other factors also have an interactive effect on the survivability and spread of ticks and their infections in host populations.

It is now evident that some tick species are spreading in parts of the world that were not previously considered to be favourable areas for ticks. The reason for this spread is not only due to the effects of climate change, but also to subtle and continuous trends of changes in the vegetation, landscape features and social habits. These changes are leading to new areas of contact between ticks, their pathogens and the interface between animals and humans. No unique pattern can explain the spread of the ticks and of the pathogens that they transmit, although in some parts of their range, climate may account for their changes in distribution. For example, *Ixodes ricinus* has spread in parts of Scandinavia as well as in the mountain ranges of the Czech Republic (Daniel *et al.*, 2003). These changes have been attributed to climate because permanent populations of *I. ricinus* are now detected in northern areas of Sweden

(Talleklint and Jaenson, 1998) and colonize a higher altitude range in the mountains.

The impact of this wide spread of *I. ricinus* is related not only to parasitism by this tick species of both humans and animals, but also to its associated pathogens and to infection of its hosts by these pathogens. *I. ricinus* is the main vector in the Palaearctic region for the prominent pathogen *Borrelia burgdorferi sensu lato* (*s.l.*), the aetiological agent of Lyme disease. While there are no adequate figures available on the importance of this disease in the countries of the European Union (as it is not a compulsorily reportable disease) it is estimated to be the most widespread infection transmitted by ticks in Europe (Rizzoli *et al.*, 2011). A recent paper suggested that this pathogen is present everywhere that the tick vector is, and that the prevalence rates of *B. burgdorferi s.l.* infection are directly dependent upon a combination of features of climate and vegetation (Estrada-Peña *et al.*, 2011). Other studies have confirmed the probable northward spread of the tick vector and of the risk of contracting Lyme disease in Scandinavia (Jaenson and Lindgren, 2011), and that the northern limit of *I. ricinus* may be predicted by some features of vegetation, which are obviously dependent upon a combination of climate factors. Lyme disease is also dependent upon a combination of reservoirs, such as rodents, birds and lizards, which are necessary for the maintenance of the pathogen population.

I. ricinus is also the only known vector in Europe for tick-borne encephalitis virus (TBEV), which causes tick-borne encephalitis (TBE) in humans. Although the ecological determinants of the spread of TBEV have been ignored, the map of reported cases of the disease has changed over the last few years. However, while there is a direct relationship between the presence of *I. ricinus* and cases of Lyme disease (or the presence of the pathogen in the tick vector), the incidence of TBE is not directly related to the abundance of tick vectors. Other authors have speculated on different theories to explain the presence of a given disease reservoir as the shape of the area where the infection is present or have evaluated the importance of a given climate pattern and its impact on the life cycle of ticks (Rogers and Randolph, 2006). It is, however, necessary to further evaluate how these factors may evolve, and how the areas where the infection is predicted to exist associate with the actual distribution of reported cases of the disease (or the records of the pathogen). A process-driven model for the life cycle of the tick *I. ricinus* would also be of great importance in providing an additional tool to understand the patterns of disease.

In the last few years, infection by the Crimean-Congo haemorrhagic fever (CCHF) virus (CCHFV) has gained much attention as a result of several serious outbreaks. CCHF is an important disease that is prevalent over wide areas of Africa, Asia, the Near East and parts of Europe (Ergönül, 2006). It has not been reported in the Americas and Australia because of the absence of the main vectors. The pathogen is mainly transmitted by the bite of ticks of the genus *Hyalomma* (and especially by *Hyalomma marginatum*), which are most prevalent in a Mediterranean-type environment, as well as in the African steppes. The disease was first diagnosed around the middle 1940s in the Crimea and then in the Congo. CCHF has not yet been diagnosed in sites or countries outside its main distribution area for 60 years. In the last few years though, this disease has re-emerged in Turkey, with more than 3000 cases, and the outbreak seems to be independent of climatic causes. Studies have hypothesized that the re-emergence of the disease is based on a series of factors that interact, mainly an abundance of ticks created by social influences and an increased availability of hosts (Ergönül, 2006).

As for other tick-borne diseases, climate trends have been commonly linked to clinical cases of CCHF. Ticks are highly sensitive to small changes in climate, and long-term, sustained and small differences in key climatic factors may drive a serious change. A framework linking the response of the tick to subtle climatic variations is, however, needed to understand the actual significant factors. An assessment of the effects of climate on the presence of human CCHF clinical cases in Turkey was carried out by Estrada-Peña *et al.* (2009). The analysis included monthly data for several climate factors. The study concluded that climate was not different between sites with active foci of the infection and sites where *Hyalomma marginatum* is common but human cases are not reported. The study also concluded that it is not possible to predict, based solely on climate factors, where new cases could appear in the reasonably near future. Climate is not the sole factor driving the virus amplification, even if climate in the autumn/winter may have a strong regulating role on the survival of tick populations (Hoogstraal, 1979). It is noteworthy that *Hyalomma* endemic areas with the mildest autumns and winters in the Mediterranean basin have historically been, and are currently, free of the infection, so the impact of a warmer climate seems to be not the only factor determining the occurrence of the disease. Efforts to link occurrence of the disease solely to an array of climate conditions in sites where *Hyalomma* ticks are absent (Randolph and Rogers, 2007) may be unrealistic. Claims, therefore, that a warmer climate influences the presence or absence of the infection (Purnak *et al.*, 2007) are unsupported when climate series are examined and compared. Bold statements such as the 'probability of hospitable climates not only extended to Spain, Italy, and northern African countries across the Mediterranean Sea, but even in large parts of North America, smaller parts of South America, and Australia' (Maltezou and Pappa, 2010) are unreliable without further knowledge of the system. Even though climatic characteristics may be consistent with sites where the disease has been reported, adequate vectors for CCHFV are known to be absent in the New World and Australia. The role of the tick vectors needs therefore to be carefully evaluated under both laboratory and field conditions, and the lack of concordance between the groups of data clarified.

There is, however, evidence that a fragmented landscape, with many small patches existing within a matrix of unsuitable tick habitat, may lead to isolated populations of both ticks and hosts, producing an amplification cycle whereby ticks feed on infected hosts (Estrada-Peña *et al.*, 2011). For CCHFV eco-epidemiology, the degree of habitat patchiness contributes to the increased contact rate among reservoir hosts, humans and ticks. It also leads to the occurrence of isolated amplification foci, with consequently higher tick exposure to humans (Estrada-Peña *et al.*, 2009). Isolated fragments surrounded by a matrix of agricultural land commonly have the poorest diversity of mammals. Species composition changes within these fragmented and smaller isolated patches, and each species differs in its ability to cross the matrix between fragments (e.g. Pires *et al.*, 2002; Lira *et al.*, 2007; Forero-Medina and Vieira, 2009). Although *Hyalomma* ticks can move horizontally, they require a vertebrate host to move over significant distances. Because these host populations are relatively isolated, there are few local movements of hosts and therefore new 'naive' animals carrying uninfected ticks do not dilute the prevalence rates in the isolated patch. However, the immune response of such isolated populations against the ticks and the pathogen may seriously decrease the population of infected ticks. These aspects of the disease have never been studied for CCHF.

It is unlikely that the recent epidemic of CCHF in Turkey is caused by a spreading wave of infected ticks through the movement of hosts, i.e. ticks infected with an 'imported' virus strain (or by a new, more infective genetic variant of the virus) spreading from central Anatolia to other regions of Turkey, as was hypothesized by Karti *et al.* (2004). Research on the epidemiology of the CCHF infection in Turkey has focused on the region of Anatolia in north-central Turkey, due to the high incidence of the disease in this area. The first clinical case of CCHF in the Thrace region (in north-western Turkey) was noticed in 2007 (Midilli *et al.*, 2009). Findings from a serosurvey on 193 human samples from three localities showed that anti-CCHFV IgM and IgG positivity rates were compatible with an epidemiological situation in which the virus has been present for a long time and is not the result of a recent invasive event from the main epidemic centre in Anatolia (Gargili *et al.*, 2011). The detection of CCHFV IgM positivity and an elevated proportion of IgG seropositivity indicated that the genetic variants Europe 1 (the clade in which viral strains from Anatolia are included) and Europe 2 are very active in the region, and that high IgG seroprevalence in humans indicates an active immune protection (Gargili *et al.*, 2011). These figures support the idea that the infection had been present in Turkey for a long time before the epidemic reports. Conditions have existed for continued cycles of amplification, with the epidemic threshold only recently being reached. However, these conditions are not the final step in a chain of recent climate trends. Such an epidemic threshold, whatever the factors behind it, has been observed simultaneously in distant regions of Turkey in only 2–4 years, where it is not the result of a moving wave of infected ticks feeding on hosts either. While landscape patchiness, and thus isolated amplification systems, seems to be behind the proximal causes of the epidemic, the factors driving the enzootic amplification cycle in Turkey remain unknown. Land use practices are also important; the agriculture in CCHF areas is based on crop rotation and land fallowing, and in a given village land is used for wheat production during half of the year, while for the remaining months it is used either for chickpea/vetch production or is left fallow. The following year the land use is rotated.

All this discussion highlights the variable impact of different biotic and abiotic factors in the delineation of the foci of pathogens transmitted by ticks. The shifting nature of these foci is far from being fully recognized, and efforts are actively being made to evaluate the main forces that make up such an intrinsically complex epidemiology.

References

Daniel, M., Danielova, V., Kriz, B., Jirsa, A. and Nozicka J. (2003) Shift of the tick *Ixodes ricinus* and tick-borne encephalitis to higher altitudes in central Europe. *European Journal of Clinical Microbiology and Infectious Disease* 22, 327–328.

Ergönül, Ö. (2006) Crimean-Congo haemorrhagic fever. *The Lancet Infectious Diseases* 6, 203–214.

Estrada-Peña, A., Naranjo, V., Acevedo-Whitehouse, K., Mangold, A.J., Kocan, K.M. and de la Fuente, J. (2009) Phylogeographic analysis reveals association of tick-borne pathogen, *Anaplasma marginale*, MSP1a sequences with ecological traits affecting tick vector performance. *BMC Biology* 7, 57.

Estrada-Peña, A., Martinez-Aviles, M. and Munoz-Reoyo, M.J. (2011) A population model to describe the distribution and seasonal dynamics of the tick *Hyalomma marginatum* in the Mediterranean Basin. *Transboundary and Emerging Diseases* 58, 213–223.

Forero-Medina, G. and Vieira, M.V. (2009) Perception of a fragmented landscape by neotropical marsupials: effects of body mass and environmental variables. *Journal of Tropical Ecology* 25, 53–62.

Gargili, A., Midilli, K., Ergönül, O., Ergin, S., Alp, H.G., Vatansever, Z., Iyisan, S., Cerit, C., Yilmaz, G., Altas, K. and Estrada-Peña, A. (2011) Crimean-Congo hemorrhagic fever in European part of Turkey: genetic analysis of the virus strains from ticks and a seroepidemiological study in humans. *Vector-Borne and Zoonotic Diseases* 11, 747–752.

Hoogstraal, H. (1979) The epidemiology of tick-borne Crimean-Congo hemorrhagic fever in Asia, Europe and Africa. *Journal of Medical Entomology* 15, 307–417.

Jaenson, T.G. and Lindgren, E. (2011) The range of *Ixodes ricinus* and the risk of contracting Lyme borreliosis will increase northwards when the vegetation period becomes longer. *Ticks and Tick Borne Diseases* 2, 44–49.

Karti, S., Odabasi, Z., Korten, V., Yilmaz, M., Sonmez, M., Caylan, R., Akdogan, E., Eren, N., Koksal, I., Ovali, E., Erickson, B.R., Vincent, M.J., Nichol, S.T., Comer, J.A., Rollin, P.E. and Ksiazek, T.G. (2004) Crimean-Congo hemorrhagic fever in Turkey. *Emerging Infectious Diseases* 19, 1379–1384.

Lira, P.K., Fernandez, F.A.S., Carlos, H.S.A. and Curzio, P.L. (2007) Use of a fragmented landscape by three species of opossum in south-eastern Brazil. *Journal of Tropical Ecology* 23, 427–435.

Maltezou, H.C. and Papa, A. (2010) Crimean-Congo hemorrhagic fever: risk for emergence of new endemic foci in Europe? *Travel Medicine and Infectious Diseases* 8, 139–143.

Midilli, K., Gargili, A., Ergonul, O., Elevli, M., Ergin, S., Turan, N., Sengöz, G., Ozturk, R. and Bakar, M. (2009) The first clinical case due to AP92 like strain of Crimean-Congo hemorrhagic fever virus and a field survey. *BMC Infectious Diseases* 9, 90.

Pires, A.S., Lira, P.K., Fernandez, F.A.S., Schittini, G.M. and Oliveira, L.C. (2002) Frequency of movements of small mammals among Atlantic coastal forest fragments in Brazil. *Biological Conservation* 108, 229–237.

Purnak T., Selvi, N.A. and Altundag, K. (2007) Global warming may increase the incidence and geographic range of Crimean-Congo hemorrhagic fever. *Medical Hypotheses* 68, 924–925.

Randolph, S.E. and Rogers, D.J. (2007) Ecology of tick-borne disease and the role of climate. In: Ergonul, O. and Whitehouse, C.A. (eds) *Crimean-Congo Hemorrhagic Fever. A Global Perspective*. Springer, Dordrecht, The Netherlands, pp. 3–13.

Rizzoli, A., Hauffe, H.C., Carpi, G., Vourc'h, G.I., Neteler, M. and Rosà, R. (2011) Lyme borreliosis in Europe. *Eurosurveillance* 16(27), pii=19906.

Rogers, D. and Randolph, S.E. (2006) Climate change and vector-borne diseases. *Advances in Parasitology* 62, 345–381.

Talleklint, L. and Jaenson, T.G.T. (1998) Increasing geographical distribution and density of *Ixodes ricinus* (Acari: Ixodidae) in Central and Northern Sweden. *Journal of Medical Entomology* 35, 521–526.

2 Basic Biology and Geographical Distribution of Tick Species Involved in the Transmission of Animal Pathogens, Including Zoonoses*

Róbert Farkas,[1] Agustín Estrada-Peña,[2] Thomas G.T. Jaenson,[3] Ilaria Pascucci[4] and Maxime Madder[5]

[1]Szent István University, Budapest, Hungary; [2]University of Zaragoza, Spain; [3]Uppsala University, Sweden; [4]Istituto Zooprofilattico Sperimentale dell'Abruzzo e del Molise "G. Caporale", Teramo, Italy; and [5]Institute of Tropical Medicine, Antwerp, Belgium and University of Pretoria, South Africa

The class Arachnida falls within the phylum Arthropoda (arthropods) and encompasses the following taxa: Acari (ticks and mites), Araneae (spiders), Opiliones (harvestmen or daddy-long-legs), Palpigradi (micro whip scorpions), Amblypygi (tailless whip scorpions), Thelyphonida (whiptailed scorpions or uropygids), Ricinulei (ricinuleids), Scorpionida (scorpions), Pseudoscorpiones (book scorpions) and Solifugae (camel spiders, sun spiders or solpugids). The Ixodida (ticks) is one of several taxa that constitute the subclass Acari (mites). Hence, ticks are mites but not all mites are ticks (Woolley, 1988).

There are three families of ticks comprising a total of at least 907 species of ticks: (i) Nutalliellidae, with a single species, *Nutalliella namaqua*; (ii) Argasidae, soft-bodied or soft ticks, with 186 species (Section 2.1 below); and (iii) Ixodidae, hard-bodied or hard ticks, with 720 species considered to have valid genus and species names (Section 2.2 below) (Barker and Murrell, 2008).

Ticks can be easily distinguished from adult insects because the tick's body is not divided into three sections; rather there is strong fusion of the body segments giving the tick body a sac-like appearance. In common with most arachnids, the adult tick and the nymph have eight legs, whereas the larva has six legs like adult insects. The tick has no distinct head, but the mouthparts and the basis capituli form the capitulum. All nymphal and adult ticks have a pair of spiracles located latero-ventrally on the abdomen. Depending on the species, a pair of simple eyes may be present.

2.1 Soft Ticks (Argasidae)

Argasid ticks, known as soft ticks, include about 180 species of four genera: *Argas*, *Ornithodoros*, *Otobius* and *Carios* (Klompen

* Róbert Farkas wrote Sections 2.1 (Introduction), 2.1.1, 2.2.2, 2.2.6; Agustín Estrada-Peña, Sections 2.1.2, 2.2.5; Thomas G.T. Jaenson, introduction of Chapter 2 Sections 2.2 (Introduction), 2.2.1; Ilaria Pascucci, Sections 2.2.3, 2.2.4; and Maxime Madder supplied the colour plates of different tick species that are cited in the text.

and Olivier, 1993). The Argasidae attain their greatest abundance in dry regions, and when they extend into humid areas they tend to select drier niches where hard ticks would not survive. This behaviour is related to their leathery cuticle, which is effective in preventing water loss. A variety of argasids occur throughout most of the tropics and subtropics of the world. Fewer species live in temperate areas and very few inhabit arctic climates. Soft ticks often live for many years. In contrast to hard ticks, they live in close proximity to their hosts: in the nests of birds, pigeon lofts, animal burrows or dens, and caves.

The body of an argasid tick has an unsclerotized, leathery cuticle with a textured surface that may be marked with grooves or folds. The many small integumental folds usually have a button-like appearance. The ticks have a flattened body margin. In nymphs and adults, the gnathosoma (mouth and feeding parts) is not visible from the dorsal view, being located ventrally. There is little sexual dimorphism. The life cycles vary greatly between the Ixodidae and the Argasidae. In the argasids, mating usually occurs off the host. Females take repeated small blood meals and lay small batches of eggs (50–500) in places where they seek shelter (e.g. cracks and crevices) after each feeding. As many as six gonotrophic cycles have been reported in some species. The six-legged larvae hatch in 1–4 weeks and, after seeking and attaching to a host, feed for 3–8 days. They then drop off and moult to the first nymphal stage. Most argasids have two or more (six or even seven in some species) nymphal stages in their life cycle, each of which must have a blood meal once before moulting to the next stage. These nymphal stages occur off the host in cracks or crevices, or beneath debris in or near the nest and burrows. In contrast to the slow-feeding hard ticks, the feeding of nymphs and adults usually lasts 1 hour or less. They are also highly resistant to starvation, which can extend their longevity even further (Hoogstraal, 1985).

Argasid ticks have a considerable economic and medical impact in many parts of the world. However, at the present time they are apparently of less importance in Europe than are the ixodid ticks. It should be noted that argasid ticks in general are xerophilic arthropods. Although in localities of extremely low relative humidity argasids may seek a somewhat more humid microhabitat, the cracks and crevices that they use seldom have a significantly high relative humidity; within this range, individual species have varying degrees of tolerance to humidity. The soft ticks are vectors of a large number of poorly known viruses (Hoogstraal, 1985) which seem of little economic interest for domestic animals. A few studies exist about those pathogens. However, perhaps the most prominent pathogen transmitted by a soft tick in the Mediterranean basin is the African swine fever (ASF) virus (ASFV). This virus has produced several epidemics in Spain and Portugal, and is transmitted by *Ornithodoros erraticus* (EFSA Panel on Health and Welfare, 2010a).

2.1.1 Genus *Argas*

Argas species are about 5–10mm long and 3–6mm wide. In general, they are oval shaped and dorsoventrally flattened. The cuticle is leathery and wrinkled with definite margins that can be seen even when the tick is engorged. The unfed adults are yellowish, but after a blood meal they appear slate grey. Eyes, when present, occur on folds lateral to the coxae. The mouthparts are on the ventral surface of the body and small. The spiracular plates, located between the coxae of legs III and IV, are relatively small and inconspicuous. Members of this genus are most commonly associated with birds and bats, although some species are found on reptiles and, to a lesser extent, on insectivorous mammals other than bats. These soft ticks are nocturnal and endophilic, and they mate in the environment. The females lay batches of 20–100 eggs in hidden places after each blood meal. The eggs hatch in about 2–3 weeks. *Argas* larvae attach to hosts and feed for about a week; they then drop to the ground to moult into a nymph. There are usually two nymphal stages. Like adults, each eight-legged nymph instar is nocturnal; they feed at night and moult in cracks and crevices to the next nymphal stage and, finally, to the adult male or female. The life cycle from eggs to adults takes about a month or longer. The adults can live up to 2 years or more without a blood meal. The larvae and nymphs may survive for a few months. The larvae and first nymphal stage are most active in early summer, the other nymphal stages in midsummer, and the adults in late summer and autumn. They overwinter in the adult stage or as eggs. Three species of *Argas* have been commonly reported in Europe, namely *Argas (Ar.) persicus*, *Ar. reflexus* and *Ar. polonicus*.

Argas persicus

The fowl or chicken tick, *Ar. persicus*, has spread widely in most parts of the world. These ticks construct large, numerous, and fairly permanent nests in the vicinity of human activity. All stages congregate on walls, in crevices or between the boards of poultry houses. Populations of the species can reach enormous numbers in poultry houses, causing irritation, loss of egg production, even high mortality because of anaemia. It is reported that larvae of *Ar. persicus* can cause fatal paralysis in young chickens. Occasionally, they parasitize small mammals and reptiles, but seldom humans. Adults feed once per month for about 2 h, but can survive for up to 4–5 years unfed. *Ar. persicus* may be the vector of the rickettsial agent *Aegyptianella pullorum*, and it is a vector of *Borrelia anserina*, the agent of fowl spirochaetosis in the Mediterranean region. The colour plate section contains two images (dorsal and ventral) of an adult male of *Ar. persicus* (Plate 1a,b).

Argas reflexus

Ar. reflexus, the European pigeon tick, originates from the Mediterranean/Near East and can be distinguished from *Ar. persicus* by its body margin. The host of *Ar. reflexus* is the wild rock pigeon, *Columba livia*, and its various domesticated forms, as well as feral town pigeons, *C. livia* var. *domestica* (Hoogstraal, 1956). *Ar. reflexus* lives in several European countries up to about 55°N (Dautel *et al.*, 1991). In central Europe, *Ar. reflexus* occurs only in or at buildings serving as pigeon breeding sites where there are microclimatically elevated temperatures, especially in summer and autumn but not necessarily in winter. Such conditions are necessary for the completion of the embryonic development. The life cycle appears to be much like that of *Ar. persicus*. All nymphal stages as well as the adults are nocturnal feeders taking blood from sleeping pigeons. Feeding is for up to 12 h and, therefore, post-larval stages might not have the opportunity to be transported to distant habitats. Heavy infestations of pigeons may cause death as a result of severe anaemia.

Ar. reflexus has a relatively long minimum generation time of 2–4 years (Dautel and Knülle, 1997). Hiding places of these soft ticks are easily found in the cracks and crevices of pigeon cotes. German scientists (Dautel *et al.*, 1994) investigated the seasonal activities of this soft tick species in Berlin. They reported that activities of all post-embryonic stages peaked from March through to early June, irrespective of whether or not hosts were available to the ticks. The immature stages displayed another peak of activity in late summer/autumn. The only survey thus far for pigeon ticks was also performed in the city of Berlin, where more than 200 infested buildings were discovered between 1989 and 1998. Most of the infestations were found in older buildings constructed before 1918 (Dautel *et al.*, 1999). Domestic chickens are apparently considerably less liable to be attacked by *Ar. reflexus* than are pigeons. In the laboratory, any available mammal may serve as host. Although the European pigeon tick is nearly always associated with pigeons, the exigencies of its domestic existence drive it to attack humans, possibly more frequently than does *Ar. persicus* (Rolla *et al.*, 2004; Spiewak *et al.*, 2006). There have been a number of reports describing human cases of a specific anti-*Ar. reflexus* allergy (Chappard *et al.*, 1996; Veraldi, 1996; Kleine-Tebbe *et al.*, 2006). Allergy to *Ar. reflexus* has to be considered in patients having allergy symptoms and living in buildings where pigeons have their nests. Severe systemic anaphylactic reactions (urticaria, angiooedema, dyspnoea,

1a

1b

2a

2b

3a

3b

Plate 1. *Argas persicus***:** adult male dorsal view (a) and ventral view (b).
Plate 2. *Argas reflexus***:** adult male dorsal view (a) and ventral view (b).
Plate 3. *Ornithodoros erraticus***:** adult (sex undermined) dorsal view (a) and ventral view (b).

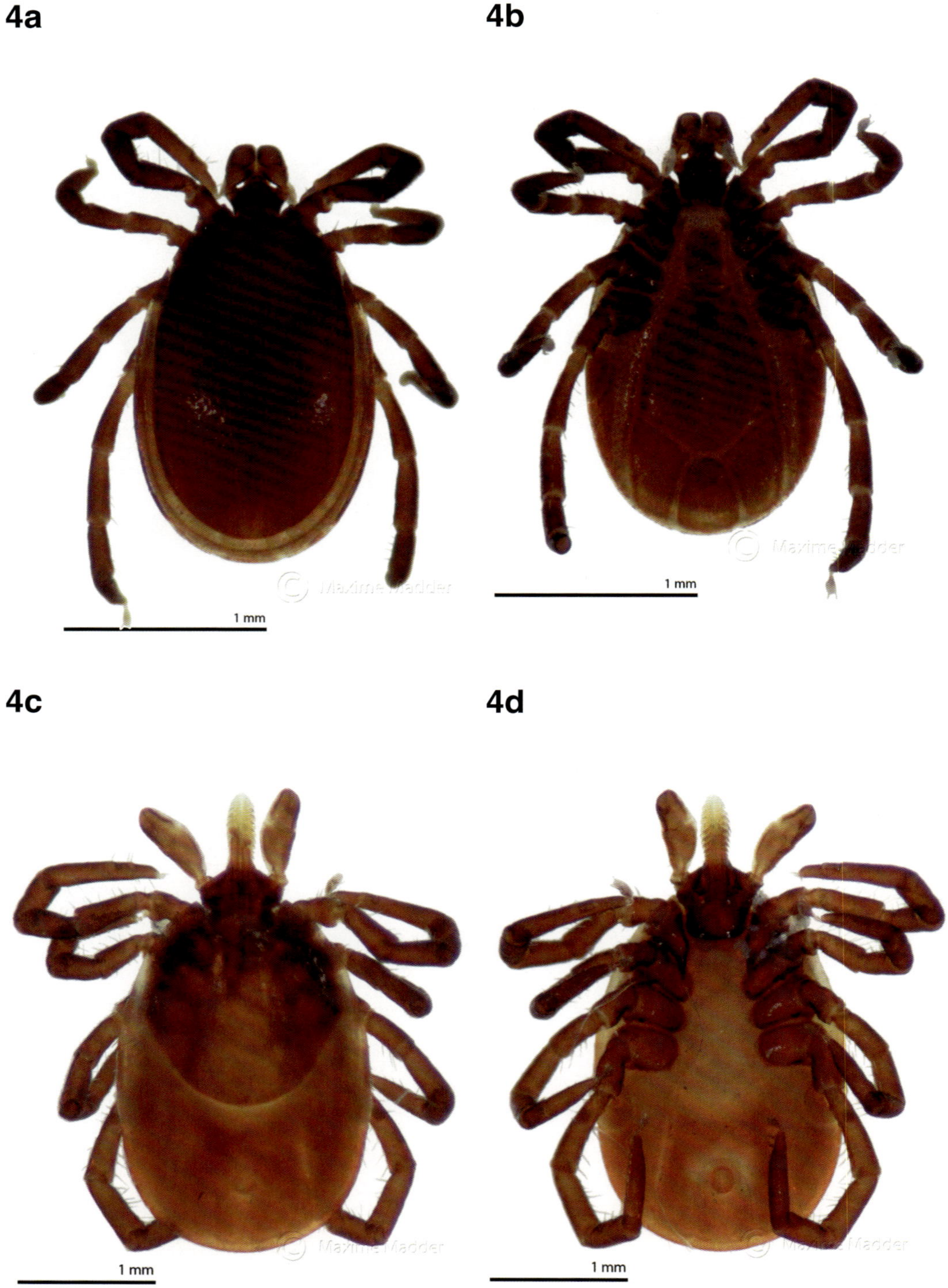

Plate 4. *Ixodes ricinus***:** adult male dorsal view (a) and ventral view (b); adult female dorsal view (c) and ventral view (d).

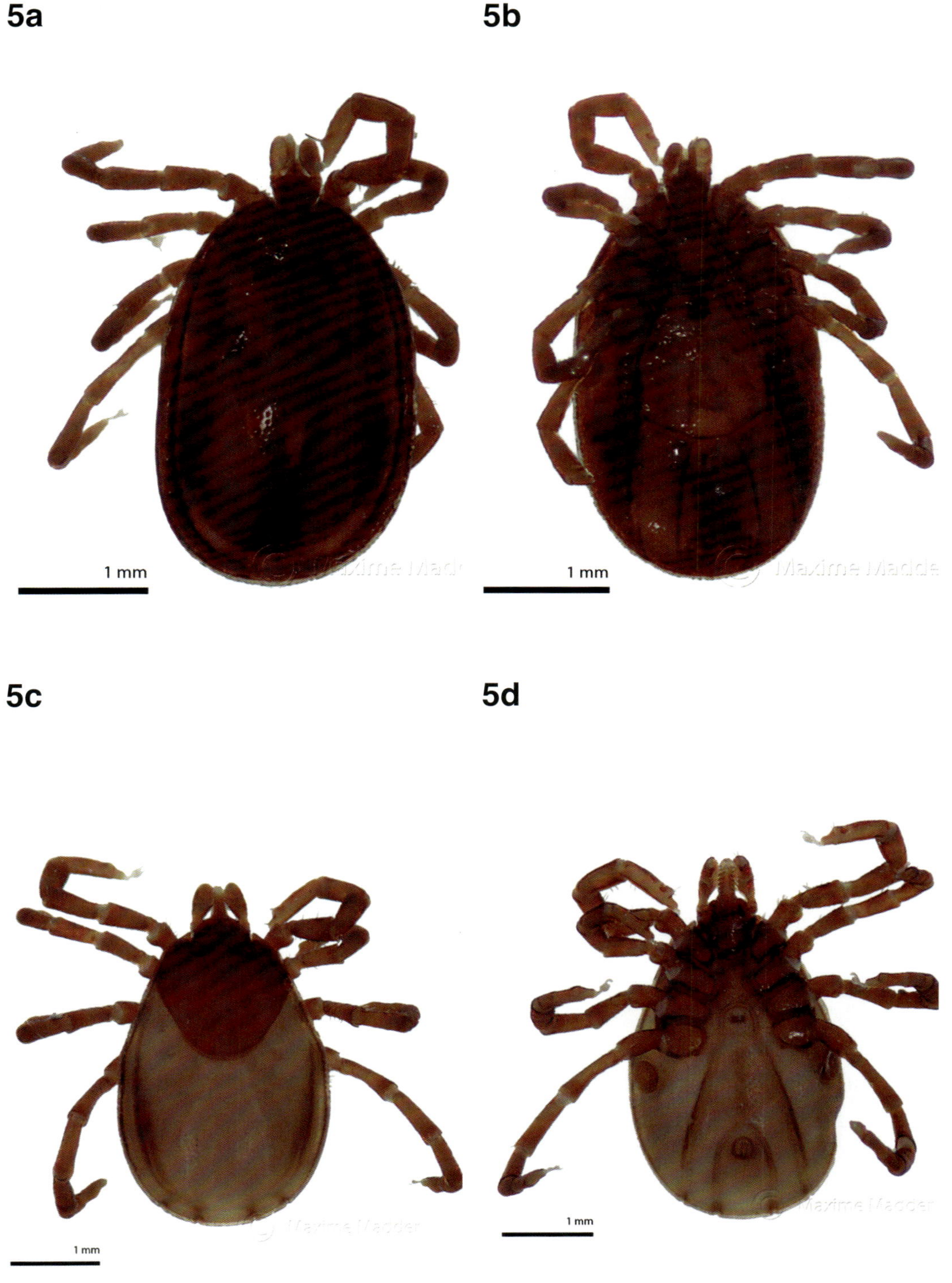

Plate 5. *Ixodes hexagonus***:** adult male dorsal view (a) and ventral view (b); adult female dorsal view (c) and ventral view (d).

6a

6b

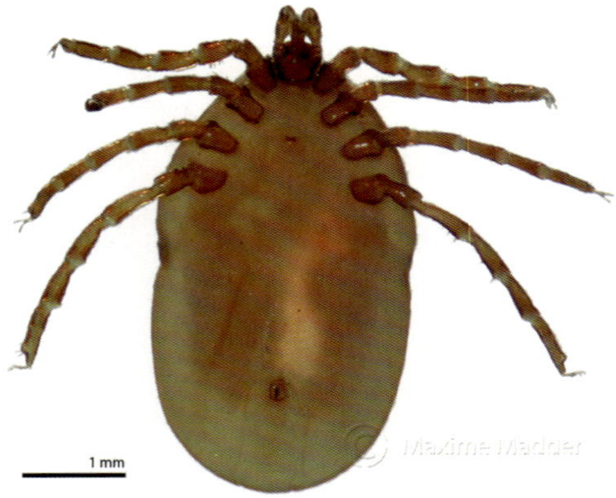

7a

7b

7c

7d

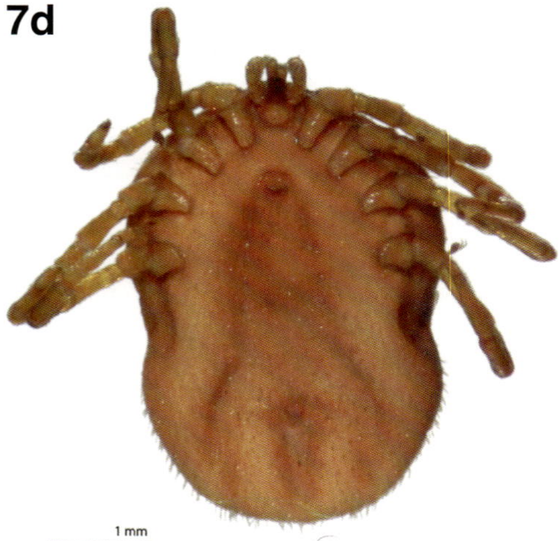

Plate 6. *Ixodes canisuga***:** adult female dorsal view (a) and ventral view (b).
Plate 7. *Ixodes uriae***:** adult male dorsal view (a) and ventral view (b); adult female dorsal view (c) and ventral view (d).

8a

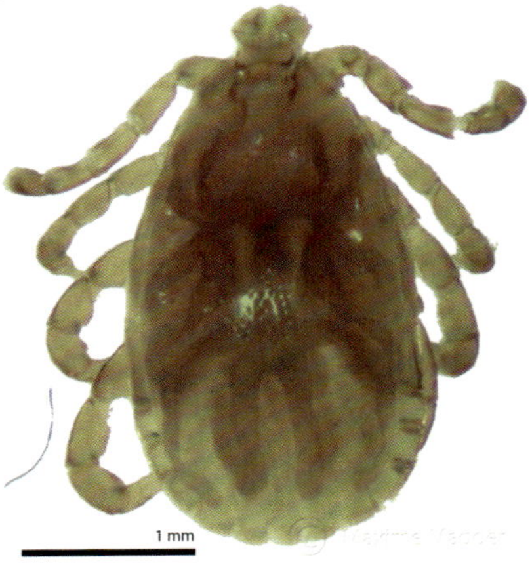

8b

8c

8d

9a

9b

Plate 8. *Haemaphysalis punctata***:** adult male dorsal view (a) and ventral view (b); adult female dorsal view (c) and ventral view (d).
Plate 9. *Haemaphysalis concinna***:** adult male dorsal view (a) and ventral view (b).

10

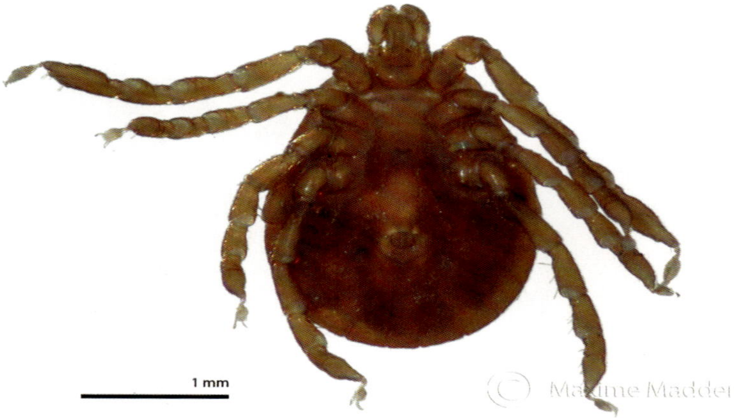

11a

11b

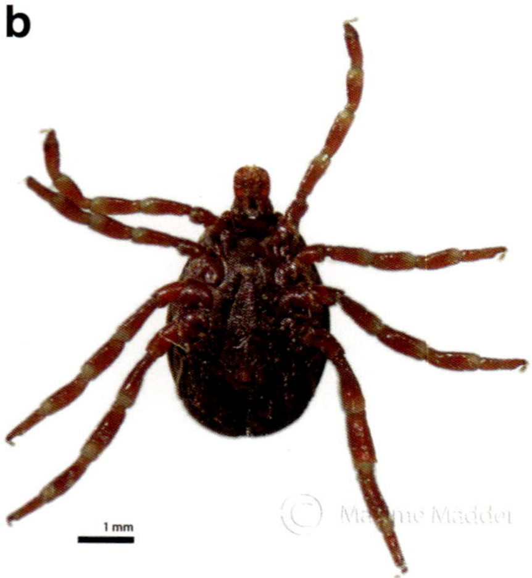

11c

11d

Plate 10. *Haemaphysalis inermis***:** adult female ventral view.
Plate 11. *Hyalomma marginatum***:** adult male dorsal view (a) and ventral view (b); adult female dorsal view (c) and ventral view (d).

12a

12b

12c

12d

Plate 12. *Hyalomma excavatum***:** adult male dorsal view (a) and ventral view (b); adult female dorsal view (c) and ventral view (d).

13a

13b

13c

13d

Plate 13. *Hyalomma scupense***:** adult male dorsal view (a) and ventral view (b); adult female dorsal view (c) and ventral view (d).

14a

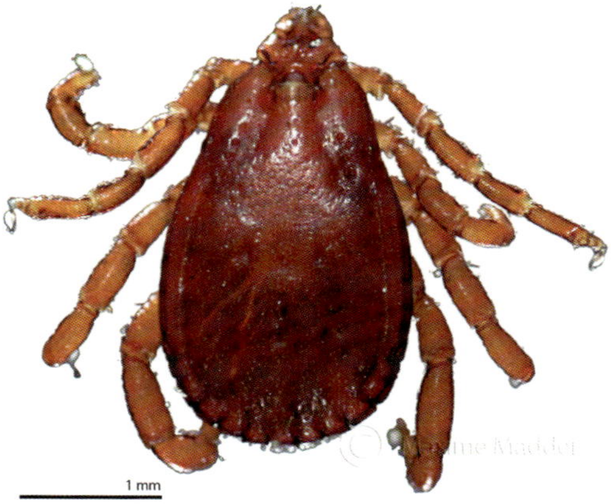

14b

14c

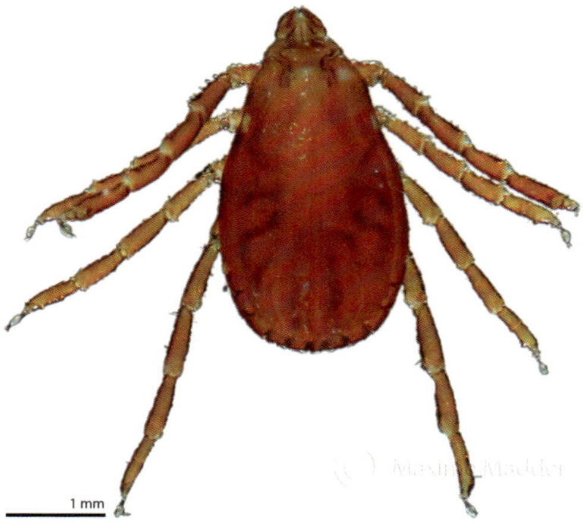

14d

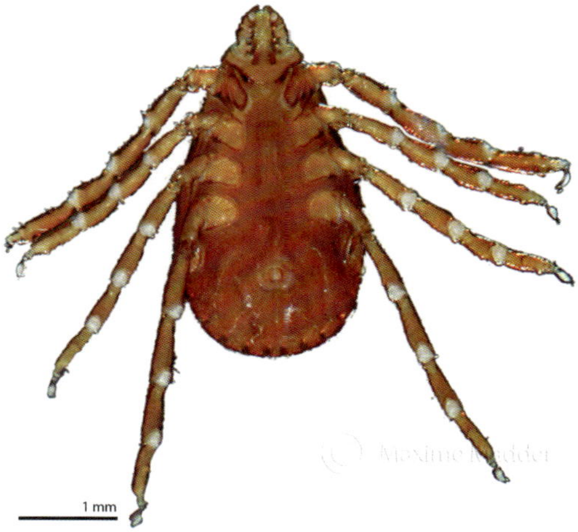

Plate 14. *Rhipicephalus sanguineus***:** adult male dorsal view (a) and ventral view (b); adult female dorsal view (c) and ventral view (d).

15a

15b

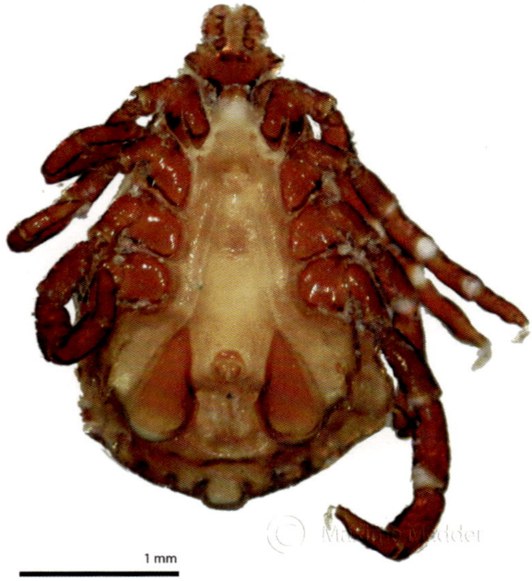

15c

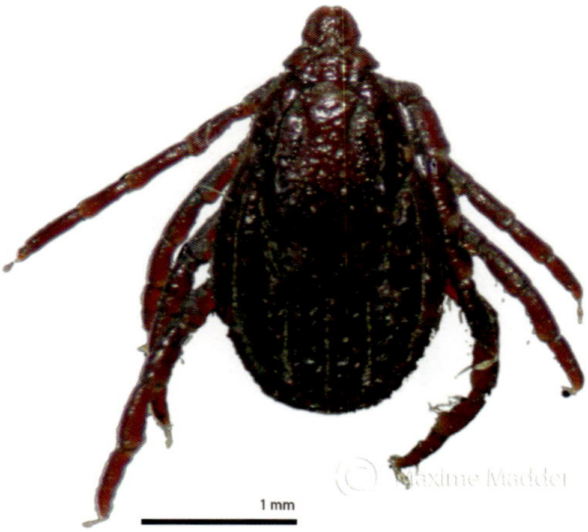

15d

Plate 15. *Rhipicephalus turanicus*, adult male dorsal view (a) and ventral view (b); adult female dorsal view (c) and ventral view (d).

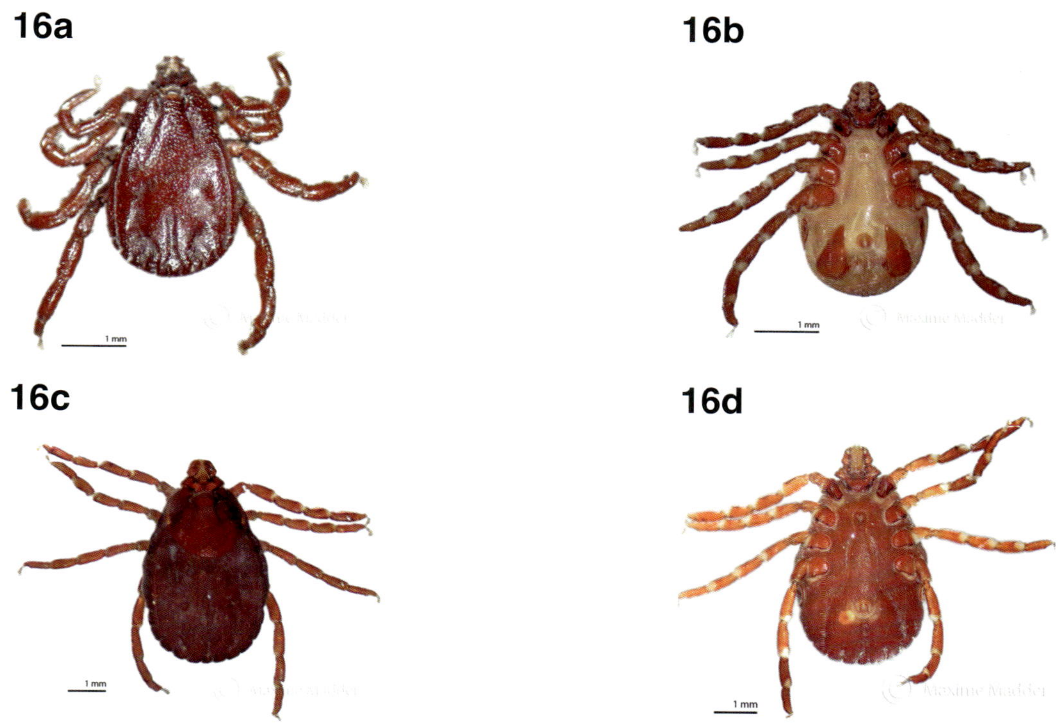

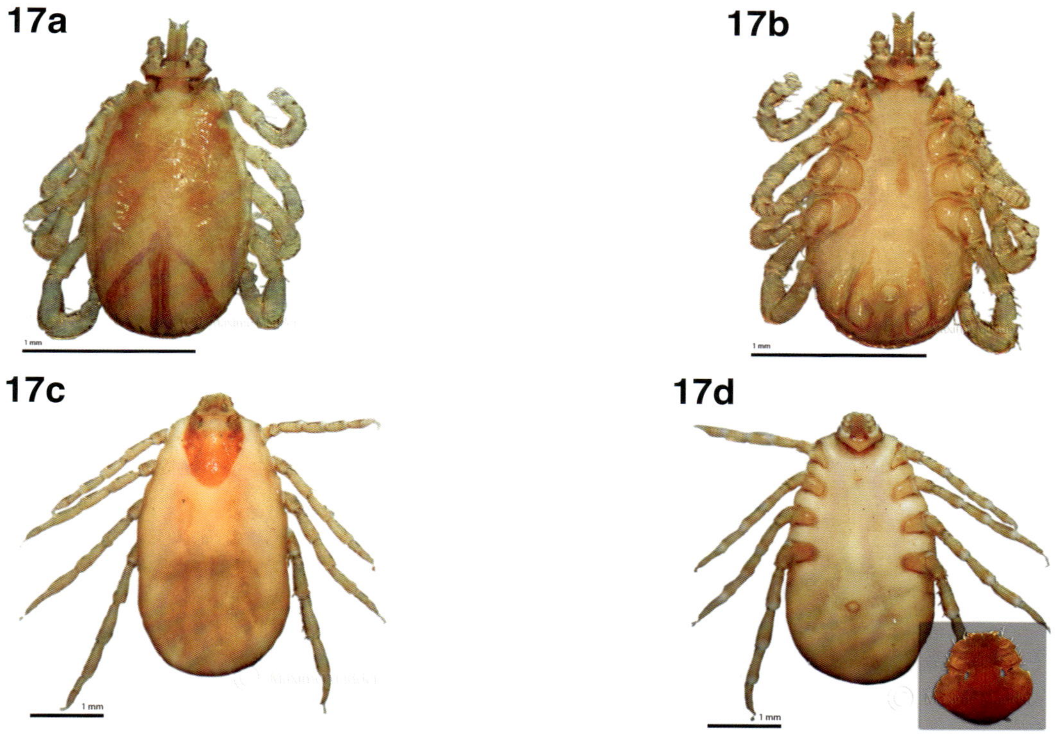

Plate 16. *Rhipicephalus bursa*: adult male dorsal view (a) and ventral view (b); adult female dorsal view (c) and ventral view (d).

Plate 17. *Rhipicephalus annulatus*: adult male dorsal view (a) and ventral view (b); adult female dorsal view (c) and ventral view (d).

18a

18b

18c

18d

19a

19b

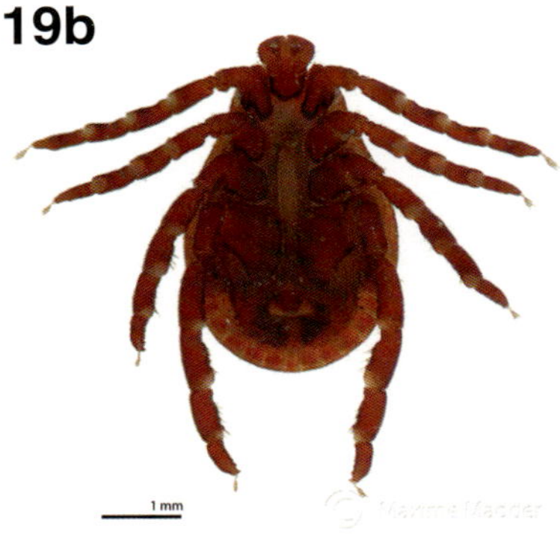

19c

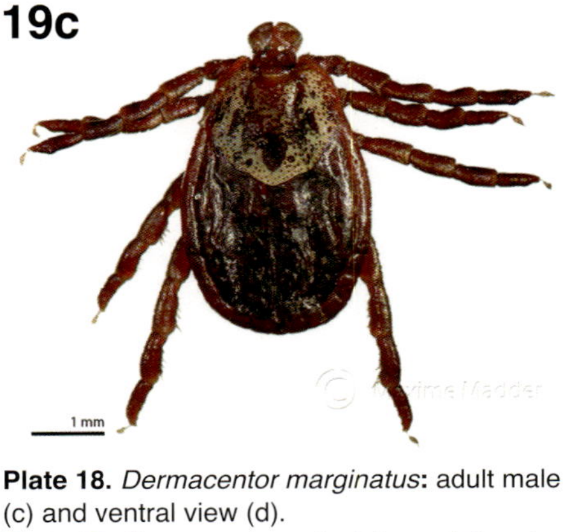

19d

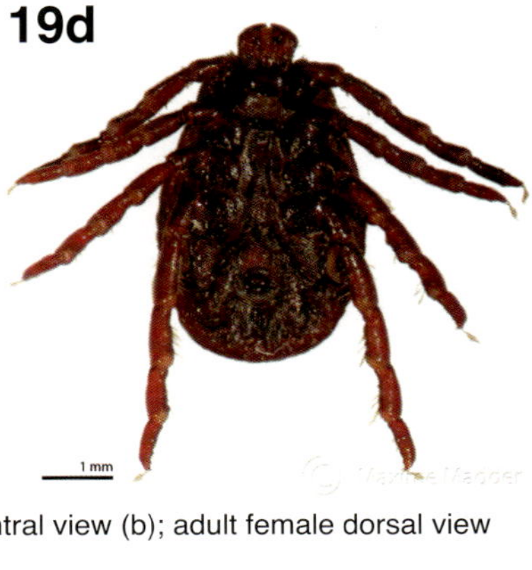

Plate 18. *Dermacentor marginatus***:** adult male dorsal view (a) and ventral view (b); adult female dorsal view (c) and ventral view (d).

Plate 19. *Dermacentor reticulatus***:** adult male dorsal view (a) and ventral view (b); adult female dorsal view (c) and ventral view (d).

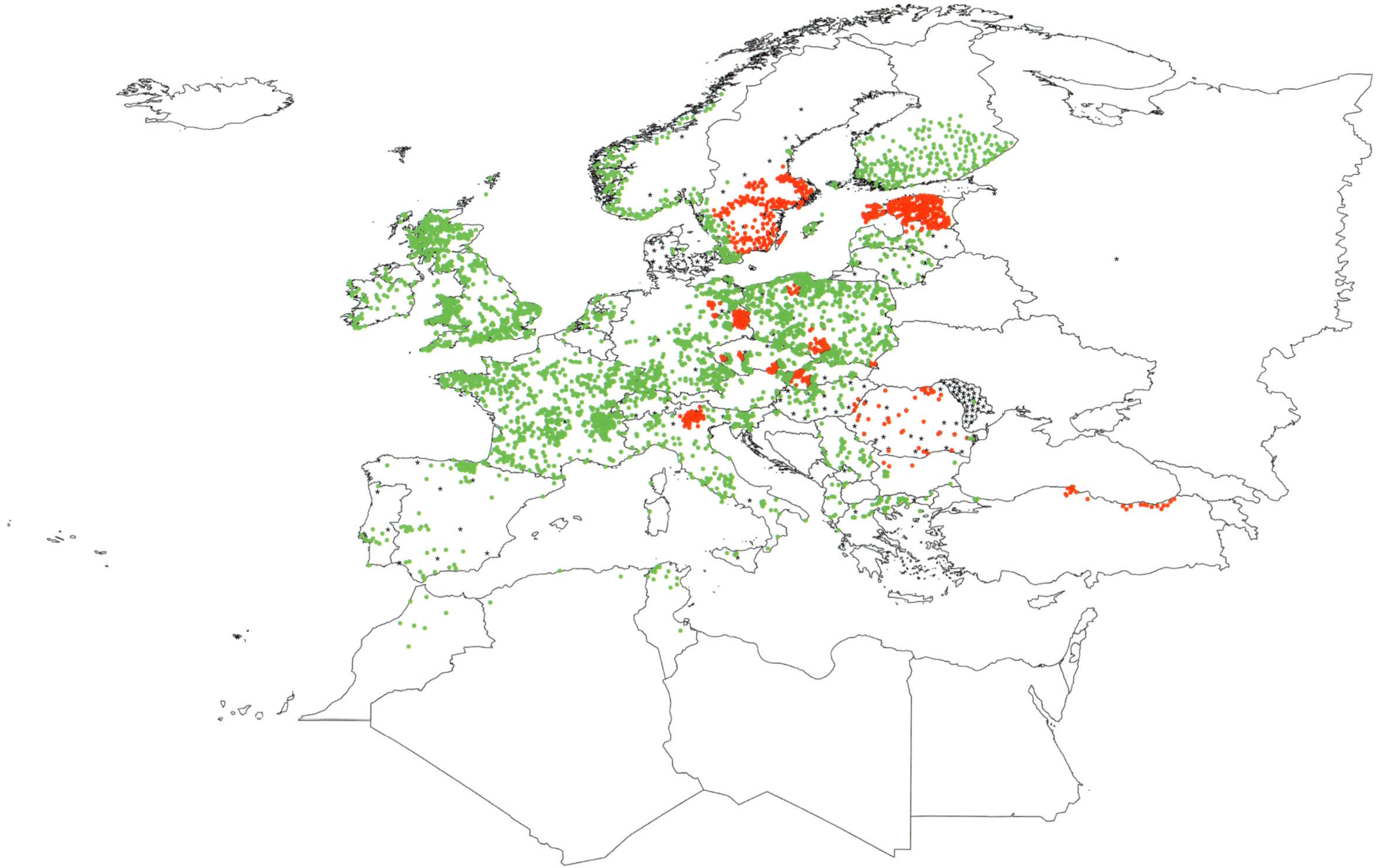

Plate 20. Reported occurrence of *Ixodes ricinus*. ⋆ Smallest administrative region or territorial unit for statistics (NUTS), data for 2000–2010; • Coordinate (latitude/longitude), data for 2000–2010; • Coordinate (latitude/longitude), historical data (before 2000).

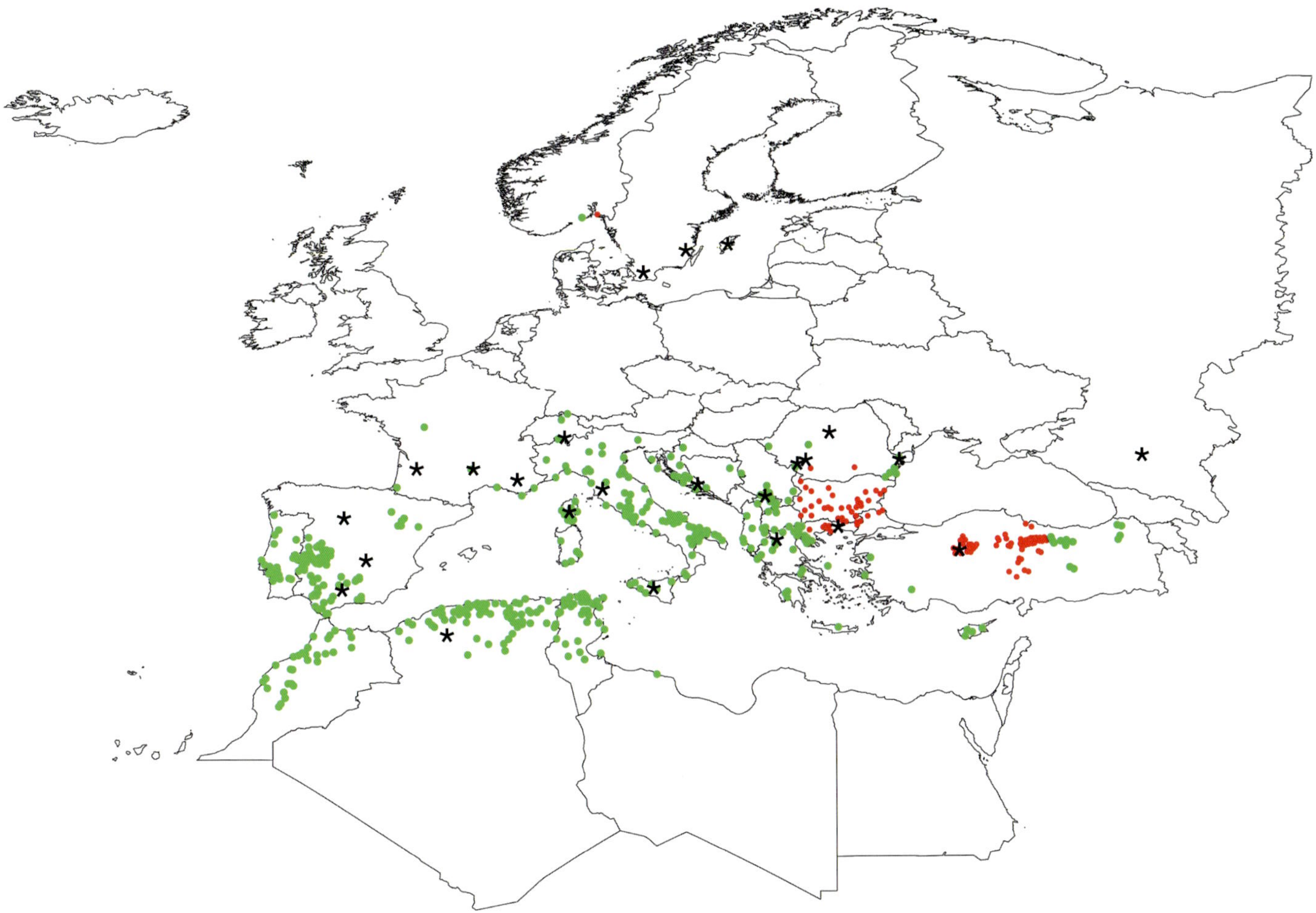

Plate 21. Reported occurrence of *Hyalomma marginatum*. ⭑ Smallest administrative region or territorial unit for statistics (NUTS), data for 2000–2010; • Coordinate (latitude/longitude), data for 2000–2010; • Coordinate (latitude/longitude), historical data (before 2000).

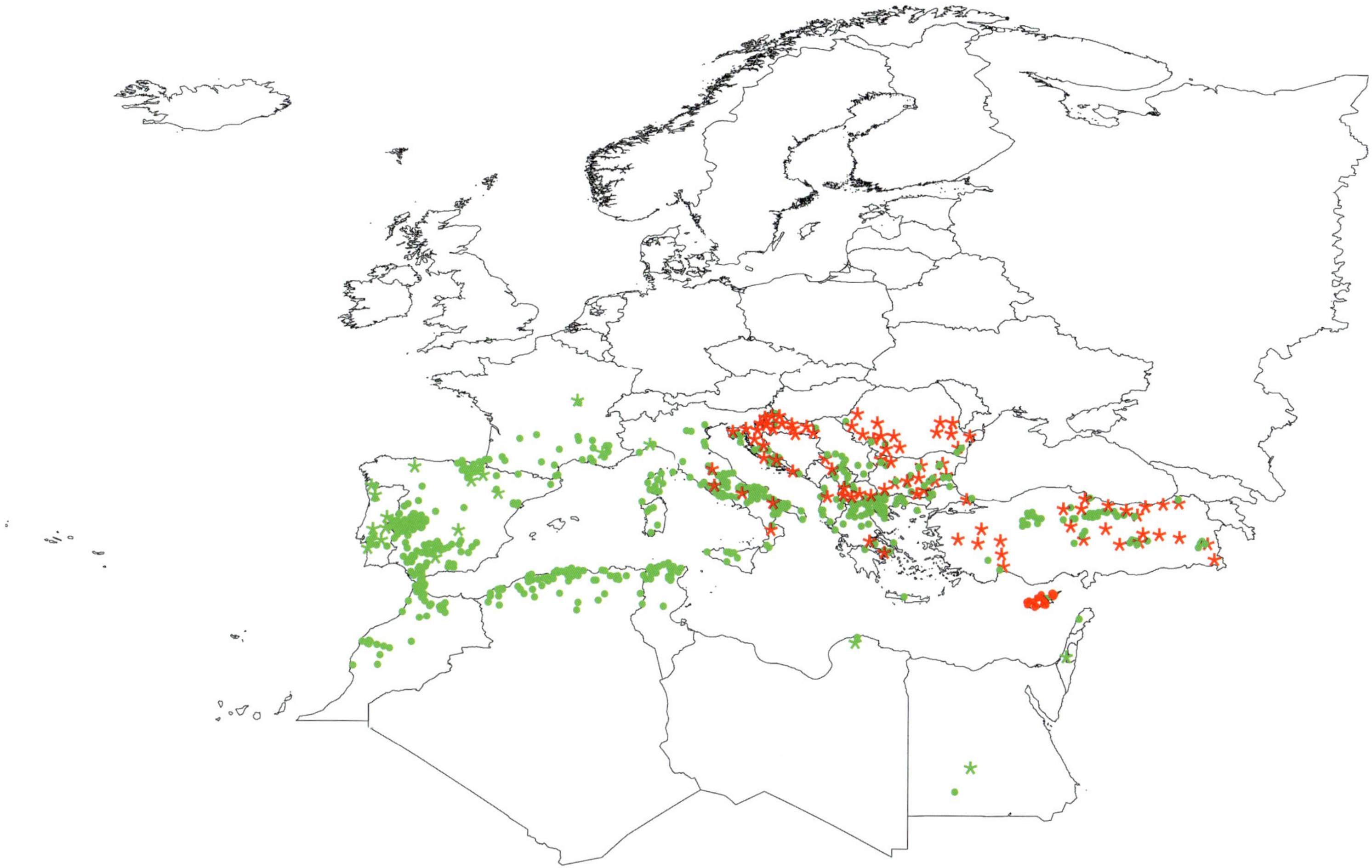

Plate 22. Reported occurrence of *Rhipicephalus bursa*. ⚹ Smallest administrative region or territorial unit for statistics (NUTS), historical data (before 2000); ⚹ Smallest administrative region or territorial unit for statistics (NUTS), data for 2000–2010; • Coordinate (latitude/longitude), historical data (before 2000); • Coordinate (latitude/longitude), data for 2000–2010.

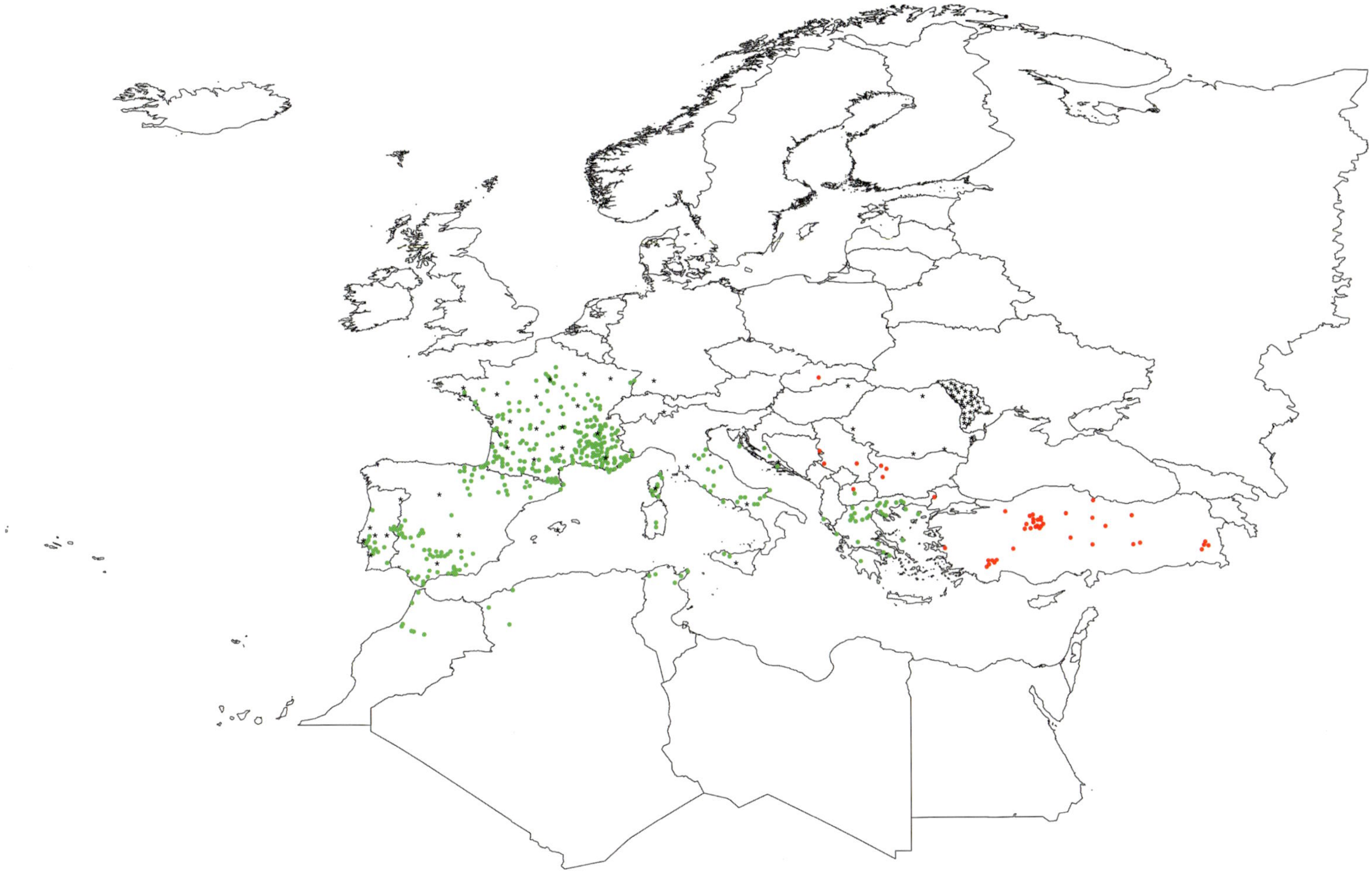

Plate 23. Reported occurrence of *Dermacentor marginatus*. ⋆ Smallest administrative region or territorial unit for statistics (NUTS), data for 2000–2010; • Coordinate (latitude/longitude), data for 2000–2010. • Coordinate (latitude/longitude), historical data (before 2000).

cardiovascular dysregulation, unconsciousness) were reported in 12 of 148 (8%) humans after bites of the soft tick *Ar. reflexus* and 146 of 148 (99%) had local reactions (Kliene-Tebbe *et al.*, 2006). Man is frequently attacked, especially in the vicinity of long unoccupied pigeon sites. The pigeon tick can withstand long periods of starvation; it may remain unfed in or near pigeon houses for many months, or even for several years. This species can also be the vector of *B. anserina*. There is no evidence that *Ar. reflexus* transmits pathogens to humans in central Europe. The colour plate section contains two photographs (dorsal and ventral) of an adult male of *Ar. reflexus* (Plate 2a,b).

Argas polonicus

Ar. polonicus is a poorly known species that seems to be restricted to domestic and wild pigeons. It has been largely confused with a related species, *Ar. vulgaris*, and it differs distinctly from *Ar. reflexus* (Siuda *et al.*, 1979). Phenotypic (involving the use of cuticular hydrocarbons) and genetic analysis also showed that the two species are different (Estrada-Peña and Dusbabek, 1992).

2.1.2 Genus *Ornithodoros*

The soft tick species within the genus *Ornithodoros* are able to transmit ASFV. All *Ornithodoros* species tested up to the present have been proved to be competent vectors for ASFV (Table 2.1). Some of these belong to the taxonomic group of *O. erraticus*, a group of species distributed mainly in the Mediterranean basin and Middle East, including Transcaucasia and parts of Russia (Morel, 1969). Sánchez-Botija (1963) demonstrated that *O. erraticus* found in pig sites can be a vector of ASFV in Europe, while

Table 2.1. Laboratory and field studies on *Ornithodoros* species to determine their vectorial capacity for African swine fever virus (ASFV) (EFSA, 2010a,b).

			Virus transmission[a]				
Species	Region	Virus presence (max. period post infection)	Trans-stadial[b]	Transovarial	Sexual	To pigs	References
O. coriaceus	North America	502 days	+ (4)	–	ND	+ 502 dpi[c]	Hess *et al.*, 1987; Endris and Hess, 1994
O. turicata	North America	23 days	ND[d]	ND	ND	+ 23 dpi	
O. puertoricensis	North America	239 days	ND	+	ND	+ 239 dpi	
O. parkeri	North America	70 days	+ (1)	ND	ND	–	
O. savignyi	Africa	106 days	+ (1)	–	ND	+ 106 dpi	Mellor and Wilkinson, 1985
O. moubata	Africa	3 years	+ (5)	+	+	+ 3 years	Kleiboeker *et al.*, 1998, 1999; Plowright, 1977; Plowright *et al.*, 1970a,b, 1974, 1994; Rennie *et al.*, 2001
O. erraticus/ *O. marocanus*	Europe	5 years	+ (5)	–	+	+ 588 dpi	Hess *et al.*, 1989; Endris *et al.*, 1992; Boinas, 1994

[a]Only the maximum lengths of time for virus presence and transmission to pigs, or horizontal or vertical transmission among tick populations, are considered; [b]number of moults; [c]days post infection; [d]ND, no data.

Plowright *et al.* (1969a,b) demonstrated that *O. moubata/porcinus* ticks associated with warthogs were also involved in virus transmission in Africa. The group of species *O. moubata/porcinus* are not found in Europe or in northern Africa. *O. erraticus* is accepted as a reservoir for ASFV with a well-established role in the maintenance of the infection in nature (Plowright *et al.*, 1970a,b). Replication of ASFV in this tick species has been demonstrated, and this allows the species to remain infectious for months and up to 5 years, posing a threat of transmitting the virus each time the ticks feed on pigs (Boinas, 1994).

The adults and nymphal stages 4 and 5 of *O. erraticus* remain infected for longer than the larvae up to nymph 3 (Boinas, 1994; Basto *et al.*, 2006). It has also been reported that the prevalence of infected *O. erraticus* and the average titres of ASFV per tick decreased over time when pigs were not present (Boinas, 1994). The maintenance of virus infection in the tick can constitute risk if the ticks are able to feed on pigs for periods of 40 weeks after an ASFV infective meal (Basto *et al.*, 2006). More recently, Vial *et al.* (2007) demonstrated ASFV DNA in *O. sonrai* collected from the field in West Africa. *O. sonrai,* although a different species, is morphologically undistinguishable from *O. erraticus*. The two species may copulate but the progeny are not fertile. At present, as there is not a solid ground test to determine which of these two tick species is which, any overlapping distribution in some areas is ignored, as well as the basic biological mechanisms of separation of the two species. It is probably that the main mechanism of isolation of the species – leading to reproductive isolation – is the simple geographical partitioning of the populations (Vial *et al.*, 2007).

Some authors consider ASFV and the *Ornithodoros* tick as co-evolving organisms. In fact, there are noticeable telomeric similarities in the genomes of ASFV and *Borrelia*, with the latter sharing the same *Ornithodoros* tick host in Africa and considered to be an original pathogen of soft ticks, suggesting that ASFV is also a primary organism of *Ornithodoros* ticks and co-adapted to its tick hosts (Hinnebusch and Barbour, 1991). This hypothesis could explain discrepancies concerning infection success rates that have been noticeable between several past surveys.

For example, De Tray (1963) reported a consistent establishment of the virus isolate 'Uganda' in specimens of the *O. moubata* group (34/35 were infected), whereas another isolate, 'Tengani', only caused persistent infection in a small proportion of ticks (2/46 were infected). More recently, Kleiboeker *et al.* (1999) compared oral and intra-haemocoelic experimental infections of *Ornithodoros* ticks collected from warthog burrows in Kruger National Park and the Northern Transvaal region of South Africa, as well as infections of ticks from Masai Mara Reserve in Kenya, by three different viruses from South Africa, Malawi and Zimbabwe; all the viruses were originally isolated from ticks collected from the field. The oral infection conducted with the isolate from Malawi was self-limiting (with a decline of virus titres and the number of ticks containing virus), while the other infections persisted. According to Kleiboeker *et al.* (1999), the cytopathology caused by the Malawi strain in infected ticks suggested the non-adaptation of the isolate to express specific genes that allow the production of large quantities of progeny virus without damaging the host cell. The reason why this virus was originally isolated from ticks could be the large opportunity for those ticks to feed on infected pigs with high viraemic titres during an ASF outbreak and the leakage of midgut contents into the haemocoel without tick mortality, instead of real adaptation of this virus isolate to the tick host (Kleiboeker *et al.*, 1999).

Dixon and Wilkinson (1988) analysed genomes of ASFV isolates collected over a 2 year period from ticks inhabiting warthog burrows in four regions of Zambia and observed additional sequences in the region close to the left-end terminus of the genome, which are not observed in domestic pig isolates. The author concluded that virus replication in ticks and warthogs may require additional host-specific genes not necessary for multiplication in domestic pigs and that the introduction of virus from tick/warthog sources into domestic pig populations would remove the selection pressure for maintaining these genes (Dixon and Wilkinson, 1988). More recently, Alfonso *et al.* (2004) and Burrage *et al.* (2004) demonstrated that the ASFV multigene family in the left variable region of the genome encodes a host range determinant required for efficient replication and generalization of infection in pigs and ticks. However, no more information is yet available on specific determinants for tick/warthog hosts, as it was previously suggested. In addition, it is unknown if ASFV is able to come back from the domestic to the sylvatic cycle, although some authors suggest that recombination processes during co-infections in ticks may exist (Plowright, 1977; Dixon and Wilkinson, 1988).

Identification

Ticks in the family Argasidae are easily separated from those in the family Ixodidae because of the presence of the mouthparts (capitulum) in a ventral position, instead of in the terminal position as in members of the family Ixodidae, and by the absence of a dorsal scutum in the adults. Furthermore, the scutum (or conscutum, in males) is very prominent in the specimens of the family Ixodidae. The species in the genus *Ornithodoros* share the features of the family, as well as the absence of a lateral suture, which separate species in the genus *Argas* from species in the genus *Ornithodoros*. Other than this feature, the mamillate structures on the dorsal surface of the genera *Argas* and *Ornithodoros* are different.

The identification of species of *Ornithodoros* is difficult because most of the species are not morphologically different. Wallis and Miller (1983) conducted electrophoretic enzyme analysis to differentiate between some of the species. In Eurasia, the adults of most species of *Ornithodoros* have a similar morphology, rendering the identification on that stage difficult, and the larvae are the most useful stages to provide an adequate determination. According to the morphological features outlined by Filippova (1966), ticks in the *O. erraticus* complex can be separated by the setal dorsal complement in the larvae, which is reduced to only two pairs of dorsal setae. Currently, however, there is no accurate way to separate the different species in the complex. Furthermore, there are no adequate DNA data for the species found in the wide distribution range of the complex, and there is strong evidence pointing to a greater diversity than has currently been reported. The colour plate section contains two photographs (dorsal and ventral) of an adult (sex undetermined) of *O. erraticus* (Plate 3a,b).

2.2 Hard Ticks (Ixodidae)

Several physiological–behavioural and morphological characteristics distinguish the hard ticks, Ixodidae from the soft ticks, Argasidae. The life cycle of the hard ticks has only one nymphal stage. A single but comparatively large blood meal is ingested by the larva, the nymph and the adult female. The ixodid female, in particular, is capable of imbibing a very large amount of blood so that the abdomen becomes enormously distended. So in general, the ixodid ticks, except the males in some species, stay attached to the host for a period of several days. The morphology of ixodids has the following characteristics: (i) an anteriorly located capitulum visible from above; (ii) behind the capitulum a hard shield, the scutum; (iii) distinct sexual dimorphism in that the dorsum of the adult ixodid male is covered completely by the scutum, but the larva, nymph and adult female have a scutum that only covers the anterior part of the dorsal abdomen.

According to Barker and Murrell (2008), the Ixodidae encompass the following genera: *Amblyomma, Anomalohimalaya, Bothriocroton, Cornupalpatum, Cosmiomma, Dermacentor, Haemaphysalis, Hyalomma, Ixodes, Margaropus, Nosomma, Rhipicentor* and *Rhipicephalus*. It may be noted that five species previously in the genus *Boophilus* have been incorporated into the genus *Rhipicephalus* (subgenus *Boophilus*).

2.2.1 Genus *Ixodes*

This is the largest genus of the family Ixodidae, the hard ticks. The ticks in this genus are so-called prostriate hard ticks, which are characterized by the anal groove encircling the anus anteriorly. The genus is present on all continents, including Antarctica. The species of most medical and veterinary importance are *I. scapularis* and *I. pacificus* in North America, *I. ricinus* in Europe, western Asia and the Mediterranean coast of North Africa, and *I. persulcatus* in north-eastern Europe and northern Asia. In Eurasia, in view of their roles as vectors of zoonotic pathogens among the natural host animals, *I. hexagonus* (present in Europe only), *I. caledonicus*, *I. frontalis*, *I. trianguliceps* and *I. uriae* may also be considered as medically and epidemiologically important tick species. Among all *Ixodes* species, approximately 20% parasitize birds and 80% parasitize mammals (Kolonin, 2009).

The host-seeking behaviour of this group of ticks can be classified as endophilic or exophilic. Endophilic ticks remain hidden close to their hosts' nest or burrows and only attach to the host when it arrives, whereas exophilic ticks seek hosts by climbing up the vegetation waiting for them to pass. This host-seeking behaviour of exophilic *Ixodes* spp. should be distinguished from that of hunting ticks, e.g. *Hyalomma* spp., which actively run towards potential hosts. Because of the difference in host-finding behaviour, only exophilic species can be collected by flagging or dragging. In the genus *Ixodes*, exophilic species such as *I. ricinus* usually mate on the host. Endophilic species such as *I. hexagonus* normally mate in the hosts' refuge (Hillyard, 1996).

Ixodes ricinus

In Europe, the most well-known tick species is *I. ricinus*; this is often called the common tick, castor bean tick or sheep tick. It is the most commonly encountered tick species in most parts of central, western and northern Europe. The distribution of this species ranges from Ireland, Britain and Scandinavia, Finland and western Russia across continental Europe southwards to the Mediterranean area into northern Africa and eastwards to Iran.

I. ricinus is a three-host tick with a very wide host range (telotropic host behaviour), which includes lizards, many species of birds and small, medium-sized and large mammals. This tick species will readily attach to humans. Jaenson *et al.* (1994) listed two lizard species, 56 bird species and 29 mammal species as hosts recorded for *I. ricinus* in Sweden alone. The colour plate section contains photographs (dorsal and ventral views) of adult male (Plate 4a,b) and female (Plate 4c,d) specimens of *I. ricinus*.

In Europe, *I. ricinus* is the vector of the agents of several zoonoses, including Lyme borreliosis (at least eight *Borrelia* genospecies have been recorded from *I. ricinus*; Rudenko *et al.*, 2011), TBE, anaplasmosis (granulocytic ehrlichiosis, tick-bite fever) due to *Anaplasma phagocytophilum* and *Rickettsia helvetica*. *I. ricinus* also transmits *Babesia divergens* (the causative agent of babesiosis in cattle and a life-threatening infection in splenectomized humans), *B. microti* and *B. venatorum*, tularaemia due to *Francisella tularensis* and the louping ill virus (LIV).

Ixodes persulcatus

The taiga tick is an extremely important vector of the TBE virus, Lyme borreliosis spirochetes and other pathogens in Finland, the Baltic States and Russia, and also eastwards to Japan. In the Baltic region and north-western Russia, *I. persulcatus* overlaps with the range of *I. ricinus* and 'replaces' *I. ricinus* eastwards through Siberia. *I. persulcatus* transmits, in general, the same pathogens as *I. ricinus*, i.e. the TBE virus, Lyme borreliosis spirochetes, *A. phagocytophilum* and *B. divergens*. Both tick species are occasionally also vectors, and possibly long-term reservoirs, of *F. tularensis*. In *I. ricinus*, it is the nymphal stage which is the most important one for transmission of pathogens to humans. In contrast, the nymphs of *I. persulcatus* rarely bite humans; therefore it is the adult females of this species which play this main role of transmitting pathogens to humans.

Ixodes hexagonus

This tick is better known as the hedgehog tick and ranges through western, central and southern Europe, including Ireland, the UK, Norway, Sweden and Poland, to Portugal, Spain, Italy, Greece, Romania and Ukraine (Kolonin, 2009). It can easily be mistaken for *I. ricinus*, but there is a pronounced hexagonal shield in female ticks. The main hosts are carnivorous mammals (Canidae, Felidae and Mustelidae), and the hedgehog, i.e. mammals with a permanent dwelling (Arthur, 1963; Jaenson *et al.*, 1994; Kolonin, 2009). It is rarely found on sheep, cattle, horses and birds (Hillyard, 1996). This tick bites man occasionally and was a pest in underground shelters in the London area during wartime (Browning, 1944). *I. hexagonus* is a competent vector of *Borrelia burgdorferi s.l.* and transfers the spirochaete both trans-stadially and transovarially (Gern *et al.*, 1991); it is also possibly a competent vector of TBEV and *Babesia microti*. The species has an endophilic feeding behaviour and remains close to the shelter of its hosts. Although relatively abundant, it is never caught by dragging or flagging. The colour plate section contains photographs (dorsal and ventral views) of adult male (Plate 5a,b) and female (Plate 5c,d) specimens of *I. hexagonus*.

Ixodes canisuga

I. canisuga Johnston 1849 (= *I. crenulatus* Koch 1844; according to Kolonin, 2009) is often confused with *I. hexagonus*, which is usually more commonly encountered. *I. canisuga* is distributed from western, northern and southern Europe through Russia into Iran, Afghanistan, India and China (Hillyard, 1996; Kolonin, 2009). All stages are parasitic on rodents and carnivores, including badgers, foxes, mustelids, dogs and cats, i.e. hosts that regularly return to a burrow or lair (Hillyard, 1996). The colour plate section contains photographs (dorsal and ventral views) of an adult female of *I. canisuga* (Plate 6a,b).

Ixodes gibbosus

The main hosts of the larvae, nymphs and adults of *I. gibbosus* are sheep and goats, but the larvae and nymphs also feed on small mammals and birds. The species is distributed from former Yugoslavia through Turkey to Israel (Kolonin, 2009).

Ixodes uriae

This is the seabird tick, which inhabits islands and mainland coasts in the subarctic and temperate regions of the northern and southern hemispheres. *I. uriae* parasitizes colony-nesting marine birds. This tick will occasionally bite people visiting seabird colonies. It is a vector of the human Lyme disease spirochaete *Borrelia garinii* (Olsen *et al.*, 1993). The colour plate section contains photographs (dorsal and ventral views) of an adult male (Plate 7a,b) and female (Plate 7c,d) of *I. uriae*.

Ixodes frontalis

Several arboviruses (Bahig, Kemerovo, TBEV) have been isolated from the bird-parasitizing tick *I. frontalis*, which is an ornithophagous ectoparasite that does not bite humans. However, it is presumably indirectly of enzootic and epidemiological importance for TBEV.

Ixodes trianguliceps

I. trianguliceps is a nidicolous ectoparasite of small mammals. It is distributed throughout most of Europe eastwards through Ukraine to Georgia, Armenia and Azerbaijan (Kolonin, 2009). Although this tick does not bite humans it may indirectly be of medical and veterinary importance in view of its presumed role as a maintenance vector of TBEV, *Borrelia afzelii* and *Babesia microti*, among its natural small mammalian hosts (rodents and insectivores).

2.2.2 Genus *Haemaphysalis*

Few of the about 160 species of *Haemaphysalis* parasitize livestock, but those that do are economically important in Eurasia, where they inhabit humid, well-vegetated habitats. The size of unfed adults ranges from 2.0 to 4.0 mm. The genus is characterized by short mouthparts and a large rear-facing spur on trochanter

I. The capitulum is short and the palps are wide. Ornamentation and eyes are absent but festoons, usually 11 in number, are present on the posterior margin. Coxa I lacks external spurs. The anal groove passes behind the anus. Nymphs of *Haemaphysalis* possess festoons but lack ornamentation and eyes (Hillyard, 1996). Species of the genus *Haemaphysalis* are three-host ticks, with the larvae and nymphs feeding on small mammals and birds. Adults infest larger domestic and wild mammals and, importantly, livestock; they are especially adapted to domestic cattle and, to a lesser extent, to sheep and goats.

Haemaphysalis punctata

The palps of the females and males are short and wide, projecting laterally beyond the basis capituli. The basis capituli of females is rectangular, about twice as broad as long, but it is only one and a half times broad as long in males. The scutum of males is longer than it is broad; in females it is about as broad. Unlike the female, the basis capituli of nymphs, in dorsal view, has sides with sharp lateral points (Hillyard, 1996). The colour plate section contains photographs (dorsal and ventral views) of an adult male (Plate 8a, b) and an adult female of *Haemaphysalis* (*Ha.*) *punctata* (Plate 8c,d). *Ha. punctata* has the largest distribution range in Eurasia. It is also known as the red sheep tick and is found in a very wide variety of habitats, providing that a high degree of humidity is present. This tick has many similarities with *I. ricinus* in its distribution, but is distinct in its physical characters from that species. This species is widely distributed throughout Europe, and it occurs in cold and wet sites in most Mediterranean countries (Estrada-Peña *et al.*, 2004). *Ha. punctata* is one of the two species of the genus *Haemaphysalis* that are likely to be found on livestock in North Africa, and it needs to be distinguished from *Ha. sulcata* (Bouattour *et al.*, 1999). It is associated with wild and domestic ruminants (Encinas Grandes, 1986; Manilla, 1998a; Cringoli *et al.*, 2002). It is frequently found where sheep, goats and cattle feed in certain open forests and shrubby pastures in the greater part of Europe, including southern Scandinavia and Britain. The species is also found on horses, deer and hares, and occasionally it attaches to humans (Hillyard, 1996). Engorgement on the host may take 6–30 days to complete and each female lays 3000–5000 eggs. Immature ticks infest birds (e.g. crows, magpies, pheasants, sparrows, blackbirds), hedgehogs, rodents and lizards. In humid climates, adults may be found feeding from October to March. Under favourable conditions the life cycle can be completed within 1 year but it usually takes 3 years.

Ha. punctata is a vector of many *Babesia* species (*B. major*, *B. bigemina*, *B. bovis*, *B. motasi*) and *Theileria* species (*T. ovis*, *T. mutans*, *T. recondite*). It is also able to transmit *F. tularensis* and *Coxiella burnetii*. Specimens of this species have been found to be infected by tick-borne encephalitis virus (TBEV) and Bhanja virus. The tick is also suspected of being able to transmit Crimean-Congo haemorrhagic fever virus (CCHFV), *Listeria monocytogenes* and *Brucella abortus*. The bite of *Ha. punctata* may cause tick paralysis (Hillyard, 1996).

Haemaphysalis concinna

Ha. concinna is widely distributed in Europe (Walter *et al.*, 1986; Sréter *et al.*, 2003). The species is restricted to temperate broadleaved and mixed forests where the environment is relatively unaltered, and the distribution is of relict type (Hoogstraal, 1972). The most important hosts for adult *Ha. concinna* are roe deer, but the ticks may occur on cattle, horses, dogs, cats and hares. The sexes are usually found attached close together and copulation occurs on the host in April. Males can be more numerous than females. The larvae and nymphs feed on deer, hedgehogs, rodents, reptiles and birds. Nymphs have also been removed from humans (Hillyard, 1996). The larvae and nymphs may be found in many months of the year; they are most numerous on small mammals in June and July. In Hungary, *Ha. concinna* showed peak activity in May (Hornok, 2009). *Ha. concinna* is known to be a carrier of *F. tularensis*. The colour plate section contains photographs (dorsal and ventral views) of an adult male of *Ha. concinna* (Plate 9a,b).

Haemaphysalis inermis

Ha. inermis, known as the winter tick, is a rather small tick. The capitulum of females differs from the normal form in the genus. The palps are relatively long and narrow, and when folded they do not project laterally beyond the relatively wide basis capituli. The scutum is broader than it is long. All coxae have a small spur. In males, the palps lack lateral projections and there are 11 well-marked festoons on the posterior margin (Hillyard, 1996). The colour plate section contains a photograph of an adult female (ventral view, Plate 10) of *Ha. inermis*.

This tick is present in some foci within many European countries (e.g. France, Poland, the Mediterranean region and the Balkan Peninsula). *Ha. inermis* is winter active and inhabits deciduous and mixed forests, grasslands and mountain slopes. The species is primarily associated with deer, with the immature stages found on small mammals, rodents and insectivores. The adults also feed on large and small domestic and wild mammals (e.g. cattle, horses, sheep, fox and hedgehogs). Humans may also be attacked (Hillyard, 1996). Copulation takes place on the host but the sexes do not attach alongside each other. The species has peak activity in November and December in Hungary (Hornok, 2009). The immature stages are rarely found in nature, primarily because they feed and drop off the host within hours – an exceptionally short period for engorgement. However, the nymph's rapid feeding contrasts with the very long time needed to develop to the next stage; nymphs may take a year or more before moulting to the adult stage. *Ha. inermis* has been noted as causing tick paralysis. In Poland it is suspected of carrying the virus that causes tick-borne encephalitis (TBE) (Hillyard, 1996).

Haemaphysalis sulcata

Ha. sulcata is one of the two species of the genus *Haemaphysalis* that are likely to be found on livestock in a wide range of climates in the Mediterranean region; it needs to be distinguished from *Ha. punctata*. In Europe, this tick species has been detected in several countries (Cyprus, Greece, Romania, former Yugoslavia, Hungary, Italy, France and Spain), always under conditions of steppe vegetation. In North Africa, the species occurs in Morocco, Algeria, Tunisia and Libya, as well as in parts of Turkey (Estrada-Peña *et al.*, 2004). Adults of the species feed on ungulates. Small ruminants, particularly sheep and goats, are the most common hosts. The preferred feeding sites for the ticks are the neck, shoulder and back. Adults are active in the autumn and winter between October and March. Immature stages feed on reptiles such as lizards, but the range of hosts of larvae and nymphs is similar to that of *Ha. punctata*. Larvae and nymphs feed on reptiles in the spring and summer (April and July). Probably because of misidentification in the past, the role in the transmission of pathogens by this species is not yet clearly defined. It is suspected of transmitting *T. annulata* and *Anaplasma ovis*. The Bhanja virus has also been isolated in *Ha. sulcata* during a survey performed in Bulgaria (Pavlov *et al.*, 1978).

Other species of *Haemaphysalis* are potentially important in Europe and the Mediterranean region. For example, examination of animals in El Arich city and El Hassanah centre in Egypt revealed *Ha. parva* on sheep, *Ha. sulcata* on goats, and *Hy. lusitanicum* on cattle (Mazyad and Khalaf, 2002). *Ha. franchinii* were also identified in small numbers in Tunisia (Bouattour *et al.*, 1999). Another important species is the rabbit tick, *Ha. leporispalustris*.

2.2.3 Genus *Hyalomma*

Hyalomma species are large ticks with pale rings on most segments of their legs. They also have long mouthparts which project to the anterior of the body and, in males, anal plate (Manilla, 1998b; Estrada-Peña *et al.*, 2004).

The hard tick genus *Hyalomma* is a complex of a few species that exhibit an almost endless variation of morphology. Its original centre of dispersal was probably Iran or southern Russia. Genetic instability may in part account for the wide morphological differences found in many specimens (Hoogstraal, 1956).

For identification purposes, there is a need to consider only the differential features not subject to change on an individual basis; these are mainly seen in replete ticks. To date, the most reliable diagnostic features are actually scanty, especially for the females.

Adult *Hyalomma* ticks share the following common features: a basis capituli that is usually rectangular, clearly wider than long; a long hypostome with a rounded tip and 3/3 denticles; long palps with the 2nd article longer than the 3rd; shiny eyes with a surrounding orbit (a circular groove surrounding the eye), which are often convex and placed in the confluence between the anterior and the posterior edges of the scutum in females and in the corresponding place in males; a dark scutum rich in punctations and grooves; festoons that are more evident in unfed ticks; long, strong and banded legs with two large spurs on the first pair of coxae; and adanal, subanal and accessory plates in the male.

In most cases, it is not possible to identify immature stages to the species level using only morphology (Manilla, 1998b).

Hyalomma ticks are often the most abundant tick parasites of livestock, including camels, in warm, arid and semiarid, generally harsh, lowland and middle altitude biotopes, and in biotopes with long dry seasons, from central and south-west Asia to southern Europe and southern Africa. Extraordinary survival factors play a large part in permitting these ticks to exist and even thrive where few or no others live.

Many *Hyalomma* species have been redescribed recently based on morphological characteristics (Apanaskevich, 2003; Apanaskevich and Horak, 2005). However, these characteristics may vary with populations, host and other environmental factors, making the ticks difficult to identify. *Hyalomma* ticks are mainly two- and three-host species (Matthysse and Colbo, 1987).

Hyalomma marginatum

Hyalomma marginatum is a large tick that can reach 18 mm length in females and 6 mm in males. Its dorsal shield is wider than long, brown in colour and rich in punctations of different size. The legs have white-yellowish rings, and the body of the replete male has subanal triangular plates that can be visible beyond the posterior edge (Manilla, 1998b). The colour plate section contains photographs (dorsal and ventral views) of adult male (Plate 11a,b) and female (Plate 11c,d) specimens of *Hy. marginatum*.

This species belongs to a hard tick species complex occurring in southern Europe, southern Asia and most of Africa. It is a characteristic species of steppe, savannah and lightly wooded hill and valley biotopes with fairly low humidity. In particular, *Hy. marginatum* inhabits the Mediterranean climatic zone, as well as steppe and foothill landscapes (south-western Palaearctic Region), from the Caspian Sea to southern Ukraine and Bulgaria, and spreads westward to Spain, Portugal, Morocco and Algeria (Manilla,1998). Sporadic reports of the species in the humid biotopes of northern Europe can be explained as introductions by migratory birds that are frequently included in the host range of the immature stages. *Hy. marginatum* is a two-host tick and the immature stages stay for a long time on the host (12–26 days) which enables them to be passively transported by migrating birds over hundreds or even thousands of kilometres (Rechav *et al.*, 1987; Manilla, 1998b).

Hy. marginatum is the main vector of CCHFV in Eurasia. The virus can also be transmitted transovarially with variable efficiency depending upon the viral strain (Zeller *et al.*, 1997). The ticks are also a vector of *Babesia caballi* and *T. equi*, and of *T. annulata* (Viseras and García-Fernández, 1999). *T. annulata* is the agent of tropical theileriosis that occurs in northern Africa, including the sub-Saharan territories, Sudan and Eritrea, southern Europe, the Near and Middle East, Central Asia, India and northern China. Its distribution reaches the Far East, where it overlaps with theileriosis caused by *T. sergenti* infection (Pipano and Shkap, 2004). *Hy. marginatum* is also involved in the maintenance of the enzootic cycle of *C. burnetii*.

Hyalomma anatolicum and *Hyalomma excavatum*

The study of the ecology, development and epizootic role of these two similar species is very difficult. Their taxonomic status is not even uniformly agreed on, and their occurrence within the same geographical area, together with the presence of hybrid forms, morphological variations and impossibility of distinguishing the larvae and nymphs, complicate this work. Ecological conditions apparently determine which form will develop in a given location (Liebish *et al.*, 1989; Apanaskevich, 2003).

A recent study that compared the morphology of each stage of the two species using several specimens belonging to different collections allowed researchers to definitely separate the two specific identities of *Hy. anatolicum* and *Hy. excavatum* (Apanaskevic and Horak, 2005).

HYALOMMA ANATOLICUM *Hy. anatolicum* is common in the southern Mediterranean area and also in eastern Africa (Sudan, Somalia and Ethiopia) and central Asia (Ahmed *et al.*, 2007; Rasulov, 2007). This is a small tick; the length and breadth of the conscutum is usually less than 4 mm and 2.3 mm in the male, and 2 mm and 1.9 mm in the female, respectively. The scutum, gnathosoma and coxae are light yellowish- or reddish-brown. Ivory coloured marbling is absent on the scutum, with large and sparse punctations in the caudal field.

The species transmits a variety of protozoan parasites trans-stadially and transovarially and is a threat to animal improvement programmes. It transmits the protozoans *T. annulata* (which causes tropical theileriosis in cattle) *T. lestoquardi* (which causes malignant ovine theileriosis), *B. caballi* and *T. equi* (which cause equine piroplasmosis). *Hy. anatolicum* is also a vector of CCHFV to humans.

HYALOMMA EXCAVATUM *Hy. excavatum* is larger than *Hy. anatolicum*. The scutum, gnathosoma and coxae are dark red- or black-brown. Ivory coloured enamelling is diffuse on the dorsal aspects of the leg segments. In the female, the cervical and lateral scutal grooves are deeper. The colour plate section contains photographs (dorsal and ventral views) of adult male (Plates 12a,b) and female (Plates 12c,d) specimens of *Hy. excavatum*.

The species is adapted to the Mediterranean and steppe climatic regions of North Africa, and to steppe climatic regions elsewhere in its wide range. *Hy. excavatum* is often less commonly found on livestock than *Hy. anatolicum*, but it has a wider geographical distribution and is found in much of North Africa from Mauritania to Egypt, and also into Sudan, Ethiopia and Eritrea. It is not found commonly south of the Sahara, but it is reported eastwards through to Iran and Turkmenistan.

The biological differences between the two species influence their vector competences. *Hy. excavatum* is questioned as a vector of *Theileria*, because its vectorial capacity is affected by its feeding habits on rodents (Manila, 1998b, Walker *et al.*, 2003).

Hyalomma scupense

Two years ago, and despite some morphological differences, the *Hyalomma* subspecies that were known as *Hy. detritum detritum* and *Hy. detritum scupense* were brought together to comprise a unique species renamed as *Hy. scupense* (Guglielmone *et al.*, 2009).

Hy. scupense is a two-host tick species that feeds on large domestic ungulates. Compared with other *Hyalomma* ticks, it is small, less punctated and does not have pale rings on its legs. The former subspecies *Hy. d. detritum* and *Hy. d. scupense* are different in their body size (*Hy. d. detritum* is generally smaller than *Hy. d. scupense*), the length of the legs, and the thickness of the tails of the spiracle plates, which are narrower in *Hy. d. detritum* than in *Hy. d. scupense* (Manilla, 1998b; Walker *et al.*, 2003). The colour plate section contains photographs (dorsal and ventral views) of adult male (Plate 13a,b) and female (Plate 13c,d) specimens of *Hy. scupense*.

This *Hyalomma* tick spreads in central Asia from northern India to the Middle East, and is also found in the southern Balkans, southern Europe and along the Mediterranean coast of Africa as far as Algeria and Morocco in the west. It is present in sub-Saharan Africa in north-central Sudan as well, perhaps having invaded from the Red Sea coast or via the Nile river valley (Manilla, 1998b).

Hy. scupense is able to transmit *T. annulata* to cattle and *T. equi* to horses, mules and donkeys. The species is also a vector of *C. burnetii*.

2.2.4 Genus *Amblyomma*

Amblyomma species are large ticks with long mouthparts and pale rings on their legs. They usually have ornate eyes and coxal spurs. Like all Ixodidae, they are protected by a rigid shield that completely covers the dorsal body surface in the male; female shields are smaller, enabling significant distension of the abdominal wall during blood feeding.

In some species, the shield is richly coloured. The long mouthparts are clearly visible on the dorsal view of the body, and these make *Amblyomma* ticks especially difficult to remove. Manual removal frequently causes serious wounds which may become secondarily infected by bacteria or screw worms (Walker *et al.*, 2003).

According to the Ixodidae evolutionary hypothesis formulated by Hoogstraal and Kim (1985), the tick genera belonging to the metastriate group originated from a group of ticks that resemble the present subfamily Amblyomminae. These ticks reached their greatest evolutionary success by simultaneously adapting themselves to the dinosaurs during the geological time from the end of the Palaeozoic to the end of the Cretaceous period. Nowadays, adults and immature stages of 37 of the around 102 known *Amblyomma* species parasitize reptiles. Moreover, reptiles, together with small mammals and birds, are often included in the host range of immature *Amblyomma* ticks that feed on mammals when they moult to become adults (Manilla, 1998).

The genus *Amblyomma* is generally confined to the tropics and subtropics. *Amblyomma* (*Am.*) *variegatum* and *Am. hebraeum* are the most important species of the genus in Africa. They occur in different geographical areas, although they are adapted to the same type of habitat. Their presence is affected by climate and the surrounding environment according to their developmental stage. Even though individuals of both species are active throughout the year, in the African ranges of distribution a certain seasonality is reported for both species. Generally, adults are more numerous during the rainy season, while larvae and nymphs start to be more active at the end of the rainy season. *Am. variegatum* and *Am. hebraeum* are morphologically similar. *Am. variegatum* adults have long mouthparts and banded legs like *Am. hebraeum*, but the colour pattern on the scutum is different and they have beady eyes. In addition, the festoons of the males are dark brown in colour.

Am. variegatum (the tropical bont tick, TBT) is the most widespread species. This tick originates from Africa and has spread widely south of the Sahara desert. It was first introduced into the Caribbean in around 1828, following the importation of cattle from Senegal, and later spread to the rest of the West Indies (Uilenberg *et al.*, 1984, Pegram *et al.*, 2005). *Am. variegatum* was similarly introduced from Africa into the Arabian Peninsula and Mauritius (Norval and Horak, 2004). It is seldom reported in the Mediterranean, but was probably introduced there by migratory birds (Albanese *et al.*, 1971, Papadopolous *et al.*, 1996).

Am. hebraeum is exclusively a southern African tick. It occurs in South Africa and in southern Mozambique. Its distribution also extends inland through the lowveld areas of Swaziland and eastern Botswana, and to southern and western Zimbabwe. The distribution of *Am. hebraeum* is affected by interspecific competition with *Am. variegatum*.

Both *Am. variegatum* and *Am. hebraeum* are the main vectors of *Ehrlichia ruminantum*, causing heartwater, a disease that is one of the main constraints to livestock production in sub-Saharan Africa. These ticks also transmit *T. velifera* and *T. mutans* to cattle. In the tropics, *Am. variegatum* is also strictly associated with dermatophilosis and the transmission of African tick bite fever (caused by *R. africae*) (Norval and Horak, 2004).

In America, *Am. maculatum* and *Am. americanum* are two of the main species of *Amblyomma* ticks. *Am. maculatum* is an important pest of livestock and is present in the Gulf coast area. It has been experimentally demonstrated to be a good vector for the agents transmitting heartwater (Uilenberg, 1982). *Am. americanum*, the lone-star tick, is abundant in the southern USA in Texas and Missouri and up the Atlantic Coast, ranging northward into New Jersey. It is also a notorious pest in Mexico as well as in Central and South America. Its distinctive feature is the silver spot on the scutum (the 'lone star'). *Am. americanum* is the vector of tularaemia, Rocky Mountain spotted fever, Q fever, and Lyme disease, and it may cause tick paralysis in humans and dogs. It is also a vector of monocytic ehrlichiosis caused by *E. chaffeensis* and *E. ewingii* in humans and dogs.

2.2.5 Genus *Rhipicephalus*

The genus *Rhipicephalus* is one of the largest of the family Ixodidae. Species in this genus have some features in common: a short hypostome and palps, a usually hexagonal basis capituli and the presence of eyes and festoons and anal plate in the males. With the exception of few species, they are inornate, hence the common name of the 'brown ticks'. This is mainly an African genus (Walker *et al.*, 2000).

Adults of most species parasitize wild and domestic artiodactyls (swine and ruminants), perissodactyls (horses) or carnivores. Immature stages feed mostly on smaller mammals; however, of those that parasitize rodents and of those that parasitize artiodactyls, a few feed on the same host as the adults. The life cycle is typically three host, but in the Mediterranean climatic zone (long, warm summer with low rainfall), *Rhipicephalus* (*Rh.*) *bursa* is a two-host species (Merck, 2010).

Many *Rhipicephalus* species have long been difficult to identify or have been incorrectly identified. Current concepts of tick phylogeny, taxonomy and nomenclature are being revised and expanded based on molecular analyses. This ongoing work is likely to expand and alter the current understanding of the phylogeny and evolution of the subfamily Rhipicephalinae. On the basis of those studies, the genus *Boophilus* has recently been included in the genus *Rhipicephalus*, but even though *Boophilus* species now belong to the same genus as *Rhipicephalus* species, the morphology and biology of the *Boophilus* species are significantly different.

The genus *Rhipicephalus* comprises 79 species, including the five species that were formerly in the genus *Boophilus*. Probably the species of *Rhipicephalus* with the largest geographical range is *Rh. sanguineus*, the 'kennel tick' or 'brown dog tick', which has travelled worldwide with domestic dogs. It is now established in buildings as far north as Canada and Scandinavia, and as far south as Australia. It can be found almost worldwide, mainly within latitudes 35°S and 50°N (Dantas-Torres, 2008). Although this species feeds primarily on dogs, it can be found on a diverse range of wild and domestic animals, including humans (Dantas-Torres, 2008). However, in urban situations everywhere within the range of distribution, dogs are virtually the only hosts of the immature stages and the adults. *Rh. sanguineus* is active throughout the year in the tropics and subtropics, but only from spring to autumn in temperate zones. Newly active adults and nymphs are frequently seen climbing walls from floor-level cracks (Merck, 2010). *Rhipicephalus* ticks are recognized worldwide as vectors of many important disease agents of animals and humans – diseases such as: Mediterranean spotted fever and other rickettsioses, Q fever, monocytic ehrlichiosis, canine babesiosis, hepatozoonosis, Nairobi sheep disease and East coast fever.

Rhipicephalus sanguineus, Rhipicephalus camicasi *and* Rhipicephalus turanicus

Rh. sanguineus is universally known as 'the kennel tick' (Manilla 1998c; Walker *et al.*, 2000; Estrada-Peña *et al.*, 2004). It is a medium sized, pale yellowish-brown or reddish-brown tick. The *Rh. sanguineus* group comprises several tick species, and the biosystematic status of the majority of them has been confused (Walker *et al.*, 2000; Estrada-Peña *et al.*, 2004). Two species of the group that closely resemble *Rh. sanguineus* are *Rh. camicasi* and *Rh. turanicus*. Females of *Rh. sanguineus* are differentiated from those of *Rh. camicasi* and *Rh. turanicus* by the genital aperture, which usually has a broad V shape in *Rh. sanguineus* but a narrow U shape in *Rh. camicasi* and *Rh. turanicus*. Both sexes of *Rh. sanguineus* have spiracle plates with tails which are narrow, and less than the width of the adjacent festoon. In *Rh. turanicus*, these tails are broad. Males of *Rh. sanguineus* do not have a depression of the cervical fields compared with the small depression that there is in *Rh. camicasi* (Walker *et al.*, 2000; Estrada-Peña *et al.*, 2004). The adults may be separated by a combination of adanal and spiracular plate characteristics, the dorsal punctation pattern (males), and by the feature of the genital aperture (females). To study the morphology of the genital aperture, it is necessary to dissect and mount it, and allow adequate cleaning to remove the chitinized structures. The colour plate section contains photographs (dorsal and ventral views) of adult male (Plate 14a,b) and female (Plate 14c,d) specimens of *Rh. sanguineus*. There are also photographs (dorsal and ventral views) of adult male (Plate 15a,b) and female (Plate 15c,d) specimens of *Rh. turanicus*.

Rhipicephalus bursa

Adults of *Rh. bursa* are large ticks. Males have very conspicuous anal plates, almost in a regular triangular shape, with a projection easily seen in the median portion and protruding to the anus (medial). Both males and females have a noticeable large dorsal punctation pattern, deeper than those observed in other species of the genus collected in the Mediterranean region. The colour plate section contains photographs (dorsal and ventral views) of adult male (Plate 16a,b) and female (Plate 16c,d) specimens of *Rh. bursa*.

Rhipicephalus annulatus (*previously known as* Boophilus)

Rhiphicephalus (Boophilus) ticks are unique in that their entire life cycle from the larva to the engorged, mated adult is confined to a single host. Females drop to the ground to oviposit. This single-host type of life cycle has numerous biological advantages. It also allows for particularly easy control by dipping infested animals. The boophilid type of life cycle eliminates danger-ridden periods between two or three different kinds of hosts, possibly in inhospitable areas and for indefinite periods. The predilection of these ticks for large domestic animals particularly favours widespread dispersal and survival, not only within a continent but also from continent to continent on imported hosts. Cattle are the chief hosts throughout the world; horses, other domestic stock, and wild antelopes and deer are less frequently attacked. Other wild animals are not commonly infested. The veterinary importance of these ticks is considerable and they are vectors of important pathogens of livestock. The only species present in the Mediterranean basin is *Rh. annulatus*. This is an important pest of domestic cattle in every kind of Mediterranean environment. Most interesting is that this tick is well adapted to xerophilic areas with periodic rains. Temperature seems to be the only restrictive factor in the dissemination of the tick towards northern latitudes. Hence it is widespread at scattered points across the Mediterranean basin. *Rh. annulatus* is the vector of some prominent pathogens, such as several species of *Babesia*, most importantly *B. bovis*, which is a potentially dangerous pathogen of cattle. The colour plate section contains photographs (dorsal and ventral views) of adult male (Plate 17a,b) and female (Plate 17c,d) specimens of *Rh. (Bo.) annulatus*.

2.2.6 Genus *Dermacentor*

The genus *Dermacentor* is small, with about 30 species of hard ticks, most of which are found in the New World. They are medium sized to large ticks, usually with ornate patterning, and resemble those of *Hyalomma*. Most species of *Dermacentor* are three-host ticks, but a few are one-host ticks. Two species, *D. marginatus* and *D. reticulatus*, are widely distributed in Europe (Hillyard, 1996).

Dermacentor marginatus

D. marginatus, also known as the 'ornate sheep tick', has an ornate scutum. It is distinguished from *D. reticulatus* by its palp which lacks a prominent rear-facing spur. The length of unfed adults is 4.6–5.8 mm; an engorged female measures up to 1.5 cm. The unfed nymph is 1.4–1.8 mm in length.

This species is one of the two representatives of the genus *Dermacentor* that infest domestic animals in the Mediterranean region. *D. marginatus* occurs widely in Europe, but has a restricted distribution in North Africa (Nosek, 1972). In the European part of its distribution, this tick species is restricted to areas with dense bush and tree cover. It is common under oak and pine vegetation. *D. marginatus* has more thermophilic requirements than other species that can occur together, such as *I. ricinus* and *Ha. punctata*. In North Africa, *D. marginatus* is restricted to the same type of habitat as *I. ricinus*. Thus it is found in the cooler and more humid parts of the Mediterranean climatic region associated with the Atlas Mountains, and is restricted to small areas of Morocco, Algeria and Tunisia (Estrada-Peña *et al.*, 2004). The ticks inhabit pastures, temperate forest and grassland, having a preference for xerophilic vegetation.

The activity of the species varies considerably according to region. Adults are active during the spring, early summer and autumn. In a study in Hungary, *D. marginatus* was found to be most numerous in February and March (Hornok, 2009). The larvae appear in June and the nymphs in July. A generation from egg to adult usually develops within a year. Adults feed on large mammals such as domesticated and wild ruminants, but also on dogs, hares, hedgehogs and humans. Wild boars *(Sus scropha)* are also important wild hosts. The larvae and nymphs feed mostly on small mammals such as rodents and birds (Nosek, 1972; Hillyard, 1996).

D. marginatus is a vector of *C. burnetii*, *R. conori*, *R. slovaca* and *R. sibirica sibirica*. *F. tularensis* may also be transmitted by this tick species (Nosek, 1972; Hillyard, 1996; Raoult *et al.*, 2002). The colour plate section contains photographs (dorsal and ventral views) of adult male (Plate 18a,b) and female (Plate 18c,d) specimens of *D. marginatus*.

Dermacentor reticulatus

D. reticulatus (syn. *D. pictus*) is also known as the 'ornate dog tick', the 'marsh tick' or the 'ornate cow tick'. This metastriate tick species has been confused with others in the genus because it has considerable morphological variability (Estrada-Peña and Estrada-Peña, 1991a,b; Zahler and Gothe, 1997). The length of unfed adults is 3.8–4.8 mm; an engorged female measures up to 1.0 cm. The unfed nymph is 1.4–1.8 mm in length. The colour plate section contains photographs (dorsal and ventral views) of adult male (Plate 19a,b) and female (Plate 19c,d) specimens of *D. reticulatus*.

The distribution area of the species extends from France and south-western England in the west to Central Asia in the east (Gilot *et al.*, 1989). *D. reticulatus* is recorded in almost all European countries but formerly was relatively rare in cold continental climates. In western and Central Europe, it does not occur north of 53–54°N latitude, for example in Scandinavia, nor in the Mediterranean climate zone. In East Europe, however, it may occur as far north as St Petersburg (60°N). Within this large distribution area, the occurrence of this species is highly focal (Dautel *et al.*, 2006). In the last two decades, several European countries have noticed changes in the distribution of *D. reticulatus* (Szymanski, 1987; Zahler and Gothe, 2001; Sréter *et al.*, 2005; Dautel *et al.*, 2006; Bullova *et al.*, 2009). The spread of *Dermacentor* ticks is thought to be linked to importation (Cochez *et al.*, 2011) and climate changes, notably the shortening of the winter period and increasing minimum temperatures. The ticks are absent from Northern Africa (Estrada-Peña *et al.*, 2004). Most important for the survival and activity of *D. reticulatus* are foci of adequate microclimate (Estrada-Peña, 2008). Therefore, the tick is abundant in open, unused habitats both in the middle and on the periphery of villages, in unused gardens, on abandoned prairies, as well as in grazing pastures and woodlands. *D. reticulatus* is absent from mountain regions, but very abundant in low-altitude hills. Favourable landscapes for the survival of the species occur where different forms of cultivation overlap and urbanization modifies landscape structures, such as is found in peri-urban colonies.

The tick has a 1 or 2 year life cycle depending on the environmental conditions. It prefers cold regions with an adequate amount of air relative humidity, so it is absent in the Mediterranean region where the relative humidity is not sufficient for its survival. The main activity period for adults is the spring, with a secondary peak in autumn, but it varies considerably according to region (Szymanski, 1987). For example, *D. reticulatus* was most numerous in September and October in Hungary (Hornok, 2009), and high numbers of questing adults were found in Belgium during winter when temperatures were just above 0°C (Madder, unpublished data). Adults parasitize larger domestic and wild mammals, cattle, horses, sheep, goats and pigs, but only occasionally bite humans. *D. reticulatus* is one of the most common tick species on dogs and wild canids wherever it exists (Gilot *et al.*, 1989; Földvári and Farkas, 2005). Copulation takes place on the host. The female feeds for 9–15 days before dropping to the ground to lay approximately 4000 eggs in sheltered spots. The larvae hatch from the eggs after 2–3 weeks and will feed for approximately 2 days on the host, then drop to the ground and moult to the nymphal stage. This stage feeds over a period of several days, drops off and moults to become an adult. The larvae and nymphs feed on a variety of small mammals, such as small rodents and carnivores, and occasionally birds (Nosek 1972; Hillyard, 1996). The immature stages are generally active from midsummer to late autumn.

D. reticulatus is known to be a vector of *Babesia* spp. (*B. canis*, *B. divergens*, *B. ovis*, *B. caballi*), *T. equi*, *A. ovis*, *R. conori* and *R. raoultii* (formerly *Rickettsia* sp. strain RpA4) (Nosek 1972; Martinod and Gilot, 1991; Hillyard, 1996; Zahler and Gothe, 2001; Dautel *et al.*, 2006). In Portugal, recent studies have confirmed the presence of *F. tularensis* in *D. reticulatus* (Lopes de Carvalho *et al.*, 2011).

References

Ahmed, J., Alp, H., Aksin, M. and Seitzer, U. (2007) Current status of ticks in Asia. *Parasitology Research* 101, 159–162.

Albanese, B., Smiraglia, C. and Lavagnino, A. (1971) Notizie sulle zecche di Sicilia con segnalazione di *Hyalomma detritum* e *Amblyomma variegatum*. *Rivista di Parassitologia* 32, 273–276.

Alfonso, P., Rivera, J., Hernaez, B., Alonso, C. and Escribano, J.M. (2004) Identification of cellular proteins modified in response to African swine fever virus infection by proteomics. *Proteomics* 4, 2037–2046.

Apanaskevich, D.A. (2003) Differentiation of closely related species *Hyalomma anatolicum* and *H. excavatum* (Acari: Ixodidae) based on a study of all life cycle stages, throughout entire geographical range. *Parazitologia* 37, 259–280.

Apanaskevich, D.A. and Horak, I. (2005) The genus *Hyalomma*. II The taxonomic status of *H. (Euhyalomma) anatolicum* Koch 1844 and *H. (Euhyalomma) excavatum* Koch 1844 with the redescription of all stages. *Acarina* 13, 181–197.

Arthur, D.R. (1963) *British Ticks*. Butterworths, London.

Barker, S.C. and Murrell, A. (2008) Systematics and evolution of ticks with a list of valid genus and species names. In: Bowman, A.S. and Nuttall, P.A. (eds) *Ticks: Biology, Disease and Control*. Cambridge University Press, Cambridge, UK, pp. 1–39.

Basto, A.P., Portugal, R.S., Nix, R.J., Cartaxeiro, C., Boinas, F., Dixon, L.K., Leitao, A. and Martins, C. (2006) Development of a nested PCR and its internal control for the detection of African swine fever virus (ASFV) in *Ornithodoros erraticus*. *Archives of Virology* 151, 819–826.

Boinas, F.S. (1994) The role of *Ornithodoros erraticus* in the epidemiology of African swine fever in Portugal. PhD Thesis, University of Reading, Reading, UK.

Bouattour, A., Darghouth, M.A. and Daoud, A. (1999) Distribution and ecology of ticks (Acari: Ixodidae) infesting livestock in Tunisia: an overview of eight years field collections. *Parassitologia* 41(Suppl. 1), 5–10.

Browning, E. (1944) On the discovery of numbers of males of *Ixodes hexagonus* Leach (Acari: Ixodidae). *Proceedings of the Linnean Society of London* 156, 96–99.

Bullova, E., Lukan, M., Stanko, M. and Petko, B. (2009) Spatial distribution of *Dermacentor reticulatus* tick in Slovakia in the beginning of the 21st century. *Veterinary Parasitology* 165, 357–360.

Burrage, T.G., Lu, Z., Neilan, J.G., Rock, D.L. and Zsak, L. (2004) African swine fever virus multigene family 360 genes affect virus replication and generalization of infection in *Ornithodoros porcinus* ticks. *Journal of Virology* 78, 2445–2453.

Chappard, C., Sainte-Laudy, J. and Emonot, A. (1996) Anaphylactic reactions caused by a pigeon's parasite (*Argas reflexus*). *Biomedical Research* 7, 11–14.

Cochez, C., Lempereur, L., Madder, M., Claerebout, E., Simons, L., De Wilde, N., Linden, A., Saegerman, C., Heyman, P. and Losson, B. (2011) Foci report of indigenous *Dermacentor reticulatus* populations in Belgium and a preliminary study on associated babesiosis pathogens. *Journal of Medical Entomology*, doi: 10.1111/j.1365-2915.2011.00998.x (early view).

Cringoli, G., Otranto, D., Testini, G., Buono, V., Di Guilio, G., Traversa, D., Lia. R., Veneziano, V. and Puccini, V. (2002) Epidemiology of bovine tick-borne diseases in southern Italy. *Veterinary Research* 33, 421–428.

Dantas-Torres, F. (2008) The brown dog tick, *Rhipicephalus sanguineus* (Latreille, 1806) (Acari: Ixodidae): from taxonomy to control. *Veterinary Parasitology* 152, 173–185.

Dautel, H. and Knülle, W. (1997) Life cycle and seasonal development of postembryonic *Argas reflexus* (Acari, Argasidae) at two thermally different sites in central Europe. *Experimental and Applied Acarology* 21, 697–712.

Dautel, H., Kahl, O. and Knülle, W. (1991) The soft tick, *Argas reflexus* (Acari, Argasidae) in urban environments and its medical significance in Berlin (West). *Journal of Applied Entomology* 11, 380–390.

Dautel, H., Kahl, O., Scheurer, S. and Knülle, W. (1994) Seasonal activities of the pigeon tick, *Argas reflexus* (Acari, Argasidae) in Berlin, Germany. *Folia Parasitologia* 41, 155–160.

Dautel, H., Scheurer, S. and Kahl, O. (1999) The pigeon tick, *Argas reflexus*: its biology, ecology, and epidemiological significance. *Zentralblatt für Bakteriologie* 289, 745–753.

Dautel, H., Dippel, C., Oehme, R., Hartelt, K. and Schettler, E. (2006) Evidence for an increased geographical distribution of *Dermacentor reticulatus* in Germany and detection of *Rickettsia* sp. RpA4. *International Journal of Medical Microbiology* 296, 149–156.

De Tray, D.E., (1963) African swine fever. *Advances in Veterinary Science* 8, 299–333.

Dixon, L.K. and Wilkinson, P.J. (1988) Genetic diversity of African swine fever virus isolates from soft ticks (*Ornithodoros moubata*) inhabiting warthog burrows in Zambia. *Journal of General Virology* 69, 2981–2993.

EFSA Panel on Animal Health and Welfare (AHAW) (2010a) Scientific opinion on African swine fever. *EFSA Journal* 8(3), 1556.

EFSA Panel on Animal Health and Welfare (AHAW) (2010b) Scientific opinion on the role of tick vectors in the epidemiology of Crimean Congo hemorrhagic fever and African swine fever in Eurasia. *EFSA Journal* 8(8), 1703.

Encinas Grandes, A. (1986) Ticks of the province of Salamance (central/NW Spain). Prevalence and parasitisation intensity in dogs and domestic ungulates. *Annales de Parasitologie Humaine et Comparée* 61, 95–107.

Endris, R.G. and Hess, W.R. (1994) Attempted transovarial and venereal transmission of African swine fever virus by the Iberian soft tick *Ornithodoros* (*Pavlovskyella*) *macrocanus* (Acari: Idodoidea: Argasidae). *Journal of Medical Entomology* 31, 373–381.

Endris, R.G., Hess, W.R. and Caiado, J.M. (1992) African swine fever virus infection in the Iberian soft tick, *Ornithodoros* (*Pavlovskyella*) *marocanus* (Acari: Argasidae). *Journal of Medical Entomology* 29, 874–878.

Estrada-Peña, A. (2008) Climate, niche, ticks, and models: what they are and how we should interpret them. *Parasitology Research* 103, 87–95.

Estrada-Peña, A. and Dusbabek, F. (1992) Cuticular hydrocarbon gas chromatography analysis of *Argas vulgaris, A. polonicus* and their hybrids. *Experimental and Applied Acarology*, 17, 365–370.

Estrada-Peña, A. and Estrada-Peña, R. (1991a) Notes on *Dermacentor* ticks (Acari: Ixodidae) (1). A proposal of nomenclature for the dorsal scutal pattern. *Acarologia* 32, 13–16.

Estrada-Peña, A. and Estrada-Peña, R. (1991b) Phenotypic variation of *Dermacentor marginatus* (Ixodidae) in the Palaearctic Region. In: Dusbabek, F. and Bukva, V. (eds) *Modern Acarology*, Volume 2. Academia, Prague and SPB Academic Publishing, The Hague, The Netherlands, pp. 155–162.

Estrada-Peña, A., Bouattour, A., Camicas, J. and Walker, L.A.R. (2004) *Ticks of Domestic Animals in the Mediterranean Region. A Guide to Identification of Species*. Produced for the European Union Project Integrated Consortium on Ticks and Tick-borne Diseases (ICTTD-2). Available on request from F.Jongejan@uu.nl.

Filippova, N.A. (1966) *Argasid Ticks (Argasidae)*. Fauna USSR 4 (3), Nauka, Zoological Institute of the Academy of Sciences of USSR, Moscow, Leningrad.

Földvári, G. and Farkas, R. (2005) Ixodid ticks attaching to dogs in Hungary. *Veterinary Parasitology* 129, 125–131.

Gern, L., Toutoungi, L.N., Hu, C.M. and Aeshlimann, A. (1991) *Ixodes* (*Pholeoixodes*) *hexagonus*, an efficient vector of *Borrelia burgdorferi* in the laboratory. *Medical and Veterinary Entomology* 5, 431–435.

Gilot, B., Pichot, J. and Doche, B. (1989) Les tiques du Massif Central (France). 1. Les Ixodidés (Acariens, Ixodoidea) parasites de carnivores et d'ongulés domestiques sur la bordure orientale du massif. *Acarologia* 30, 191–207.

Guglielmone, A.A., Robbins, R.G., Apanaskevich, D.A., Petney, T.N., Estrada-Peña, A. and Horak, I.G. (2009) Comments on controversial tick (Acari: Ixodida) species names and species described or resurrected from 2003 to 2008. *Experimental and Applied Acarology* 48, 311–327.

Hess, W.R., Endris, R.G., Haslett, T.M., Monahan, M.J. and McCoy, J.P. (1987) Potential arthropod vectors of African swine fever virus in North America and the Caribbean basin. *Veterinary Parasitology* 26, 145–155.

Hess, W.R., Endris, R.G., Lousa, A. and Caiado, J.M. (1989) Clearance of African swine fever virus from infected tick (Acari) colonies. *Journal of Medical Entomology* 26, 314–317.

Hillyard, P. (1996) *Ticks of North-West Europe*. Natural History Museum, London.

Hinnebusch, J. and Barbour, A.G. (1991) Linear plasmids of *Borrelia burgdorferi* have a telomeric structure and sequence similar to those of a eukaryotic virus. *Journal of Bacteriology* 173, 7233–7239.

Hoogstraal, H. (1956) Ticks of the Sudan (with special reference to Equatoria Province and with preliminary reviews of the genera *Boophilus, Margaropus* and *Hyalomma*). In: *African Ixodoidea I*. Research Report NM 005 050.29.07, Department of the Navy, Bureau of Medicine and Surgery, Washington, D.C., pp. 436–451.

Hoogstraal, H. (1972) The influence of human activities on tick distribution, density and diseases. *Wadomosci Parazytologiczne* 18, 501–511.

Hoogstraal, H. (1985) Argasid and nuttalliellid ticks as parasites and vectors. *Advances in Parasitology* 24, 135–238.

Hoogstraal, H. and Kim, K.C. (1985) Tick and mammal coevolution, with emphasis on *Haemaphysalis*. In: Kim, K.C. (ed.) *Coevolution of Parasitic Arthropods and Mammals*. John Wiley, New York, pp. 505–568.

Hornok, S. (2009) Allochronic seasonal peak activities of *Dermacentor* and *Haemaphysalis* spp. under continental climate in Hungary. *Veterinary Parasitology* 163, 366–369.

Jaenson, T.G.T., Tälleklint, L., Lundqvist, L., Olsen, B., Chirico, J. and Mejlon, H. (1994) Geographical distribution, host associations, and vector roles of ticks (Acari: Ixodidae, Argasidae) in Sweden. *Journal of Medical Entomology* 31, 240–256.

Kleiboeker, S.B., Burrage, T.G., Scoles, G.A., Fish, D. and Rock, D.L. (1998) African swine fever virus infection in the argasid host, *Ornithodoros porcinus porcinus*. *Journal of Virology* 72, 1711–1724.

Kleiboeker, S.B., Scoles, G.A., Burrage, T.G. and Sur, J.H. (1999) African swine fever virus replication in the midgut epithelium is required for infection of *Ornithodoros* ticks. *Journal of Virology* 73, 8587–8598.

Kleine-Tebbe, J., Heinatz, A., Gräser, I., Dautel, H., Hansen, G.N., Kespohl, S., Rihs, H.P., Raulf-Heimsoth, M., Vater, G., Rytter, M. and Haustein, U.F. (2006) Bites of the European pigeon tick (*Argas reflexus*): risk of IgE-mediated sensitizations and anaphylactic reactions. *Journal of Allergy and Clinical Immunology* 117, 190–195.

Klompen, J.S.H. and Oliver J.H. (1993) Systematic relationships in the soft ticks (Acari: Ixodida: Argasidae). *Systematic Entomology* 18, 313–331.

Kolonin, G.V. (2009) Fauna of ixodid ticks of the world (Acari, Ixodidae). Kolonon, G.V., Moscow. Available at: http://www.kolonin.org (accessed 26 June 2012).

Liebisch, A., Rahaman, M.S. and Hoogstaal, H. (1989) Tick fauna of Egypt with special reference to studies on *Hyalomma anatolicum anatolicum*, the natural vector of cattle theileriosis. In: Channabasavanna, G.P. and Viraktamat, C.A. (eds) *Progress in Acarology, Vol. I*. Oxford and IBH Publishing, Oxford, UK, pp. 53–59.

Lopes de Carvalho, I., Santos, N., Soares, T., Zé-Zé, L. and Sofia Núncio, M. (2011) *Francisella*-like endosymbiont in *Dermacentor reticulatus* collected in Portugal. *Vector-Borne and Zoonotic Diseases* 11(2), 185–188. doi:10.1089/vbz.2010.0014.

Manilla, G. (1998a) Subfam. Haemaphysalinae. In: *Acari, Ixodida – Fauna d'Italia, Volume 36*. Edizioni Calderini, Bologna. Italy, pp. 133–174.

Manilla, G. (1998b) Subfam. Hyalomminae. *In: Acari, Ixodida – Fauna d'Italia, Volume 36*. Edizioni Calderini, Bologna, Italy, pp. 174–200.
Manilla, G. (1998c) Subfam. Rhipicephalinae. In: *Acari, Ixodida – Fauna d'Italia, Volume 36*. Edizioni Calderini, Bologna, Italy, pp. 200–242.
Martinod, S. and Gilot, B. (1991) Epidemiology of canine babesiosis in relation to the activity of *Dermacentor reticulatus* in southern Jura (France). *Experimental and Applied Acarology* 11, 215–222.
Matthysse, I.G. and Colbo, M.I. (1987) *The Ixodid Ticks of Uganda Together with Species Pertinent to Uganda Because of their Present Known Distribution*. Entomological Society of America, College Park, Maryland.
Mazyad, S.A.M. and Khalaf, S.A.A. (2002) Studies on *Theileria* and *Babesia* infecting live and slaughtered animals in Al-Arish and El-Hasanah, North Sinai governorate, Egypt. *Journal of the Egyptian Society of Parasitology* 32, 601–610.
Mellor, P.S. and Wilkinson, P.J. (1985) Experimental transmission of African swine fever virus by *Ornithodoros savinyi* (Audouin). *Research in Veterinary Science* 39, 353–356.
Merck (2010) *Haemaphysalis* spp. In: *The Merck Veterinary Manual*. Merck Sharp and Dohme, Whitehouse Station, New Jersey. Available at: http://www.merckvetmanual.com/mvm/index.jsp?cfile=htm/bc/72110.htm (accessed 14 March 2010).
Morel, P. (1969) *Ticks from Africa and the Mediterranean Basin* [in French]. CD-ROM, CIRAD, Montpellier, France.
Norval, R.A.I. and Horak, I.G. (2004) Vectors: ticks. In: Coetzer, J.A.W. and Tustin, R.C. (eds) *Infectious Diseases of Livestock, Volume 1*, 2nd edn. Oxford University Press Southern Africa, Cape Town, South Africa, pp. 3–42.
Nosek, J. (1972) The ecology and public health importance of *Dermacentor marginatus* and *D. reticulatus* ticks in Central Europe. *Folia Parasitologica* 19, 93–102.
Olsen, B., Jaenson, T.G.T., Noppa, L., Bunikis, J. and Berström, S. (1993) A Lyme borreliosis cycle in seabirds and *Ixodes uriae* ticks. *Nature* 362, 340–342.
Papadopoulus, B., Morel, P.C. and Aeschliman, A. (1996) Ticks of domestic animals in the Macedonia region of Greece. *Veterinary Parasitology* 63, 25–40.
Pavlov, P., Rosicky, B., Hubalek, Z., Daniel, M., Bardos, V., Minar, J. and Juricova, Z. (1978) Isolation of Banjia virus from ticks of the genus *Haemaphysalis* in southeast Bulgaria. *Folia Parasitologica* (Praga) 25, 67–73.
Pegram, R.G., Rota, A., Onkelinx, R., Wilson, D.D., Bartlette, P., Nisbett, B.S., Swanston, G., Vanterpool, P. and de Castro, J.J. (2005) Eradicating the tropical bont tick from the Caribbean. *World Animal Review* 87-1996/2. Food and Agriculture Organization, Rome Available at: http://www.fao.org/ag/aga/agap/frg/feedback/war/W2650T/w2650t06.htm (accessed 7 July 2007).
Pipano, E. and Shkap, V. (2004) *Theileria annulata* theileriosis. In: Coetzer, J.A.W. and Tustin, R.C. (eds) *Infectious Diseases of Livestock, Volume 1*, 2nd edn. Oxford University Press Southern Africa, Cape Town, South Africa, pp. 486–497.
Plowright, W. (1977) Vector transmission of African swine fever virus. In: *Agricultural Research Seminar on Hog Cholera/Classical Swine Fever and African Swine Fever, Tierarztliche Hochschule, Hannover, 6–11 September 1976*, pp. 575–587.
Plowright, W., Parker, J. and Peirce, M.A. (1969a) African swine fever virus in ticks (*Ornithodoros moubata*, Murray) collected from animal burrows in Tanzania. *Nature* 221, 1071–1073.
Plowright, W., Parker, J. and Pierce, M.A. (1969b) The epizootiology of African swine fever in Africa. *Veterinary Record* 85, 668–674.
Plowright, W., Perry, C.T. and Pierce, M.A. (1970a). Transovarial infection with African swine fever virus in the argasid tick, *Ornithodoros moubata porcinus*, Walton. *Research in Veterinary Science* 2, 582–584.
Plowright, W., Perry, C.T., Peirce, M.A. and Parker, J. (1970b) Experimental infection of the argasid tick, *Ornithodoros moubata porcinus*, with African swine fever virus. *Archiv für die Gesamte Virusforsch* 31, 33–50.
Plowright, W., Perry, C.T. and Greig, A. (1974) Sexual transmission of African swine fever virus in the tick, *Ornithodoros moubata porcinus*, Walton. *Research in Veterinary Science* 17, 106–13.
Plowright, W., Thomson, G.R. and Neser, J.A. (1994) African swine fever. In: Coetzer, J.A.W., Thomson, G.R. and Tustin, R.C. (eds) *Infectious Diseases of Livestock: With Special Reference to Southern Africa, Volume 1*. Oxford University Press Southern Africa, Cape Town, South Africa, pp. 567–599.
Raoult, D., Lakos, A., Fenollar, F., Beytout, J., Brouqui, P. and Fournier, P.E. (2002) Spotless rickettsiosis caused by *Rickettsia slovaca* and associated with *Dermacentor* ticks. *Clinical Infectious Diseases* 34, 1331–1336.

Rasulov, I. (2007) Ticks status in Central Asia with a special emphasis on Uzbekistan. *Parasitology Research* 101(Suppl. 2), 183–186.

Rechav, Y., Zeederberg, M.E. and Zeller, D.A. (1987) Dynamics of African tick (Acari: Ixodoidea) populations in a natural Crimean-Congo hemorrhagic fever focus. *Journal of Medical Entomology* 24, 575–583.

Rennie, L., Wilkinson, P.J. and Mellor, P.S. (2001) Transovarial transmission of African swine fever virus in the argasid tick *Ornithodoros moubata*. *Medical and Veterinary Entomology* 15, 140–146.

Rolla, G., Nebiolo, F., Marsico, P., Guida, G., Bigo, P., Riva, G. and Zanotta, S. (2004) Allergy to pigeon tick (*Argas reflexus*): demonstration of specific IgE-binding components. Case report. *International Archives of Allergy and Immunology* 135, 293–295.

Rudenko, N., Golovchenko, M., Grubhoffer, L. and Oliver, J.H. Jr (2011) Updates on *Borrelia burgdorferi sensu lato* complex with respect to public health. *Ticks and Tick-borne Diseases* 2, 123–128.

Sánchez-Botija, C. (1963) Reservoirs of ASFV: a study of the ASFV in arthropods by means of haemadsorption. *Bulletin de l'Office Internationale des Epizooties* 60, 895–899.

Siuda, K., Hoogstraal, H., Clifford, C.M. and Wassef, H.Y. (1979) Observations on the subgenus *Argas* (Ixodoidea: Argasidae: *Argas*). 17. *Argas (A.) polonicus* sp. n. parasitizing domestic pigeons in Kraków, Poland. *Journal of Parasitology* 65, 170–181.

Spiewak, R., Lundberg, M., Johansson, G. and Buczek, A. (2006) Allergy to pigeon tick (*Argas reflexus*) in Upper Silesia, Poland. *Annals of Agricultural and Environmental Medicine* 13, 107–112.

Sréter, T., Széll, Z. and Varga, I. (2003): Ectoparasite infestations of red foxes (*Vulpes vulpes*) in Hungary. Short communication, *Veterinary Parasitology* 115, 349–354.

Sréter, T., Széll, Z. and Varga, I. (2005) Spatial distribution of *Dermacentor reticulatus* and *Ixodes ricinus* in Hungary: evidence for change? Short communication. *Veterinary Parasitology* 128, 347–351.

Szymanski, S. (1987) Seasonal activity of *Dermacentor reticulatus* (Fabricius, 1794). (Acarina, Ixodidae) in Poland. *Acta Parasitologia Polonica* 29, 247–255.

Uilenberg, G. (1982) Experimental transmission of *Cowdria ruminantium* by the Gulf Coast tick *Amblyomma maculatum:* danger of introducing heartwater and benign African theileriasis onto the American mainland. *American Journal of Veterinary Research* 43, 1279–1282.

Uilenberg, G., Barré, N., Camus, E., Burridge, M.J. and Garris, G.I. (1984) Heartwater in the Caribbean. *Preventive Veteterinary Medicine* 2, 255–267.

Veraldi, S. (1996) Acute urticaria caused by pigeon ticks (*Argas reflexus*). *International Journal of Dermatology* 35, 34–35.

Vial, L., Wieland, B., Jori, F., Etter, E., Dixon, L. and Roger, F. (2007) African swine fever virus DNA in soft ticks, Senegal. *Emerging Infectious Diseases Journal* 13, 1928–1931.

Viseras, J., and García-Fernández, P. (1999) Studies on theileriosis in southern Spain. *Parassitologia* 41 (Suppl. 1), 111–115.

Walker, A., Bouattour, A., Camicas, J.L., Estrada-Peña, A., Horak, I.G., Latif, A.A., Pegram, R.G. and Preston, P.M. (2003) *Ticks of Domestic Animals in Africa: A Guide to Identification of Species* (revised edn 2007). Produced for the European Union Project Integrated Consortium on Ticks and Tick-borne Diseases (ICTTD-2). Copyright University of Edinburgh, Edinburgh, UK, 2003. Published by Bioscience Reports, Edinburgh, UK, 2007. Available at: http://www.alanrwalker.com/index/cms-filesystem-action/tickguide-africa-web-08.pdf (accessed 27 June 2012).

Walker, J.B., Keirans, J.E. and Horak, I.G. (2000) *The Genus* Rhipicephalus *(Acari, Ixodidae): A Guide to the Brown Ticks of the World*. Cambridge University Press, Cambridge, UK.

Wallis, G.P. and Miller, B.R. (1983) Electrophoretic analysis of the ticks *Ornithodoros* (*Pavlovskyella*) *erraticus* and *O.* (*P.*) *sonrai* (Acari: Argasidae). *Journal of Medical Entomology* 20, 570–571.

Walter, G., Kock, D. and Liebisch, A. (1986) Beitrag zur Zecken-Fauna der Bundesrepublik Deutschland (Arachnida: Acarina: Ixodidae). *Senckenbergiana Biologica* 67, 199–206.

Woolley, T.A. (1988) *Acarology: Mites and Human Welfare*. John Wiley, New York.

Zahler, M. and Gothe, R. (1997) Evidence for the reproductive isolation of *Dermacentor marginatus* and *Dermacentor reticulatus* (Acari: Ixodidae) ticks based on cross-breeding, morphology and molecular studies. *Experimental and Applied Acarology* 21, 685–696.

Zahler, M. and Gothe, R. (2001) A new endemic focus of the bont tick *Dermacentor reticulatus* in Bavaria – risk of further endemic spreading of canine babesiosis. *Tierarztliche-Praxis-Ausgabe-K, Kleintiere/Heimtiere* 29, 121–123.

Zeller, H., Cornet, J., Diop, A. and Camicas, J. (1997) Crimean-Congo hemorrhagic fever in ticks (Acari: Ixodidae) and ruminants: field observations of an epizootic in Bandia, Senegal (1989–1992). *Journal of Medical Entomology* 34, 511–551.

3 Factors Influencing the Spread and Distribution of Ticks

Maxime Madder[1] and Ilaria Pascucci[2]
[1]*Institute of Tropical Medicine, Antwerp, Belgium and University of Pretoria, South Africa; and* [2]*Istituto Zooprofilattico Sperimentale dell'Abruzzo e del Molise "G. Caporale", Teramo, Italy*

3.1 Climate Changes

Long-time series of historical data related to climatic factors and the geographical distribution of ticks and tick-borne pathogens are required to determine the impact of climate changes on the spread of these vectors and the pathogens they carry. To our knowledge, few such attempts have been made, mainly because of the lack of data collection over a sufficiently long period of time to allow for comparison in relation to existing climatic data. One of the most recent and long-term studies has been described by Jaenson *et al.* (2012). In this article it was shown that deer abundance and climate change seem to be the most important factors explaining the spread of *Ixodes ricinus* northwards in Sweden during the last 30 years. It is suggested that the milder climate has made it possible for roe deer to expand northwards and into northern Sweden. As the roe deer is, generally, the most important host of adults of *I. ricinus* in Sweden the expansion of this tick host has favoured the subsequent expansion of the tick into the same northern area (Jaenson *et al.*, 2012).

It is possible, however, to affirm that despite some evidence it is difficult to implicate climate change as the main cause for an increase in tick-borne diseases. In order to predict future tick-borne scenarios, climate change models are required that take into account the dynamic biological processes involved in vector abundance and pathogen transmission which affect the complex ecology and epidemiology of tick-borne diseases such as Crimean-Congo haemorrhagic fever (CCHF), tick-borne encephalitis (TBE) and Lyme borreliosis. According to Gray *et al.* (2009), climate changes have influenced the survival of ticks (direct effect). Empirical proofs have been obtained that climate may be in part responsible for shaping the prevalence rates of *Borrelia burgdorferi s.l.* in questing nymphal *I. ricinus* (Estrada-Peña *et al.*, 2011). However, studies on the dynamics of CCHF cases in Turkey did not show any influence of climate when sites where the disease had been reported were compared with sites where the disease was absent (Estrada-Peña *et al.*, 2010).

Changes in climate and the length of the different seasons directly affect tick survival, activity and development, but there is limited evidence that rising temperature results in a greater abundance of ticks simply by increasing rates of development. Effects of the climate on the geographical distribution of a tick species are generally more evident along the border of its distribution, affecting the geographic range potentially suitable for the tick. A study by Jaenson and Lindgren (2011) has shown that a warmer climate in Northern Europe has extended the vegetation period which has allowed ticks to be active for a longer time during the year; this seems to have enlarged the geographic range available for the tick. Then again, changes in development rates make tick cohorts available to different diapause windows (largely determined by day length), thus changing the patterns of seasonal activity and significantly altering generation length. Probably the most interesting tick species to work on and to explore the diversity of seasonal patterns is *I. ricinus*, the most abundant tick in Europe and an important pest of both animals and humans. However, a process-driven model, which is necessary to study such diverse seasonality patterns, is still not available.

The indirect effects of climate change are more complicated to evaluate. Climate changes can indirectly alter survival of the ticks by modifying the type of vegetation and, as mentioned above, the host abundance. For instance, climate will affect the vegetation, which in many ways may influence the number of infected ticks. In addition, dense humid vegetation will, for some tick species, be beneficial to tick survival and could lead to an increase in tick numbers as well. Such vegetation may also benefit pathogen-reservoir hosts such as rodents and shrews, and increase the survival and abundance of tick-maintenance hosts such as deer (Jaenson *et al.*, 2012). All these factors may boost the density of infected ticks, but the direct impact on animal and human infection rates is unknown and has not yet been evaluated for any of the many pathogens that the ticks can transmit.

The influence of mast years on rodent abundance and consequently on tick density is one example of how vegetation will indirectly affect the density of *Borrelia*-infected *Ixodes* ticks (Ogden *et al.*, 2005). After mast years, rodent populations significantly increase as a result of an extended breeding season and higher survival during winter. This is due to an increase in food supply and consequently in body fat. The higher host population density affects the tick density in the following year and possibly also the risk of disease transmission (Tack *et al.*, 2010).

Climate change may also influence the risk of infection by affecting human activities; for example, the long-term use of land, e.g. for agriculture or tourism, could be affected by climate change. Weather patterns also have an effect by influencing short-term human behaviour so that tick-bite risk may be linked to the seasonality of certain human activities such as hiking, hunting, and berry and mushroom picking. Generally climate effects are more easily noticeable close to the geographical distribution limits of both vector and pathogen, but the magnitude of the effects of climate change in an endemic area is the result of the interaction of many other parameters associated with socio-economics, human migration and settlement, ecosystems and biodiversity, bird migrations, land use and land cover, human culture and behaviour, and immunity in the human and animal populations (Gray *et al.*, 2009).

3.2 Changes in Land Use

Changes in land use often affect the host–tick relationship and consequently increase the transmission of tick-borne pathogens. For instance, during World War II, after the occupation of Crimea (1941–1944), normal agricultural activities were disrupted and the common sport of hunting European hares was abandoned. When Soviet troops reoccupied the hilly Crimean steppes in 1944, the hares had become excessively abundant and neglected pastures were overgrown with weeds. European hares are a reservoir of CCHF virus (CCHFV) and one of the primary hosts of immature *Hyalomma*, and with the expansion of the hare population came the first documented CCHF outbreak of the modern era (Ergönül, 2006). A similar explanation was suggested for an outbreak in Turkey as hunting and pasturing had been abandoned in the fields in the affected region from between 1995 and 2001 because of terrorist activities (Ergönül, 2006). In this period, the numbers of small mammals (e.g. hares) and wild animals (e.g. wild boars) increased, as did the population of *Hyalomma* ticks. When the fields became available again for hunting and pasturing, cattle, sheep and humans were exposed to virus-carrying ticks (Ergönül, 2006).

3.3 Movement of Animals

3.3.1 Migratory birds

The introduction and spread of tick species by migratory birds are documented for those tick species which are ornithophilic in the larval and nymphal stages. *Hyalomma* (*Hy.*) *marginatum* and *Hy. rufipes*, and to a lesser extent some species of *Amblyomma*, belong to this group of ticks. Their potential spread northwards and the establishment of permanent populations in the north are therefore of great importance, particularly as the immature stages of *Hyalomma* are frequently found on migratory birds flying northward to temperate Europe (e.g. Jaenson *et al.*, 1994; Molin *et al.*, 2011). The dispersal of ticks by birds may be restricted to a short distance during local post-breeding flights or over extremely long distances during migration flights. In studies conducted on birds migrating through Egypt between 1955 and 1973, it was discovered that birds migrating from Eurasia to Africa carried tick species that were characteristic of the fauna of Europe and Asia (Hoogstraal, 1963). As another example, the introduction of *Am. variegatum* in southern Italy and Greece by migratory birds has been sporadically reported (Albanese *et al.*, 1971; Papadopoulus *et al.*, 1996).

Hy. rufipes is the most common tick found on birds migrating northwards from sub-Saharan Africa to Eurasia in the spring (Hoogstraal, 1963; Linthicum and Bailey, 1994). The risk of diffusion of CCHFV into areas where the virus does not occur is linked to the possibility of infected *Hyalomma* ticks reaching and becoming established in such 'uninfected areas'. This probability is greater for ticks associated with migratory birds (Manilla, 1998a). Hoogstraal (1979) reported that many bird species are responsible for the intracontinental and intercontinental dissemination of ticks associated with CCHFV. While this fact is not the most important factor in the introduction of CCHFV to Europe, it has nevertheless been hypothesized that the outbreak of CCHF in Turkey in 2002 could have been caused by birds carrying CCHFV-infected ticks from the Balkans (Karti *et al.*, 2004). The occurrence of this disease in Turkey is under discussion, taking into consideration the available epidemiological data. It currently seems impossible that migrating birds from the Balkans could have introduced the disease into Turkey and have it spread in an interval of weeks. The current opinion is that the virus has been there many years and that some change/s favoured its contact with humans so that disease incidence increased (Ozdarendeli *et al.*, 2010).

3.3.2 Importation of exotic and wildlife species

There are many examples of exotic ticks that have been introduced into countries by the importation of exotic animals. Some reptile-associated African *Amblyomma* tick species were found on the American mainland, mainly as a result of the importation of tick-infested reptiles from Africa (i.e. infested by *Am. marmoreum* or *Am. sparsum*). Also, in Poland, *Am. latum* and *Am. transversale* were found on pythons and *Am. exornatum*, *Am. flavomaculatum*, *Am. latum* and *Am. nuttalli* on monitor lizards, both reptiles imported from Africa (Nowak, 2010). In addition, in the Netherlands, *Am. flavomaculatum* was collected from an African savannah monitor lizard (Nijhof *et al.*, 2007), and other tick species such as *Am. marmoreum*, *Am. dissimile* and *Hy. rufipes* have occasionally been collected from imported tortoises and snakes (Bronswijk *et al.*, 1979; Jongejan, 2001). Similarly, a constant threat for the introduction of heartwater into the American mainland exists mainly by the importation of livestock from the Caribbean Islands infested with *Am. variegatum* ticks, but also through the introduction of lightly infested wild ungulates and tick-infested reptiles from Africa, as mentioned above (Burridge *et al.*, 2002).

3.3.3 Movement of domestic animals

Ticks associated with pets are able to reach and establish themselves in areas far from their origins. One of the most typical examples is *Rhipicephalus* (*Rh.*) *sanguineus*, universally known as the 'kennel tick' (Manilla,

1998b; Walker *et al.*, 2000; Estrada-Peña *et al.*, 2004). This is a monotropic three-host tick strictly associated with the domestic dog. *Rh. sanguineus* is probably one of the most widely distributed tick species in the world. Travellers with dogs have brought this important vector species permanently into the latitudes approximately between 50°N and 30°S (Walker *et al.*, 2000). Furthermore, *Dermacentor reticulatus* (the ornate dog tick, marsh tick or cattle tick) may have been introduced into north-western Europe in a similar way, in addition to the routes described previously.

D. reticulatus is a tick species that appears to be expanding its range in north-western Europe. A study performed in Germany by Dautel *et al.* (2006) in 2003 and 2004 showed that the distribution and abundance of *D. reticulatus* has increased on deer and vegetation. Populations have also been found over the last few years in the Netherlands and Belgium (Nijhof *et al.*, 2007; Cochez *et al.*, 2011). In the Netherlands, several populations have been found in freshwater tidal marshes, mainly in the south of the country; these were most likely established after the introduction of southern European cattle breeds. Ponies were exported to Belgium to manage vegetation, and these ponies then introduced this tick species to that country. In Belgium, *D. reticulatus* has been found on tick (flag or drag) cloth dragged in different nature reserves and peri-urban areas, and on dogs. Most of these areas are frequented by dogs, possibly explaining further spread of *D. reticulatus*, as no large mammals are present there.

The introduction of *Am. variegatum* into the Caribbean area is an example of how the movement of domestic animals can spread exotic ticks. The initial introduction of *Am. variegatum* into Guadeloupe and probably also into most of the West Indies islands was due to the importation of infested N'Dama cattle in the 1800s, or to tick-infested litter used during cattle transportation by ships from Senegal to the West Indies (Barré *et al.*, 1987), or from East Africa to the Indian Ocean islands. Cattle transportation between islands also explains the introduction of the tick at least into Marie-Galante, Antigua and Martinique. But from the late 1960s, the rapid increase in the number of infested islands in the Caribbean area was linked to the presence and spread of the cattle egret in the West Indies. This cattle-associated African bird species was also introduced into the Caribbean area (Corn *et al.*, 1993; Barré *et al.*, 1995). The reinfestation of some countries, for example Puerto Rico, after a successful eradication campaign might also be due to the movement of this bird (Bokma and Shaw, 1993). *Am. variegatum* is thus the only African vector of cowdriosis that has established itself successfully outside the continent (Walker and Olwage, 1987). It remains a threat in areas where climatic conditions are suitable for its establishment, as it has already succeeded in colonizing new territories through the movement of migratory birds which cannot be controlled.

Rh. (Boophilus) microplus, the pantropical blue tick, is without a doubt one of the most important vectors of livestock pathogens in the world, mainly because of its large geographical distribution and its ability to transmit both Asiatic and African redwater babesiosis caused by *Babesia bovis* and *B. bigemina* (Coetzer and Tustin, 2004).

This tick species originates from Asia, but during the latter half of the 19th century it, and the babesiae it transmits, was spread by cattle transported to Australia, Madagascar, South Africa, Latin America, Mexico and the USA (Hoogstraal, 1956; Temeyer *et al.*, 2004). *Rh. microplus* was eradicated from the USA in 1943 through the costly Cattle Fever Tick Eradication Program which started in 1906. The most resistant strain studied to date has been shown to survive the dipping procedure used at the USA–Mexico border as part of an importation barrier to prevent the re-entry of *Rh. microplus* into the USA (Temeyer *et al.*, 2004).

Until recently, *Rh. microplus* was not present in West Africa, but this tick species has unfortunately been introduced into the Ivory Coast (Madder *et al.*, 2007, 2011) and Benin (Madder *et al.*, 2012) as a result of uncontrolled cattle imports for cattle improvement programmes. Because of the high adaptability of ticks and, more specifically, of *Rh. microplus*, the risk of the introduction of this species into the Mediterranean area and the establishment of local populations seems possible.

References

Albanese B., Smiraglia C. and Lavagnino A. (1971) Notizie sulle zecche di Sicilia con segnalazione di *Hyalomma detritum* e *Amblyomma variegatum*. *Rivista di Parassitologia* 32, 273–276.

Barré, N., Uilenberg, G., Morel, P.C. and Camus, E. (1987) Danger of introducing heartwater onto the American mainland: potential role of indigenous and exotic *Amblyomma* ticks. *Onderstepoort Journal of Veterinary Research* 54, 405–417.

Barré, N., Garris, G. and Camus, E. (1995) Propagation of the tick *Amblyomma variegatum* in the Caribbean. *Scientific and Technical Review/Revue Science et Technique* (World Organisation for Animal Health, OIE) 14, 841–855.

Bokma, B.H. and Shaw, J.L. (1993) Eradication of a new focus of *Amblyomma variegatum* in Puerto Rico. *Revue d'Elevage et de Médecine Vétérinaire des Pays Tropicaux* 46, 355–358.

Bronswijk, J.E.M.H. van, Rijntjes, R.H. and Garben, A.F.M. (1979) *De Teken (Ixodidae) van de Beneluxlanden*. Wetenschappelijke Mededeling van de Koninklijke Nederlandse Natuurhistorische Vereniging, Utrecht, The Netherlands.

Burridge, M.J., Simmons, L.A., Peter, T.F. and Mahan, S.M. (2002) Increasing risks of introduction of heartwater onto the American mainland associated with animal movements. *Annals of the New York Academy of Sciences* 969, 269–74.

Cochez, C., Lempereur, L., Madder, M., Claerebout, E., Simons, L., De Wilde, N., Linden, A., Saegerman, C., Heyman, P. and Losson, B. (2011) Foci report of indigenous *Dermacentor reticulatus* populations in Belgium and a preliminary study on associated babesiosis pathogens. *Journal of Medical Entomology*, doi: 10.1111/j.1365-2915.2011.00998.x (early view).

Coetzer, J.A.W. and Tustin, R.C. (eds) (2004) *Infectious Diseases of Livestock*, 2nd edn. Oxford University Press Southern Africa, Cape Town, South Africa.

Corn, J.L., Barré, N., Thiebot, B., Creekmore, T.E., Garris, G.I. and Nettles, V.F. (1993) Potential role of cattle egrets, *Bubulcus ibis* (Ciconiformes: Ardeidae), in the dissemination of *Amblyomma variegatum* (Acari: Ixodidae) in the Eastern Caribbean. *Journal of Medical Entomology* 30, 1029–1037.

Dautel, H., Dippel, C., Oehme, R., Hartelt, K. and Schettler, E. (2006) Evidence for an increased geographical distribution of *Dermacentor reticulatus* in Germany and detection of *Rickettsia* sp. RpA4. *International Journal of Medical Microbiology* 296, 149–156.

Ergönül, Ö. (2006) Crimean-Congo haemorrhagic fever. *The Lancet Infectious Diseases* 6, 203–214.

Estrada-Peña, A., Bouattour, A., Camicas, J. and Walker, L.A.R. (2004) *Ticks of Domestic Animals in the Mediterranean Region. A Guide to Identification of Species*. Produced for the European Union Project Integrated Consortium on Ticks and Tick-borne Diseases (ICTTD-2). Available on request from F.Jongejan@uu.nl.

Estrada-Peña, A., Zatansever, Z., Gargili, A. and Ergönül, Ö. (2010) The trend towards habitat fragmentation is the key factor driving the spread of Crimean-Congo haemorrhagic fever. *Epidemiology and Infection* 138, 1194–1203.

Estrada-Peña, A., Ortega, C., Sánchez, N., Desimone, L., Sudre, B., Suk, J.E. and Semenza, J.C. (2011) Correlation of *Borrelia burgdorferi sensu lato* prevalence in questing *Ixodes ricinus* ticks with specific abiotic traits in the western palearctic. *Applied Environmental Microbiology* 77, 3838–3845.

Gray, J.S., Dautel, H., Estrada-Peña, A., Kahl, A. and Lindgren, E. (2009) Effects of climate change on ticks and tick-borne diseases in Europe. *Interdisciplinary Perspectives on Infectious Diseases* 2009, Article ID 593232, doi:10.1155/2009/593232.

Hoogstraal, H. (1956) Ticks of the Sudan (with special reference to Equatoria Province and with preliminary reviews of the genera *Boophilus*, *Margaropus* and *Hyalomma*). In: *African Ixodoidea I*. Research Report NM 005 050.29.07, Department of the Navy, Bureau of Medicine and Surgery, Washington, D.C., pp. 436–451.

Hoogstraal, H. (1963) Ticks on birds migrating from Europe and Asia to Africa 1959–61. WHO Bulletin.

Hoogstraal, H. (1979) The epidemiology of tick-borne Crimean-Congo hemorrhagic fever in Asia, Europe and Africa. *Journal of Medical Entomology*15, 307–417.

Jaenson, T.G.T. and Lindgren, E. (2011) The range of *Ixodes ricinus* and the risk of contracting Lyme borreliosis will increase northwards when the vegetation period becomes longer. *Ticks and Tick-Borne Diseases* 2, 44–49.

Jaenson, T.G.T., Tälleklint, L., Lundqvist, L., Olsen, B., Chirico, J. and Mejlon, H. (1994) Geographical distribution, host associations, and vector roles of ticks (Acari: Ixodidae, Argasidae) in Sweden. *Journal of Medical Entomology* 31, 240–256.

Jaenson, T.G.T., Jaenson, D.G.E., Eisen, L., Petersson, E. and Lindgren, E. (2012) Changes in the geographical distribution and abundance of the tick *Ixodes ricinus* during the past 30 years in Sweden. *Parasites and Vectors* 5, 8, doi:10.1186/1756-3305-5-8.

Jongejan, F. (2001) Teken en door teken overgedragen ziekten. *Diergeneeskundig Memorandum* 48, 1–51.

Karti, S., Odabasi, Z., Korten, V., Yilmaz, M., Sonmez, M., Caylan, R., Akdogan, E., Eren, N., Koksal, I., Ovali, E., Erickson, B.R., Vincent, M.J., Nichol, S.T., Comer, J.A., Rollin, P.E. and Ksiazek, T.G. (2004) Crimean-Congo hemorrhagic fever in Turkey. *Emerging Infectious Diseases* 19, 1379–1384.

Linthicum, K.J. and Bailey, C.L. (1994) Ecology of Crimean-Congo hemorrhagic fever. In: Sonenshine, D.E. and Mather, N. (eds) *Ecological Dynamics of Tick-borne Zoonoses*. Oxford University Press, New York, pp. 392–437.

Madder, M., Thys, E., Geysen, D., Baudoux, C. and Horak, I. (2007) *Boophilus microplus* ticks found in West Africa. *Experimental and Applied Acarology* 43, 233–234.

Madder, M., Thys, E., Achi, L., Touré, A., and De Deken, R. (2011) *Rhipicephalus (Boophilus) microplus*: a most successful invasive tick species in West-Africa. *Experimental and Applied Acarology* 53, 139–145.

Madder, M., Adehan, S., De Deken, R., Adehan, R. and Lokossou, R. (2012) New foci of *Rhipicephalus microplus* in West Africa. *Experimental and Applied Acarology* 56, 385–390.

Manilla, G. (1998a) Subfam. Hyalomminae. In: *Acari, Ixodida – Fauna d'Italia, Volume 36*. Edizioni Calderini, Bologna, Italy, pp. 174–200.

Manilla, G. (1988b) Subfam. Rhipicephalinae. In: *Acari, Ixodida – Fauna d'Italia, Volume 36*. Edizioni Calderini, Bologna, Italy, pp. 200–242.

Molin, Y., Lindeborg, M., Nyström, F., Madder, M., Hjelm, E., Olsen, B., Jaenson, T. and Ehrenborg, C. (2011) Migratory birds, ticks, and *Bartonella*. *Infection Ecology and Epidemiology* 1, doi: 10.3402/iee.v1i0.5997.

Nijhof, A.M., Bodaan, C., Postigo, M., Nieuwenhuijs, H., Opsteegh, M., Franssen, L., Jebbink, F. and Jongejan, F. (2007) Ticks and associated pathogens collected from domestic animals in the Netherlands. *Vector-borne and Zoonotic Diseases* 7, 585–595.

Nowak, M. (2010) Parasitation and localisation of ticks (Acari: Ixodidae) on exotic reptiles imported into Poland. *Annals of Agricultural and Environmental Medicine* 17, 237–242.

Ogden, N.H., Bigras-Poulin, M., O'Callaghan, C.J., Barker, I.K., Lindsay, L.R., Maarouf, A., Smoyer-Tomic, K.E., Waltner-Toews, D. and Charron, D. (2005) A dynamic population model to investigate effects of climate on geographic range and seasonality of the tick *Ixodes scapularis*. *International Journal for Parasitology* 35, 375–389.

Ozdarendeli, A., Canakoglu, N., Berber, E., Aydin, K., Tonbak, S., Ertek, M., Buzgan, T., Bolat, Y., Aktas, M. and Kalkan, A. (2010) The complete genome analysis of Crimean-Congo hemorrhagic fever virus isolated in Turkey. *Virus Research* 147, 288–293.

Papadopoulus B., Morel P.C. and Aeschliman A. (1996) Ticks of domestic animals in the Macedonia region of Greece. *Veterinary Parasitology* 63, 25–40.

Tack, W., Madder, M. and Verheyen, K. (2010) De impact van mastjaren op knaagdieren, teken en de ziekte van Lyme. *Antenna* 4(3), 29–31.

Temeyer, K.B., Pound, J.M., Miller, J.A., Chen, A.C., Pruett, J.H. Jr, Guerrero, F., Davey, R.B., Untalan, P.M., Lohmeyer, K.H., Li, A.Y., Miller, R. and George, J.E. (2004) Organophosphate resistance in Mexican strains of *Boophilus microplus*: a major threat to the U.S. cattle industry. *Southern Association of Agricultural Scientists Bulletin of Biochemistry and Biotechnology* 17, 43–51.

Walker, J.B. and Olwage, A. (1987) The tick vectors of *Cowdria ruminantium* (Ixodoidea, Ixodidae, genus *Amblyomma*) and their distribution. *Onderstepoort Journal of Veterinary Research* 54, 353–379.

Walker, J.B., Keirans, J.E. and Horak, I.G. (2000) *The Genus* Rhipicephalus *(Acari, Ixodidae): A Guide to the Brown Ticks of the World*. Cambridge University Press, Cambridge, UK.

4 Tick-borne Infections (Including Zoonoses) in Europe and the Mediterranean Basin*

Frank Koenen,[1] Ilaria Pascucci,[2] Thomas G.T. Jaenson,[3] Maxime Madder,[4] Rita de Sousa,[5] Agustín Estrada-Peña,[6] Róbert Farkas[7] and Mo Salman[8]

[1]Veterinary and Agrochemical Research Centre, CODA-CERVA, Brussels, Belgium; [2]Istituto Zooprofilattico Sperimentale dell'Abruzzo e del Molise "G. Caporale", Teramo, Italy; [3]Uppsala University, Sweden; [4]Institute of Tropical Medicine, Antwerp, Belgium and University of Pretoria, South Africa; [5]National Institute of Health Doutor Ricardo Jorge, IP (INSA), Águas de Moura, Portugal; [6]University of Zaragoza, Spain; [7]Szent István University, Budapest, Hungary; and [8]Colorado State University, USA

4.1 African Swine Fever

African swine fever (ASF) is a viral swine disease caused by the African swine fever virus (ASFV), an icosahedral complex DNA virus that is a unique member of the *Asfarviridae* family. It affects only porcine species, those of all breeds and ages. The disease was first described in Kenya by Montgomery in 1921 when the virus spread from infected warthogs (*Phacochoerus aethiopicus*) to domestic pigs (*Sus scrofa*) and resulted in a 100% case fatality rate. The disease is currently present in Africa, mainly in countries located south of the Sahara, and in most cases it is endemic. In Europe, ASF is still endemic in Sardinia. More recently, in 2007, ASFV spread to the Transcaucasian countries (TCC) and the Russian Federation (RF) (Sánchez-Vizcaíno, 2006).

Pigs are the only domestic animal species that are naturally infected by ASFV. Wild boar have also been identified as susceptible to ASFV infection, with clinical signs and case fatality rates similar to those observed in domestic pigs in Spain, Portugal and Sardinia (Italy) and, experimentally, in feral pigs in Florida (Sánchez-Vizcaíno, 2006). Wild boar and feral pigs can transmit the virus directly to domestic swine. In Africa, it has been observed that ASFV induces an inapparent infection in three species of wild pigs (warthogs, bush pigs and red river hogs), while the role played by the giant forest hog has not yet been clarified (Jori and Bastos, 2009).

At present, no treatment or effective vaccine against ASFV is available. Since 1963, when the first live attenuated vaccine was used in Portugal, many efforts have been made in this area, but with unsatisfactory results. As no vaccine for ASFV currently exists, the prevention of this disease in

* Frank Koenen wrote Section 4.1; Ilaria Pascucci, Section 4.2; Thomas G.T. Jaenson, Sections 4.3, 4.7, 4.11.2; Maxime Madder, Sections 4.4, 4.5; Rita de Sousa, Section 4.6; Agustín Estrada-Peña, Section 4.8; Róbert Farkas, Sections 4.9, 4.10, 4.11.3, 4.11.4; Mo Salman, Section 4.11.1.

disease-free areas depends on preventing the introduction of the virus (Sánchez-Vizcaíno, 2006).

ASFV is spread among domestic pigs via the oral–nasal route. However, it has also been demonstrated that the virus can be infectious by a number of other routes, including tick bites, experimental inoculation via cutaneous scarification, and by the intramuscular, intravenous, subcutaneous and intraperitoneal routes. The infection usually commences in the monocytes and macrophages of the tonsils and mandibular lymph nodes. From there it spreads through the draining lymph nodes and blood to the target organs (lymph nodes, bone marrow, spleen, lung, liver and kidney) which are the principal sites of secondary replication.

The clinical signs of ASF can resemble a variety of other swine haemorrhagic diseases and can easily be confused with classical swine fever (hog cholera) and erysipelas. Laboratory tests are necessary to confirm the diagnosis. ASF can also present different clinical signs that depend primarily upon the virulence of the virus, infectious dose and mode of infection, with a range of clinical forms varying from acute to subclinical and chronic.

Some species of soft ticks have proved to be ASFV reservoirs and vectors, such as *Ornithodoros moubata* and *O. porcinus* in Africa and *O. erraticus* in the Iberian Peninsula (Spain and Portugal). In *O. moubata*, transovarial and trans-stadial ASFV transmission have been described; in *O. erraticus*, only trans-stadial transmission has been demonstrated. Other soft tick species that are widely distributed in North and South America have been identified as harbouring and transmitting ASFV, and in the experimental setting, *O. savignyi*, present in Africa, has been shown to transmit ASFV to domestic pigs (see Table 2.1).

In Africa, ASFV is maintained by a cycle of infection between wild pigs and soft ticks. In some of these wild pigs, ASFV infection is characterized by low levels of virus in the tissues and low or undetectable levels of viraemia; however, these levels of virus are sufficient to infect soft tick vectors and for tick transmission of the virus to domestic pigs. This cycle of the virus makes ASF very difficult to eradicate in Africa. In Sardinia, where ASF is still present, wild boars are as susceptible as domestic pigs. No ticks from the *O. erraticus* complex have been found in Sardinia.

Experience of past outbreaks of ASF outside Africa has shown that the introduction of ASFV into a non-infected pig population within a free region is most often linked to entry through international ports or airports. The infected material is typically garbage containing uncooked pork that is used for pig feeding (Sánchez-Vizcaíno, 2006). Once ASFV is established in domestic swine, infected animals are the most important source of virus dissemination to susceptible pigs. In Europe, ASFV was introduced for the first time in 1957 into Portugal through waste from international flights. Although this first outbreak was rapidly eradicated, in 1960 the virus entered Europe again in Lisbon (Portugal) and spread through the rest of Portugal and Spain, where ASFV remained endemic until 1995. During this period, some other outbreaks occurred in other European countries, affecting Andorra (1975), Belgium (1985), France (1964, 1967 and 1974), Malta (1978), the Netherlands (1986) and Italy (1967, 1969 and 1993), including the island of Sardinia, where ASF has remained endemic since 1978. All of these virus introductions were also linked to swill feeding.

In Europe, several epidemiological paths are known to be able to maintain ASFV in domestic pig populations and this complicates the control of the disease. The main routes of transmission are: swill, domestic pig and wild boar interactions, and pig–tick interactions (Sánchez-Vizcaíno, 2006). In 2010, a scientific opinion issued by the EFSA (European Food Safety Authority) Panel on Animal Health and Animal Welfare (2010a) on ASF contained an assessment of the risk of introduction of this virus into the European Union (EU), especially from the Caucasus (Wieland *et al.*, 2011). Another scientific report from the EFSA Panel on Animal Health and Animal Welfare (2010c) discusses the role of tick vectors in the epidemiology of Crimean-Congo haemorrhagic fever (CCHF) and ASF in Eurasia, and contains geographic distribution maps of the tick vector and ASFV.

4.1.1 Virus–tick interaction

Several factors can influence the vector competence of soft tick species for ASFV. Some authors consider ASFV and the *Ornithodoros* tick as co-evolving organisms. Actually, noticeable telomeric similarities in the genomes of ASFV and *Borrelia* (the latter shares the same *Ornithodoros* tick host in Africa and is considered to be an original pathogen of soft ticks) suggest that ASFV is also a primary organism of *Ornithodoros* ticks and that it coadapts to its tick hosts (Hinnebusch and Barbour, 1991). This hypothesis could explain the noticeable discrepancies concerning infection success rates that have been reported in several past surveys. For example, De Tray (1963) reported consistent establishment of the virus isolate 'Uganda' in specimens of the *O. moubata* group (34/35 were infected) whereas another isolate, 'Tengani', caused persistent infection only in a small proportion of ticks (2/46 were infected). More recently, Kleiboeker *et al.* (1999) compared oral and intra-haemocoelic experimental infections of *Ornithodoros* ticks collected from warthog burrows in Kruger National Park and the Northern Transvaal region of South Africa, as well as infections of ticks from Masai Mara Reserve in Kenya. These researchers used three different viruses from South Africa, Malawi and Zimbabwe, all originally isolated from ticks collected in the field. The oral infection using the isolate from Malawi was self-limiting (decline of virus titres and number of ticks containing virus), while the others persisted. According to Kleiboeker *et al.* (1999), the cytopathology caused by the Malawi strain in infected ticks suggested the non-adaptation of the isolate to express specific genes that allow the production of large quantities of progeny virus without damaging the host cell. The reason why this virus was originally isolated from ticks could be the large opportunity for those ticks to feed on infected pigs with high viraemic titres during an ASF outbreak and the leakage of midgut contents into the haemocoel without tick mortality, instead of real adaptation of this virus isolate to the tick host (Kleiboeker *et al.*, 1999). Dixon and Wilkinson (1988) suggested that virus replication in ticks and warthogs may require additional host-specific genes that are not necessary for multiplication in domestic pigs, and that the introduction of virus from tick/warthog sources into domestic pig populations would remove the selection pressure for maintaining these genes. However, no more information is currently available on specific determinants for tick/warthog hosts, as was previously suggested. In addition, it is unknown whether ASFV is able to come back from the domestic to the sylvatic cycle, although some authors suggest that recombination processes during co-infections in ticks may exist (Plowright, 1977; Dixon and Wilkinson, 1988).

4.1.2 Genetic diversification of ASFV

By sequencing the C-terminal end of the *p72* gene, Bastos *et al.* (2003) and Lubisi *et al.* (2005) observed higher genetic variations in genotypes directly isolated from *Ornithodoros* ticks and warthogs, or in genotypes circulating in East and southern Africa where the sylvatic cycle plays a crucial part in the epidemiology of ASF. Some other genotypes were only found in domestic pigs and presented low genetic divergence (Lubisi *et al.*, 2005). In Madagascar, using concatenated sequences of the *p22* and *p32* genes, Michaud (Michaud *et al.*, 2007) detected relatively high genetic divergence between Malagasy virus isolates collected on domestic pigs from 1998 and 2003, compared with that observed on West African and European isolates since the 1970s. In Madagascar, it has been suspected that the introduced virus was adapted to local bush pigs and *Ornithodoros* ticks, leading to its accelerated diversification. Such diversification phenomena as reported in several African countries has been previously analysed by Dixon and Wilkinson (1988) on Zambian virus isolates from *Ornithodoros* ticks. A considerable genetic diversity was observed between virus isolates from ticks collected from the same regions and even from the same warthog burrows. This diversity resulted from peculiar point mutations all along the length of the genome, instead of insertions/deletions in the region close to the left-hand terminus of the genome usually observed for host selection (Dixon and Wilkinson, 1988). In this case, ticks would be able to enhance the diversification of ASFV and the emergence of new virulent isolates to domestic pigs. However, no information is yet available on the location and the expression of this genetic diversification. In addition, this process does not seem compatible with the persistence of ASFV in Iberian *Ornithodoros* ticks and the observed genetic homogeneity of ASFV in Europe.

4.2 Crimean-Congo Haemorrhagic Fever

CCHF is a tick-borne zoonotic infection that has public health concern in several regions of the world including Africa, the Middle East, the Balkans region, Greece, Turkey and western Asia. The infection is caused by a virus belonging to the genus *Nairovirus* (family *Bunyaviridae*) and is transmitted by several species of hard (ixodid) ticks, particularly by those belonging to the *Hyalomma* genus (Horak *et al.*, 2001).

Phylogenetic analyses performed on S-, M- and L-RNA segments of the Crimean-Congo haemorrhagic fever virus (CCHFV) showed virus strains grouped in seven different clades (Deyde *et al.*, 2006). Three clades are distributed in Africa, two in Europe and the other two in Asia. Despite the potential for dispersal of the virus in Africa and Eurasia, it appears that circulation of the virus is largely confined to within two specific regions in these continents, and corresponds to the distribution and dispersal of tick vectors of the virus (Burt and Swanepoel, 2005; Paweska, 2007). It has also been concluded that viral strains have a latitudinal relationship in which there is not much interchange of the strains between different latitudes (i.e. Africa and Europe).

The virus may be maintained in tick populations during inter-epizootic periods through several mechanisms, such as transstadial and transovarial transmission, and non-viraemic transmission of ticks aggregated on the same host (co-feeding). Outbreaks usually take place during the peak activity periods of *Hyalomma* ticks, coinciding with the hot and dry season (Swanepoel, 2006).

There are a large number of potential vertebrate host reservoirs for CCHFV, reflecting the diverse feeding preferences of the immature and adult tick vectors. Antibodies against the virus have been detected in domestic and wild animals, including hares, hedgehogs, rodents, bats, and large mammals such as giraffes and rhinoceroses. The most important source of virus transmission is immature ticks of the *Hyalomma* genus which have fed on the blood of viraemic small vertebrates. Once infected, the tick remains infected throughout its life (transstadial transmission), and the mature tick may transmit the virus to large vertebrates such as livestock. Domestic ruminants such as cattle, sheep and goats will have viraemia for approximately a week after becoming infected. They may be the source of infection for humans during slaughtering or veterinary procedures such as castration. In Appendix 1, Table 1.24 contains data on the serological surveillance of CCHF in domestic animals. The level of viraemia in birds is usually low and unnoticeable, even though migratory birds may play role in the epidemiology of the virus by disseminating infected *Hyalomma* ticks. Tick-infested birds migrating from Russia across the Black Sea were suggested as a link to introduction of the virus in Turkey (Karti *et al.*, 2004). Nevertheless, many social and environmental factors affect CCHF occurrence in Turkey. The epidemiological features of the disease are still under discussion.

The most common cause of infection in humans is a bite from an infected tick. Infection can also occur through direct contact with blood or tissues from infected humans or livestock. The highly pathogenic nature of the virus occasionally results in serious nosocomial outbreaks (Swanepoel, 1995; Aradaib *et al.*, 2010; Elata *et al.*, 2011; Naderi *et al.*, 2011). Clinical infection in humans is initially manifested as an acute febrile illness that can be followed by a fatal haemorrhagic syndrome with case-fatality rates of up to 50% (Swanepoel *et al.*, 1987). Misdiagnosis is frequently due to the nonspecific clinical signs (Fisgin *et al.*, 2010). The diagnosis is based on serology or viral RNA detection by molecular techniques. During human infection, viral genomes are present in saliva and urine with viral loads similar to those in blood (Bodur *et al.*, 2010). Two animal models have been recently established by using genetically defective mice in which the interferon response had been altered (Bente *et al.*, 2010; Bereczky *et al.*, 2010). These animal models will be beneficial for the development of treatments and vaccines (Keshtkar-Jahromi *et al.*, 2011).

Outbreaks in South Africa arose among slaughterhouse operators during the slaughter of ostriches which were heavily infested with ticks. The infection occurred when the infected ticks on the carcasses were squashed during skinning (Swanepoel, 1998). Nevertheless, meat from butchered animals does not pose a risk; in this substrate the CCHFV is quickly inactivated by a drop in pH, as occurs during the maturation process that the meat undergoes after slaughter. A scientific report from the EFSA Panel and Animal Health and Welfare (2010c) addresses the role of tick vectors in the epidemiology of CCHF and ASF in Eurasia.

4.3 Tick-borne Encephalitis Group

4.3.1 Tick-borne encephalitis

Tick-borne encephalitis (TBE) is one of the most important and serious human infections occurring in Europe and many parts of Asia. The aetiological agent, tick-borne encephalitis virus (TBEV), is a member of the genus *Flavivirus* of the family *Flaviviridae*. TBEV is believed to cause at least 11,000 human cases of encephalitis in Russia and more than 3000 cases in the rest of Europe annually (Donoso Mantke *et al.*, 2008). Related viruses are louping ill virus (LIV), Langat virus (LGTV) and Powassan virus (POWV), which may also cause human encephalitis, and Omsk haemorrhagic fever virus (OHFV), Kyasanur Forest disease virus (KFDV) and Alkhurma virus (ALKV), which cause serious haemorrhagic fevers rather than encephalitis (Gritsun *et al.*, 2003).

Three subtypes of the TBEV are recognized: the Western or Central European subtype, including the Kumlinge virus on Åland in Finland; the Siberian subtype; and the Far Eastern subtype. Recently, the Siberian subtype was recognized as a human pathogen in western Finland where populations of the vector, *Ixodes persulcatus*, were also recorded.

I. ricinus is the main tick vector involved in TBEV infections in Europe. All its stages can attack humans although the nymphs are the most important ones as vectors of the virus to humans. This is due to several factors: the unfed larvae are usually not infected but may become infected while taking their first blood meal on a viraemic host or, which is more important, by co-feeding with infective nymphs; the nymphs are far more abundant in nature than the adult ticks; and the colourful adult females are relatively large (3.5 mm) compared with the dull-coloured smaller (1.5 mm) nymphs and, therefore, are more easily detected and removed when encountered on the human body. In TBE foci in central and northern Europe, the infection prevalence of TBEV in nymphs ranges around 0.1–0.5%, with that in adults about 0.3–6.0%. The infection prevalence in adult females of *I. persulcatus*, which is the main vector stage for the Eastern TBEV, tends to be much higher (up to 40%) than in *I. ricinus* (Labuda and Nuttall, 2008). Co-circulations of both Western TBEV and Eastern TBEV occur in some foci in the Baltic States, where the distributions of the two tick species overlap. The Far Eastern subtype has been discovered not only in Siberia but also in some European localities (Chausov *et al.*, 2010). In Estonia, all three human-pathogenic subtypes of TBEV have been found in the same areas (Golovljova *et al.*, 2004). Apart from the two main vector species, *I. ricinus* and *I. persulcatus*, several other tick species, including *I. hexagonus*, *I. arboricola*, *Haemaphysalis* (*Ha.*) *concinna*, *Ha. inermis* and *Ha. punctata* are competent but secondary vectors (Labuda and Nuttall, 2008).

Until not long ago, it was believed that viraemic small rodents, particularly the bank vole, *Clethrionomys glareolus*, and the field mouse, *Apodemus flavicollis*, and insectivores were the principal reservoirs of TBEV that infected the vectors. However, the viraemia in these rodents is usually of short duration (2 days). For the maintenance of TBEV in *I. ricinus* populations, co-feeding transmission between infective nymphs and susceptible larvae feeding very close to and on the same small rodent is now considered to be much more important than transmission via viraemic small mammals (Labuda and Nuttall, 2008).

Many cases of TBE in humans are unrecognized and without clinical signs or symptoms. In some cases, however, the clinical syndrome of TBE disease is severe, with a life-threatening neurological syndrome and high case fatality (5 to 35%) in its Eastern form, mainly in Russia. In contrast, the case fatality in Western Europe, mainly in Central and Northern Europe, is usually comparatively low (approximately 1%) with nearly all deaths confined to patients above 60 years of age. Patients infected with the Siberian subtype may suffer from a milder but often more chronic disease compared with the disease caused by the Far Eastern subtype.

The incidence of TBE usually fluctuates from year to year, but an increased incidence has been noted in some countries (Danielova *et al.*, 2006; Lindquist and Vapalahti, 2008), and new TBE foci seem to have appeared, especially in the last decade. This is presumably the result of a complex interaction of factors such as the changing climate affecting the vector both directly and indirectly by affecting the plant and host communities, sociopolitical changes and technological factors, e.g., better diagnostic methods and increased awareness (Donoso Mantke *et al.*, 2008; Telford and Goethert, 2008).

4.3.2 Louping ill

Louping ill virus (LIV), also known as ovine encephalitis/encephalomyelitis virus, is closely related to TBEV and is the only member of this virus complex present in the British Isles, where the vector is *I. ricinus* (Reid, 1988). Louping ill (LI) is endemic in sheep-farming areas of northern England, Scotland, Wales, Ireland and Norway. Several tick hosts, such as the red grouse, willow grouse, field vole and deer become viraemic when infected with LIV. The viraemia, however, is usually too low to be infective to feeding tick larvae. In contrast to these wild hosts, sheep and red grouse consistently develop viraemia sufficient to infect tick larvae and amplify the virus. Occasionally, horses, cattle and goats develop viraemias sufficiently high to be infective to tick larvae (Reid, 1988). Mountain hares, *Lepus timidus*, may be maintenance hosts for LIV by non-viraemic transmission between co-feeding ticks, as experimentally demonstrated by Jones *et al.* (1997). One of the main assumptions for the transmission of a vector-borne pathogen is usually that feeding by the vector is the sole or main route of host infection. Gilbert *et al.* (2004), however, demonstrated experimentally a transmission route whereby an important tick host, the red grouse (*Lagopus lagopus scoticus*), became infected with LIV after eating infected *I. ricinus* ticks. These authors estimated from field observations conducted in Scotland that this mode of infection could account for 73–98% of all virus infections in wild red grouse in their first season. Certainly, this way of transmission has potential implications for the understanding of other vector-borne pathogens in which hosts may ingest vectors through foraging or grooming.

LI is principally a disease of sheep and red grouse, and less commonly of cattle, other domesticated animals and birds. Dogs, and particularly sheepdogs and hunting dogs in endemic areas, are occasionally infected. Clinical signs include fever, ataxia, trembling, salivation, coma and death. The virus can cause severe encephalitis in humans; about 35 cases have been recorded. Most of these cases were due to accidents while handling the virus in the laboratory

(Labuda and Nuttal, 2008). Definitive diagnosis is based upon the isolation and identification of the virus (Reid, 1988; Lobetti, 2007). LIV infection in sheep is exacerbated by co-infection with *Anaplasma phagocytophilum* (Reid, 1988).

Experimental and trans-stadial transmission have been reported in *Rhipicephalus* (*Rh.*) *appendiculatus* and *Hyalomma* (*Hy.*) *anatolicum*, but there is no evidence that they are natural vectors or that any other tick species except *I. ricinus* play any significant role in the epidemiology of the disease (Reid, 1988).

Although *I. ricinus* is the primary vector and virus reservoir, because of the trans-stadial – but presumably not transovarial – transmission, the vector efficiency of this species is relatively restricted. Even when virions are acquired by the feeding larvae, only a few of the nymphs become infected. Like the prevalence of TBEV in the *I. ricinus* population, the prevalence of LIV in *I. ricinus* is also low.

4.3.3 Other viruses related to TBE

Infections of domesticated animals similar to TBE and LI also occur in other European countries (Spain, Bulgaria, Greece and Turkey). Greek goat encephalitis virus (GGEV), which was isolated from the brain of a newborn goat with neurological symptoms, is currently classified in the TBEV group. The vector of GGEV has not yet been specifically identified but it is considered as likely to be *I. ricinus*. A study during 2003–2006 in goat and sheep farming rural areas of northern Greece suggested the presence of TBEV in two pools of *I. ricinus* ticks. Sequence analysis showed that the virus was GGEV. These virus-positive ticks were detected in regions where a high prevalence of TBE antibodies was present in humans. TBEV is considered not to be endemic in Greece, so most probably the seroprevalence of TBE antibodies in humans is due to cross-reactivity to GGEV (Grard *et al.*, 2007). The Turkish subtype (Turkish sheep encephalitis virus) is more closely related to LIV and should be reclassified.

4.4 Anaplasmoses

Alphaproteobacteria of the order *Rickettsiales* are obligate intracellular organisms with a wide range of eukaryotic hosts. There are two well-characterized families: *Anaplasmataceae* and *Rickettsiaceae*. In the family *Anaplasmataceae*, four genera have been identified: *Anaplasma*, *Ehrlichia*, *Wolbachia* and *Neorickettsia*. Several species of the genus *Anaplasma* pose severe threats to livestock and human health. The main *Anaplasma* species responsible for animal infections or zoonosis in Europe and the Mediterranean basin are listed in Table 4.1. Anaplasmosis was formerly known as gall sickness, a disease of ruminants caused by intraerythrocytic bacteria. As a result of a taxonomic reorganization of the order *Rickettsiales* (Dumler *et al.*, 2001), some species of the genus *Ehrlichia* (*E. equi*, *E. phagocytophila* and *Ehrlichia* spp. causing human granulocytic ehrlichiosis – HGE) were renamed as *A. phagocytophilum* in the genus *Anaplasma*. This species is the aetiological agent of human and animal granulocytic anaplasmosis. Similarly, *E. bovis* and *E. platys* are now known as *A. bovis* and *A. platys*. These latter, recently added, species all invade blood cells other than erythrocytes in their respective mammalian hosts.

Infection of domestic and wild animals and humans with these organisms may lead to a clinical disease collectively called anaplasmosis that manifests as a febrile systemic illness with haematological abnormalities and lymphadenopathy (Rikihisa, 2006).

4.4.1 *Anaplasma phagocytophilum*

A. phagocytophilum, formerly known as *E. phagocytophila* and *E. equi*, and the agent of HGE, is a commonly found bacterium causing tick-borne fever (TBF) in sheep; pasture fever in cattle and wild ruminants (deer, bison and wild goats); human granulocytic anaplasmosis in humans; equine granulocytic anaplasmosis in horses; and canine granulocytic anaplasmosis in dogs (Strle, 2004; Rymaszewska and Grenda, 2008). It was first recognized in Scotland in 1932 and is now identified in most other European

Table 4.1. Anaplasmoses in Europe and in the Mediterranean basin transmitted by hard ticks.

Anaplasma spp.	Disease	Host range	Tick involved (in Europe and the Mediterranean basin)
A. phagocytophilum	Tick-borne fever (TBF) Pasture fever Human granulocytic anaplasmosis	Sheep Cattle, wild ruminants Humans	*I. ricinus, Ha. punctata, I. persulcatus, I. trianguliceps, Rh. sanguineus*
	Equine granulocytic anaplasmosis	Horses, llamas, rodents	
	Canine granulocytic anaplasmosis	Dogs	*I. ricinus, Ha. punctata, I. persulcatus, I. trianguliceps, Rh. sanguineus*
A. marginale	Bovine anaplasmosis	Ruminants	*I. ricinus, I. persulcatus, Rh. sanguineus, Rh. bursa, Rh. annulatus*
A. centrale	Bovine anaplasmosis	Cattle	*I. ricinus, I. persulcatus, Rh. sanguineus, Rh. bursa, Rh. annulatus*
A. bovis	Bovine mononuclear or agranulocytic anaplasmosis	Cattle, small mammals	*Hy. excavatum, Rh. sanguineus, Rh. turanicus*
A. ovis	Ovine anaplasmosis	Goats, sheep, cattle	*Rh. bursa, Rh. sanguineus*
A. platys	Canine infectious cyclic thrombocytopenia	Dogs	*Rh. sanguineus, Rh. turanicus*

A., *Anaplasma*; *Ha.*, *Haemaphysalis*; *Hy.*, *Hyalomma*; *I.*, *Ixodes*; *Rh.*, *Rhipicephalus*.

countries. Apart from domestic ruminants, free-living ruminants such as feral goats, and red, fallow and roe deer tested positive for *A. phagocytophilum*. Not only a tick-ruminant cycle, but also a rodent-tick cycle, is believed to maintain the TBF variants. The wood mouse (*Apodemus sylvaticus*), yellow-necked mouse (*A. flavicollis*), field vole (*Microtus agrestis*) and bank vole (*Myodes glareolus*) are found to be competent reservoirs of infection (Barandika *et al.*, 2007). The bacterium infects granulocytic leucocytes (neutrophils, eosinophils and basophils), monocytes and tissue macrophages.

TBF in sheep and pasture fever in cattle are characterized by fever, neutropenia, lymphopenia, thrombocytopenia and general immunosuppression (Woldehiwet, 2006). Human granulocytic anaplasmosis is a multisystemic disease that occurs more in adults than in children, especially in persons above the age of 60 years. The disease is characterized by acute fever, headache, myalgia, nausea and lethargy, similar to symptoms of the common flu. In particular, immunocompromised patients are at high risk. Meningoencephalitis, respiratory distress, shock and opportunistic infections are occasional complications. In Europe, no fatal cases have been reported, but the mortality rate in the USA has been shown to be between 7 and 10% (reviewed by Bakken and Dumler, 2008 and Rymaszewska and Grenda, 2008). Equine granulocytic anaplasmosis occurs in horses as their natural host but also in llamas and rodents. It is generally a benign disease in these animals, yet fulminating cases have been described. Mortality is low, and the disease is always acute, never chronic.

The main vector of *A. phagocytophilum* is *I. ricinus* (Strle, 2004), and the prevalence of infection varies among regions and with the development stage of the tick (Stuen, 2007). For example, in unfed nymphs the infection rate varied between 0.25 and 25% (Walker *et al.*, 2001). The survival of the parasite is believed to be over a year while ticks are awaiting a new host. Only trans-stadial transmission occurs.

The transmission of *A. phagocytophilum* has also been associated with other tick species, such as *Ha. punctata*, in areas of the UK where *I. ricinus* was not present (MacLeod, 1936). It has also been linked with *I. persulcatus*, *I. trianguliceps* and *Rh. sanguineus* (Alekseev *et al.*, 1998; Ogden *et al.*, 1998; Alberti *et al.*, 2005). The role of the latter species as a vector is not yet determined, as only one *Rh. sanguineus* was found to be positive; this tick was removed from a dog showing clinical signs of tick-borne disease (TBD).

4.4.2 *Anaplasma marginale*

Bovine anaplasmosis, caused by *A. marginale*, was formerly known as gall sickness, and is a disease that affects domestic and wild ruminants (water buffalo, bison, African antelopes and mule deer). The disease is characterized by fever, anaemia, weight loss, reduction of milk production and in pregnant females, abortion; it may lead to death (Rymaszewska and Grenda, 2008; Kocan *et al.*, 2010). *A. marginale* is present in tropical and subtropical regions although this bacterial species is frequently detected in Europe (Sicily, Hungary and Spain) (de la Fuente *et al.*, 2005; Naranjo *et al.*, 2006; Hornok *et al.*, 2007b; Torina *et al.*, 2007, 2008). It is an obligate intracellular species invading erythrocytes mostly in ruminants, both domestic and wild, i.e. calves, water buffalo, bison, African antelopes and mule deer.

In cattle, the disease in cattle causes considerable losses to dairy and beef industries worldwide. Trans-stadial transmission of the bacteria is effected by ticks of approximately 20 species; in Europe this mainly involves *I. ricinus*, *I. persulcatus*, *Rh. sanguineus*, *Rh. bursa* and *Rh.* (*Boophilus*) *annulatus* (Kocan *et al.*, 2004). Calves under the age of 6 months have innate resistance and will not develop clinical anaplasmosis, no matter the immune status of the mothers. Thereafter, the risk for serious diseases increases with age, unless sufficient contact in the first months of life allowed for the development of immunity. Hence, cattle reared in endemic regions develop a naturally acquired immunity, quite often without passing through a stage of clinical disease, as endemic stability means that all calves need to come into contact with the disease, reservoirs and stable vector populations.

Wild ruminants (antelopes, buffalo, deer, eland) can function as reservoirs of *A. marginale* and the infection can be maintained in game reserves (deer-to-deer transmission) without bovine intervention being necessary (Potgieter and Stoltsz, 2004).

It appears that *A. marginale* is often introduced into a herd by ticks, but subsequently mechanical transmission (transmission by insects or by veterinary interventions) may become more important. Mechanical transmission occurs via the contaminated mouthparts of biting flies, but can only be achieved within a few minutes after the initial bite, although the pathogen can remain viable and infective in arthropods for several days after ingestion (Ewing, 1981; Hornok *et al.*, 2008). Horseflies (*Tabanus* spp.) and stable flies (*Stomoxys* spp.), and to a lesser extent mosquitoes (*Psorophora* spp.), transmit *A. marginale* and also *A. centrale*.

4.4.3 *Anaplasma centrale*

A. centrale is considered as a separate species or subspecies of *A. marginale*, and is also an intraerythrocytic tick-borne pathogen that causes mild infections in cattle. A cross-immunity between the two bacteria exists, and because of its mild virulence this naturally attenuated strain has been used for more than 100 years in live-blood vaccines to protect cattle from the more virulent *A. marginale* (Potgieter and Stoltsz, 2004). These vaccines are mainly used in Africa, Australia, Latin America and Israel (Rymaszewska and Grenda, 2008).

Not much is known about the epidemiology of *A. centrale*; only few strains have been characterized. In Europe, the pathogen has been detected mainly in Italy (George *et al.*, 2001; Carelli *et al.*, 2008).

The first case of bovine anaplasmosis caused by *A. centrale* in Europe was reported in 2008 in Italy (Carelli *et al.*, 2008). The case involved a naturally occurring infection in a dairy cow, and the clinical signs were typical of acute anaplasmosis caused by *A. marginale*. Molecular analysis of the Italian strains linked these to the African *A. centrale* strains (Carelli *et al.*, 2008).

Recently, a complete genome sequence of *A. centrale* was compared with that of virulent *A. marginale* (Herndon *et al.*, 2010) with the aim of identifying possible outer membrane protein candidates for the development of a safer inactivated vaccine.

4.4.4 *Anaplasma bovis*

A. bovis is the aetiological agent of bovine mononuclear or agranulocytic anaplasmosis, a disease occurring mainly in cattle and small mammals (Goethert and Telford, 2003). Goats appear to be resistant. *A. bovis* infects the monocytes of the peripheral blood and the macrophages of the reticuloendothelial system. Infection may occur with limited or no clinical signs. The disease is characterized by weakness, weight loss, fever, enlargement of the prescapular lymph nodes, paleness of the mucous membranes and mucous nasal secretion (Uilenberg, 1997). This disease has been reported in Italy (Georges *et al.*, 2001) and Israel (Harrus *et al.*, 2011), but is most commonly present in South America, West, Central and southern Africa, and the Indian subcontinent. The transmission of the disease is transstadial by known vectors: *Amblyomma* (*Am.*) *variegatum*, *Rh. appendiculatus* and *Hy. excavatum* (Coetzer and Tustin, 2004). The pathogen has recently been detected in Israel in unfed *Rh. sanguineus* and *Rh. turanicus* adults collected from vegetation (Harrus *et al.*, 2011).

4.4.5 *Anaplasma ovis*

A. ovis mainly infects wild ruminants and small ruminants like sheep and goats, and is prevalent worldwide (Rymaszewska and Grenda, 2008). In Europe, it has been detected in Italy, Hungary and Turkey (de la Fuente *et al.*, 2002, 2005; Christova *et al.*, 2003; Hornok *et al.*, 2007b; Aktas *et al.*, 2009). This bacterium also infects erythrocytes, but in general anaplasmosis due to *A. ovis* in small ruminants is a benign infection with low morbidity and mortality. Goats are normally more susceptible than sheep or cattle. The biological vector of *A. ovis* in the Mediterranean basin is *Rh. bursa* (Friedhoff, 1997), but possibly also *Rh. sanguineus. A. ovis* was identified molecularly in the salivary glands of two *Rh. sanguineus* ticks collected from sheep in Turkey (Aktas *et al.*, 2009), although the authors also suggest that *D. marginatus* could play a role in the transmission because of the high prevalence of ticks found in the affected flock. In contrast to *A. marginale*, vaccination with *A. centrale* does not protect against *A. ovis*.

4.4.6 *Anaplasma platys*

Canine anaplasmosis or canine infectious cyclic thrombocytopenia (CICT) is caused by *A. platys*. The bacterium multiplies in platelets, but infected dogs may remain asymptomatic (Harvey *et al.*, 1978). Clinical signs are usually not observed, apart from occasional haemorrhages after trauma or surgery. The infection has been detected worldwide, and in Europe cases have been reported in Spain, France, Greece and Italy (Sainz *et al.*, 1999; Sparagano *et al.*, 2003; Mylonakis *et al.*, 2004; Alberti and Sparagano, 2006; Torina *et al.*, 2008; Yabsley *et al.*, 2008). Serological tests may cross-react with other *Anaplasma* and *in vivo* tests may be inaccurate because of low bacteraemias (de la Fuente *et al.*, 2006).

Cases of importation of both the infectious agent and the vector have been reported in dogs visiting the Mediterranean region (Heyman *et al.*, 2007; Nijhof *et al.*, 2007). The implicated vector seems to be *Rh. sanguineus* (Inokuma *et al.*, 2000; Sanogo *et al.*, 2003; Sparagano *et al.*, 2003), a cosmopolitan tick species that also transmits *E. canis*, although ticks of *Rh. sanguineus* fed on experimentally infected dogs were not able to infect naive dogs in the adult stage (Simpson *et al.*, 1991).

Similarly, as mentioned for *A. bovis*, unfed *Rh. turanicus* adults collected from the vegetation also have been shown to harbour *A. platys* (Harrus *et al.*, 2011).

4.5 Ehrlichioses

4.5.1 *Ehrlichia canis*

E. canis is a tick-transmitted obligate intracellular Gram-negative bacterium; in dogs it infects monocytes and causes classical canine monocytic ehrlichiosis. It resides as a colony within a membrane-lined intracellular vacuole or morula. Canine monocytic ehrlichiosis is also known by other names, such as tracker dog disease, tropical canine pancytopenia, canine haemorrhagic fever and canine typhus. The disease has been known since 1935 as a disease of dogs and other canids (jackal, wolf, fox, coyote) (Donatein and Lestoquard, 1935), but its importance was not seriously considered until 1968 when an epizootic occurred in Vietnam among military working dogs.

The bacterium is transmitted transstadially by the kennel tick (or brown dog tick), *Rh. sanguineus*, and is widespread in tropical and temperate areas of the world (from 50°N to 35°S). Its distribution has expanded with the distribution of its vector and is maintained in nature by persistent infections of wild and domestic canids (Groves *et al.*, 1975). In Europe, *E. canis* is restricted to the Mediterranean and Balkan countries (Hornok *et al.*, 2010), although several cases of canine ehrlichiosis have been detected in more northern regions in dogs with a history of travelling (Hirsch and Pantchev, 2008).

The disease has three clinicopathological phases: acute, subclinical and chronic (Skotarczak, 2003). German Shepherd dogs and their crosses are particularly prone to more severe signs of disease, and infections in this breed are associated with a poorer prognosis (Raoult and Parola, 2007).

Several cases of human infections with *E. canis* have been reported (Maeda *et al.*, 1987; Sambri *et al.*, 2004; Perez *et al.*, 2006). Clinical signs of canine monocytic ehrlichiosis in humans are very similar to those of human monocytic ehrlichiosis, a zoonosis present in the USA and caused by the closely related *E. chaffeensis* (Perez *et al.*, 2006).

4.5.2 *Ehrlichia ruminantium*

E. ruminantium, previously known as *Cowdria ruminantium*, causes heartwater or cowdriosis (other names are black gall sickness, mad gall sickness, infectious exudative pericarditis or malignant rickettsiosis of ruminants). Heartwater is an infectious tick-transmitted disease of ruminants, i.e. bovines, sheep, goats and various wild species. Transmission is related to the *Amblyomma* vector. In Africa, at least ten species are capable of transmission; the most important are *Am. variegatum* and *Am. hebraeum*, and the adults of both species parasitize cattle. The most widely distributed *Amblyomma* species in Africa is *Am. variegatum*, and this even spreads outside the continent. *Am. hebraeum* is the most important vector in southern Africa.

Heartwater occurs in sub-Saharan Africa and in several African islands, the islands in the Indian Ocean and several Caribbean islands, i.e. Guadeloupe and Antigua. In continental America, it has not been observed, in spite of the presence of potential vectors (Coetzer and Tustin, 2004). The control of *Am. variegatum* in the Caribbean has been a daunting task. Although the ticks had been eradicated for a number of years, the islands became infested again after the import of immature stages on migrating cattle egrets (Corn *et al.*, 1993).

Am. variegatum sporadically occurs in the Mediterranean basin, most probably imported on migratory birds (Papadopoulos *et al.*, 1996). According to a predictive GIS (geographical information system) model using temperature and land use, the survival of *Am. variegatum* and *Am. hebraeum* in this region would be best suited to Sardinia, Sicily and the south-western part of the Italian peninsula (Pascucci *et al.*, 2007). This poses a possible risk for the introduction of *E. ruminantium* into the Mediterranean region.

4.6 Rickettsioses

Rickettsial diseases are among the oldest known arthropod-borne diseases affecting human health throughout the ages. These diseases exist primarily in endemic and enzootic foci and occasionally give rise to sporadic or seasonal outbreaks, causing illness and death worldwide (Maxey, 1899). Evidence suggests that rickettsiae and *Rickettsia*-like organisms have evolved and survived as obligate intracellular bacteria, cultivating long and well-established relationships with arthropods (lice, mites, fleas and ticks) and vertebrate hosts (Azad and Beard, 1998). The vectors of rickettsiae do include lice, mites and fleas, but most *Rickettsia* species are associated with ticks, which are both their vectors and reservoirs. Ticks become infected with rickettsiae mainly by feeding on rickettsaemic host animals, particularly young rodents, and by the passage of rickettsiae via eggs to the progeny of infected female ticks. Trans-stadial persistence from larvae to nymphs and to adults is also important for maintaining the infection so that transovarial transmission is successful (Burgdorfer and Brinton, 1975). Geographic distribution and activity of infected ticks are important determinants in the epidemiology of tick-borne rickettsioses. Humans are accidental hosts that become infected when ticks containing virulent rickettsiae in their salivary glands take a blood meal and inject the rickettsiae into the feeding site. Currently, there are 12 tick-transmitted species of rickettsiae that are known to cause disease in man. Eight are present in Europe and the Mediterranean basin, and the majority have been described in the last 25 years. This boom in new or rediscovery of *Rickettsia* species in recent years is undoubtedly related to the development of cell culture systems and the advent of molecular genetics technology. Likewise, the detection and identification of long-known pathogens in patients from new and distinct geographic regions has refuelled interest in these agents. These emerging infections often have diverse presentations that underscore the need for increased awareness among clinicians.

4.6.1 Rickettsiae as bacteria

The genus *Rickettsia* is classified in the *Alphaproteobacteria* subdivision, order *Rickettsiales* (which also includes the families *Anaplasmataceae* and *Holosporaceae*), family *Rickettsiaceae* (Dumler *et al.*, 2001). This family includes the genus *Rickettsia* along with *Orientia* (Tamura *et al.*, 1995).

Rickettsiae are coccobacillary Gram-negative, obligate intracellular bacteria that reside free in the cytosol and occasionally in the nucleus of host cells (Walker, 2006). This lifestyle within a highly specialized niche, the eukaryotic cell, has given rise to unique adaptations such as the reduction of bacterial metabolism and the exploitation of host metabolites. The ability to transport substrates that are present in the host cell cytosol but rarely available in the extracellular milieu is likely to have contributed to the evolution of the small-sized rickettsial genome (1.11–1.12 Mb). Many of the *de novo* biosynthetic pathways characteristic of free-living bacteria are no longer present in rickettsiae (Andersson and Kurland, 1998).

The rickettsiae are traditionally divided into two antigenically distinct groups based on their lipopolysaccharide (LPS): the typhus group (TG) that includes two species, *R. prowazekii*, the agent of epidemic typhus transmitted by the human body louse (*Pediculus humanus corporis*), and *R. typhi*, the agent of murine typhus transmitted by the rat flea; and the spotted fever group (SFG) that includes the majority of rickettsial species, which are transmitted by different arthropod vectors.

Different phylogenetic studies using the sequencing of 16S rDNA (Roux and Raoult, 1995) and diverse rickettsial genes – *glt*A (Roux *et al.*, 1997), *omp*A (Fournier *et al.*, 1998), *omp*B (Roux and Raoult, 2000), *sca*2 (Sekeyova *et al.*, 2001), *sca*4 (Ngwamidiba *et al.*, 2005) and *sca*1 (Ngwamidiba *et al.*, 2006) – have shown distinct groups of *Rickettsia* species, but some authors still believe that it is difficult to define the SFG species. This is because some species of *Rickettsia* are so closely related that they could be considered to be strains of a single species (Walker, 2007). For example, the sequence analysis of 16S rDNA for *R. conorii*,

R. sibirica, *R. africae*, *R. parkeri* and many others has shown less than 0.5% of divergence between them (Roux and Raoult, 1995).

The diagnosis of rickettsial disease is most often confirmed by serological tests because culture, rickettsial DNA detection by PCR and other methods require specialized laboratories, and also because serum is usually the most commonly available sample. Indirect immunofluorescent assay remains the gold standard serological test and is the mostly widely used technique for diagnosis of rickettsial diseases (Brouqui *et al.*, 2004). However, owing to the presence of shared protein and LPS antigens among spotted fever group rickettsiae, the use of serological methods to distinguish between infections due to closely related rickettsiae is extremely difficult. Western blot (WB) and cross adsorption (CA) immunoassays can sometimes assist in the differentiation of SFG or TG rickettsiae; however, interpretation should be undertaken cautiously because some strains are so closely related that they cannot be distinguished (Teysseire and Raoult, 1992). The advent of molecular methods based on PCR has though enabled the development of specific and rapid tools for the detection and identification of *Rickettsia* species. It should be noted that the sensitivity of PCR in skin biopsy (eschar) is higher than in blood. Culture of the agent is the ultimate criterion to confirm the diagnosis and to identify the species of *Rickettsia* from the patient's blood (Gouriet *et al.*, 2005; de Sousa *et al.*, 2008b). This technique can reach a good level of success if the sample is collected and preserved under appropriate conditions until isolation procedures begin.

The main *Rickettsia* species responsible for animal infections or zoonoses in Europe and the Mediterranean basin are listed in Table 4.2.

4.6.2 *Rickettsia conorii*

R. conorii, the causative agent of Mediterranean spotted fever (MSF), is the most frequently isolated rickettsia and has

Table 4.2. Rickettsioses in Europe and in the Mediterranean basin transmitted by hard ticks.

Rickettsia spp.	Disease	Tick involved (based on publications review)
R. conorii Malish	Mediterranean spotted fever (MSF)	*Rh. sanguineus*, *Rh. turanicus*
R. conorii Israeli spotted fever strain	Israeli spotted fever (ISF)	
R. conorii Astrakhan strain	Astrakhan fever (AF)	*Rh. pumilio*, *Rh. sanguineus*
R. sibirica mongolitimonae strain	Lymphangitis-associated rickettsiosis (LAR)	*Hy. anatolicum*, *Hy. excavatum*, *Rh. pusillus*
R. slovaca	Tick-borne lymphadenopathy (TIBOLA)/*Dermacentor*-borne necrosis erythema lymphadenopathy (DEBONEL)	*D. marginatus*, *D. reticulatus*, *Ha. inermis*, *Ha. punctata*
R. massiliae	Unnamed	*Rh. sanguineus*, *Rh. turanicus*
R. aeschlimannii	Unnamed	*D. reticulatus*, *Ha. punctata*, *Ha. inermis*, *Hy. detritum* (syn. *Hy. scupense*), *Hy. marginatum*, *Rh. bursa*, *Rh. sanguineus*, *Rh. turanicus*
R. helvetica	Unnamed	*D. reticulatus*, *I. hexagonus*, *I. ricinus*, *I. ventalloi*
R. monacensis	Unnamed	*Ha. punctata*, *I. ricinus*

D., *Dermacentor*; *Ha.*, *Haemaphysalis*; *Hy.*, *Hyalomma*; *I.*, *Ixodes*; *Rh.*, *Rhipicephalus*.

the widest geographical distribution of the SFG rickettsial species. *Rh. sanguineus*, commonly known as the brown dog tick, is the main vector and reservoir for *R. conorii* strains in the Mediterranean area, former USSR, northern Africa and India (Rehacek and Tarasevich, 1988). In some situations, mainly in anthropozoonotic and domestic cycles, dogs are the main host for feeding of all the stages of *Rh. sanguineus* (Gilot, 1984). Dogs are transient reservoirs because they have a short-lived rickettsaemia after infection, but do not seem to be a reservoir for *R. conorii*. One study performed in Portugal also showed that dogs can present with febrile illness related to infection with *R. conorii* Malish and *R. conorii* Israeli spotted fever strains (Alexandre *et al.*, 2011).

MSF was described for the first time by Conor and Brusch in Tunisia in 1910, and is widely distributed in the Old World, being endemic in southern Europe, Africa, the Middle East, India and Pakistan. Case reports and isolations of *R. conorii* from patients in the Mediterranean area continue to be described, including recent cases in Croatia (Sardelic *et al.*, 2003), Turkey (Kuloglu *et al.*, 2004) and Greece (Psaroulaki *et al.*, 2005a). Sporadic cases have also occurred in non-endemic countries in northern and central Europe, such as Belgium (Lambert *et al.*, 1984), Switzerland (Chamot *et al.*, 1987), Sweden (Vene, 1989) and Germany (McDonald *et al.*, 1988), as well as in Japan (Yoshikawa, *et al.*, 2005), the USA (Anderson *et al.*, 1981) and the UK (Chai *et al.*, 2008). Most of these cases were attributed to the introduction of imported vectors or travellers who acquired the infection in endemic areas (Walker, 2003; Menn *et al.*, 2010).

In some countries, the incidence of MSF is unknown, and most of the knowledge is based on the human seroprevalence of antibodies to SFG rickettsiae. Inasmuch as the reporting of cases of MSF is not obligatory in all endemic countries, it is difficult to compare the incidence of the disease in different regions. However, it seems that in the last few decades there has been an increase in reported MSF cases in Portugal, Italy, Spain, France and Israel (Walker and Fishbein, 1991); this is likely to be due to climate changes influencing tick activity. In Portugal, the national incidence rate during the period 1989–2003 was $8.9/10^5$ inhabitants per year. During that same period the incidence of MSF in the Bragança and Beja districts of Portugal reached rates of $56.8/10^5$ inhabitants and $47.4/10^5$ inhabitants, respectively (de Sousa *et al.*, 2003; de Sousa and Bacellar, 2004). A study by de Sousa *et al.* (2006a), showed that the average winter temperatures in Portugal over the last 10 years have been the warmest on record and this, coupled with low rainfall, possibly played a role in doubling the number of confirmed MSF cases during 2000–2005.

In Italy, during a 5 year period from 1998 to 2002, a national incidence rate of $1.6/10^5$ inhabitants was reported. However, the same period of time in Sicily, accounting for 51.4% of all clinical cases, had an incidence rate of $9.3/10^5$ inhabitants (Ciceroni *et al.*, 2006). In France, a prospective study in the south of Corsica showed a higher incidence of $48/10^5$ inhabitants compared with other regions (Raoult *et al.*, 1985). Otero *et al.* (1982), estimated that in Israel the annual incidence of MSF is $6.2/10^5$ inhabitants and that the highest incidences of the disease occur in the western coastal area and the southern Negev Desert.

MSF is seasonal, and in most of European countries the cases are encountered in late spring and summer (81–88%), peaking in July and August when immature stages of the tick predominate. As the larvae and nymphs of the associated ticks are small and their attachment to the body during feeding is painless, they are more difficult to detect and so more likely to transmit MSF organisms (Raoult and Roux, 1997).

The disease is characterized by a generalized endothelial infection of the microvasculature and the main clinical features are due to the injury of blood vessels. The histopathological phenomenon of vasculitis can involve all the organs, not only the skin, and it has particularly serious manifestations when the lungs and brain are affected (Valbuena and Walker, 2009). The incubation period ranges from 3 to 7 days after the tick bite, but it can be longer. The onset of MSF is generally abrupt and the disease is characterized by fever, a maculopapular rash involving the entire body, including the palms and soles, and the presence of an eschar at the site of the tick bite. Occasionally, the eschar is not found and is seen rarely in multiples. Recently, clinical signs and symptoms were reported in a group of 71 Portuguese patients with confirmed diagnosis either by PCR or isolation of *R. conorii* Malish. The main clinical signs were fever (94%), maculopapular rash (94%), eschar (60%), myalgias (84%), headache (78%), asthenia (95%) and anorexia (73%). Purpuric or petechial rashes were present in 6% of patients with severe forms; this was indicative of a bad prognosis (de Sousa *et al.*, 2008b).

R. conorii has always been considered to produce a less severe disease than *R. rickettsii*. However, severe forms of MSF have been reported in 6% of patients, and a case fatality rate of 1.4–13% was reported for hospitalized patients in France, Israel, Spain, Algeria (Oran) and Portugal (Ruiz-Beltran *et al.*, 1985; Walker *et al.*, 1987; Amaro *et al.*, 2003; de Sousa *et al.*, 2008b). Moreover, in southern Portugal the fatality rate in 1997 reached 32.3% in hospitalized patients (de Sousa *et al.*, 2003). The co-morbidity condition of diabetes mellitus was identified as the risk factor for a fatal outcome. Statistical analysis of a representative sample of Portuguese patients with MSF also showed that alcoholism increases the risk for a fatal outcome. Other underlying conditions, such as heart failure, increasing age and glucose-6-phosphate dehydrogenase (G6PD) deficiency can be also be implicated in severe illness (Walker, 1990; Regev-Yochay *et al.*, 2000).

It is still unclear whether the more severe and fatal cases are simply related to host factors or whether there are differences in the virulence of the bacteria causing the disease.

4.6.3 *Rickettsia conorii* Israeli spotted fever strain

Israeli spotted fever (ISF) was described for the first time in 1946 in Israel, and some years later the aetiological agent was isolated from *Rh. sanguineus* ticks and from a patient (Goldwasser *et al.*, 1974). The disease that initially seemed restricted to that area is actually widespread in the Mediterranean basin, and has been detected in *Rh. sanguineus* collected in Portugal and Sicily (Giammanco *et al.*, 2003; de Sousa *et al.*, 2007). Several clinical cases were described in those same countries, and more recently patients have also been reported in Tunisia and Libya (Bacellar *et al.*, 1999; Giammanco *et al.*, 2005; de Sousa *et al.*, 2008b; Znazen *et al.*, 2011). In general, clinical and laboratory data are similar to those found in patients infected with *R. conorii* strains. However, studies in Portuguese and Sicilian patients infected with the *R. conorii* ISF strain showed that only 39% and 40% of patients had eschars; this is less frequent than what is reported by patients infected with the *R. conorii* Malish strain (de Sousa *et al.*, 2005; Giammanco *et al.*, 2005). In Israel, eschars have been described in only 4% of cases, and some clinical series reported the total absence of an eschar (Gross and Yagupsky, 1987; Wolach *et al.*, 1989). A prospective study conducted in Portugal during 1994–2006 identified and compared two groups of patients infected by Malish and ISF strains, confirmed either by isolation or DNA detection by PCR (de Sousa *et al.*, 2008b). Although the comparison of the clinical manifestations of MSF caused by different strains revealed a tremendous overlap, some differences were found. Patients infected with ISF had a recognized tick bite, but significantly fewer reported presence of an eschar compared with patients infected by the Malish strain. Also, significantly higher percentages of ISF patients had nausea, vomiting and increased levels of total bilirubin, γ-glutamyl transferase and alkaline phosphatase. The most important and statistically significant finding documented in the study was that the ISF strain was associated with a higher number of fatal cases. Severe forms and fatal cases were also described in patients in Sicily and Tunisia, and also in travellers that visited endemic areas (de Sousa *et al.*, 2003, 2008b; Giammanco *et al.*, 2005; Boillat *et al.*, 2008; Chai *et al.*, 2008). The differences in virulence between strains are not yet understood and comparative analysis of genomes will probably lead to better comprehension.

4.6.4 *Rickettsia conorii* Astrakhan strain

Astrakhan fever (AF) was first reported in the 1970s in patients living in rural areas of Astrakhan, a region of Russia located on the Caspian Sea (Tarasevich and Mediannikov, 2006). The appearance of the clinical cases coincided with the construction of a petrochemical complex that was releasing enormous quantities of CO_2 into the atmosphere. It seems that the CO_2 attracted ticks to this area and somehow increased the probability of exposure of the population to the vector (Tarasevich and Mediannikov, 2006). Later, the disease was serologically diagnosed, and the *R. conorii* Astrakhan strain was isolated from humans and *Rh. pumilio* ticks (Tarasevich *et al.*, 1991; Eremeeva *et al.*, 1994). Since 1983, more than 2000 cases of the disease have been registered in a small area of the Astrakhan region (Tarasevich *et al.*, 1991). Infection caused by the *R. conorii* Astrakhan strain showed similar clinical manifestations to those of the other strains of *R. conorii*; however, eschars have been reported in only 23% of patients (Tarasevich *et al.*, 1991). *R. conorii* Astrakhan strain has recently been isolated from a patient in Chad and from *Rh. sanguineus* ticks in Kosovo (Fournier *et al.*, 2003a,b).

4.6.5 *Rickettsia sibirica mongolitimonae* strain

R. sibirica mongolitimonae strain – initially named strain HA-91 – was originally isolated from a *Hy. asiaticum* tick collected in the Alashian region of Inner Mongolia in 1991 (Yu *et al.*, 1993). In Europe and the Mediterranean basin, the *R. sibirica mongolitimonae* strain was detected in *Hy. anatolicum* from Greece, *Hyalomma* sp. from Israel and *Rh. pusillus* from Portugal and Spain (Psaroulaki *et al.*, 2005b; de Sousa *et al.*, 2006a; Toledo *et al.*, 2009b). In 1996, the first human case of infection caused by this *Rickettsia* strain was described in southern France. The new strain was isolated from the blood and skin of a patient admitted in March to the Hospital La Timone in Marseille (Raoult *et al.*, 1996). The patient had no travel history, and disease manifestations similar to MSF, but the unusual aspect of the case was its occurrence in March when MSF is rarely reported. Subsequently, other human cases were described in France, and the diagnosis was confirmed by rickettsial isolation and/or PCR detection of the agent in eschars (Fournier *et al.*, 2005). Clinical cases with isolation or DNA detection of the agent were also reported in Greece, Portugal and Spain, and in French travellers who visited Algeria and Egypt (Fournier *et al.*, 2005; Psaroulaki *et al.*, 2005b; de Sousa *et al.*, 2006b; Aguirrebengoa *et al.*, 2008; Scolovschi *et al.*, 2010). In France, most of the reported cases caused by the *R. sibirica mongolitimonae* strain have occurred in the spring; this is in contrast with the majority of Portuguese cases that occurred in summer during the MSF season. The occurrence of these cases in different months could be related to the differences in seasonal activity and population dynamics of the different vectors. The clinical presentation of *R. sibirica mongolitimonae* infection has included fever, a rare or diffuse maculopapular rash and eschars; 50% of patients also have a lymphangitis expanding from the inoculation eschar to the draining lymph node (de Sousa *et al.*, 2008a). The latter feature has led to the name given to this disease: lymphangitis-associated rickettsiosis (LAR) (Fournier *et al.*, 2005).

4.6.6 *Rickettsia slovaca*

R. slovaca was first isolated in 1968 from a *Dermacentor marginatus* tick in Slovakia (Rehacek, 1984). In subsequent years, suspected cases of infection caused by *R. slovaca* were reported in patients from Hungary and Slovakia (Mittermayer *et al.*, 1980; Rehacek, 1984; Raoult *et al.*, 2002). However, the first proven case of *R. slovaca* infection was reported only in 1997 in a French patient who presented with an eschar on the scalp and enlarged cervical lymph nodes after receiving a bite from a *Dermacentor* tick (Raoult *et al.*, 1997). The lymphadenopathy present in most of the patients that have been seen infected with *R. slovaca* has led to this clinical syndrome being named tick-borne lymphadenopathy (TIBOLA). Spanish researchers have also coined the name *Dermacentor*-borne necrosis erythema lymphadenopathy (DEBONEL) (Lakos, 1997; Raoult *et al.*, 1997). *R. slovaca* has been identified in *D. marginatus* and *D. reticulatus* in most European countries. The prevalence rates of *R. slovaca* infection that have been found in *Dermacentor* ticks in Europe range from 21% in Hungary to higher rates in Spain (40.6%), Portugal (41.5%) and Switzerland (45.4%) (Beati *et al.*, 1994; Bacellar *et al.*, 1995; Lakos and Raoult, 1999; Oteo *et al.*, 2006; Parola *et al.*, 2009; Milhano *et al.*, 2010).

The number of *R. slovaca* infections in Europe is still under-evaluated; the majority of patients have been reported from France, Hungary, Spain, Slovakia and Italy (Lakos, 2002; Raoult *et al.*, 2002; Ibarra *et al.*, 2006; Selmi *et al.*, 2008; Parola *et al.*, 2009). The epidemiological and clinical findings on *R. slovaca* infections among patients in France, Hungary, Spain and Italy showed that the infection occurred mainly during the colder months of the year, mostly from October to April, in accordance with the density and activity of *Dermacentor* ticks. It seems that children and women had a higher risk of infection, and patients were more frequently bitten on the scalp (68–100%). Fever is present in 12–67% of the patients; rash is rare (14–23%) compared with other rickettsioses. The enlargement of lymph nodes has been reported in almost all of the patients (74–100%) (Lakos, 2002; Raoult *et al.*, 2002; Ibarra *et al.*, 2006; Selmi *et al.*, 2008, Parola *et al.*, 2009). No complications were observed, but 21–52% of patients developed a localized alopecia at the site of the tick bite and around 37% suffered from persistent asthenia. Only 50% of the patients develop detectable antibodies, which may reflect the fact that this disease is a localized infection.

R. raoultii (formerly genotypes RpA4, DnS14, DnS28), the other aetiological agent of TIBOLA/DEBONEL, was described for the first time in *Rh. pumilio* collected in the Astrakhan region and in *D. nutallii* in Siberia. Later, this *Rickettsia* was detected in *Dermacentor* ticks in other European countries (Rydkina *et al.*, 1999). The pathogenicity of the species has been suggested by the amplification of its DNA from the blood and skin biopsy samples of patients with a clinical picture of *R. slovaca*-like infection. This *Rickettsia* species was also found in *Dermacentor* removed from Spanish and French patients with cases of TIBOLA/DEBONEL (Ibarra *et al.*, 2006; Parola *et al.*, 2009). Nevertheless, *R. slovaca* remains the main aetiological agent responsible for the majority of TIBOLA/DEBONEL cases.

4.6.7 *Rickettsia massiliae*

R. massiliae was isolated in 1992 from ticks in France near Marseille (Beati and Raoult, 1993). Subsequently, this rickettsia has been detected by molecular methods and isolated in several countries in Europe (Scolovschi *et al.*, 2010). *R. massiliae* has been found mainly in ticks from the *Rhipicephalus* genus: *Rh. sanguineus* and *Rh. turanicus*. The first human case was described in a Sicilian patient who was admitted at Palermo hospital in 1985. However, it was not until 2005 that the isolate discovered 20 years before was characterized and identified as *R. massiliae* (Vitale *et al.*, 2006). The patient presented with an eschar on his right ankle, and a maculopapular rash involving the palms and soles, revealing similar manifestations to those found with other rickettsioses. He recovered completely after receiving tetracycline. Recently, another case of *R. massiliae* infection was diagnosed in a patient from Argentina. The diagnosis was confirmed by PCR in a Spanish laboratory (García-García *et al.*, 2010).

4.6.8 *Rickettsia aeschlimannii*

R. aeschlimannii was first characterized in 1997 from *Hy. marginatum* ticks from Morocco (Beati *et al.*, 1997). Later, this rickettsia was also described from *Hy. marginatum* ticks from Portugal, France, Spain, Croatia, Algeria, Italy and Egypt (Beati *et al.*, 1997; Bacellar, 1999; Punda-Polic *et al.*, 2002; Fernández-Soto *et al.*, 2003). However, until now the two described human cases caused by *R. aeschlimannii* were only diagnosed in Africa. The first case was in a French traveller who had visited Morocco, and the second case was detected in a South African patient who was bitten by a *Rh. appendiculatus* tick (Pretorius and Birtles, 2002). Symptoms exhibited by the patients were similar to those of MSF, and infection by *R. aeschlimannii* was confirmed by PCR amplification of rickettsial DNA from serum and skin biopsy (Pretorius and Birtles, 2002; Raoult *et al.*, 2002). Recent studies in Algeria, Corsica, Greece, Spain, Germany and the European part of Russia also revealed that other tick species can harbour *R. aeschlimannii*, such as: *Hy. anatolicum*, *Hy. detritum* (syn. *Hy. scupense*), *Hy. rufipes*, *D. reticulatus*, *I. ricinus*, *Ha. punctata*, *Ha. inermis*, *Rh. bursa*, *Rh. sanguineus* and *Rh. turanicus* (Fernández-Soto *et al.*, 2003, Matsumoto *et al.*, 2004; Psaroulaki *et al.*, 2006; Mokrani *et al.*, 2008; Bitam *et al.*, 2009; Shypnov *et al.*, 2009).

4.6.9 *Rickettsia helvetica*

R. helvetica was detected for the first time in Swiss *I. ricinus* in 1979, and confirmed as a new member of the SFG *Rickettsia* in 1993 (Beati *et al.*, 1993). *R. helvetica* has been detected and isolated from *I. ricinus* in many European countries, including Austria, Bulgaria, Denmark, Eastern Ukraine, France, Germany, Hungary, Italy, Moldova, the Netherlands, Poland, Portugal, Slovenia, Spain, Sweden and the UK (Bacellar *et al.* 1995; Parola *et al.*, 1998; Nilsson *et al.*, 1999; Beninati *et al.*, 2002; Christova *et al.*, 2003; Prosenc *et al.*, 2003; Fernández-Soto *et al.*, 2004; Sréter-Lancz *et al.*, 2005; Nijhof *et al.*, 2007; Skarphedinsson *et al.*, 2007; Blaschitz *et al.*, 2008; Chmielewski *et al.*, 2009; Movila *et al.*, 2009; Pluta *et al.*, 2010; Tijsse-Klassen *et al.*, 2011). In Portugal and Spain, the organism has been also detected in *I. ventalloi* parasitizing birds (Santos-Silva *et al.* 2006; Movila *et al.*, 2011). The prevalence of *R. helvetica* in ticks from different countries has been found to range from 2.8% in Poland to 91.4% in the south of Germany (Chmielewski *et al.*, 2009; Silaghi *et al.*, 2011). Detection of co-infections with *R. helvetica* and *Borrelia burgdorferi s.l.* in the same vector (*I. ricinus*) has gained attention for possible exacerbation of the illness arising from *R. helvetica* infection or Lyme borreliosis (Fernández- Soto *et al.*, 2004; Milhano *et al.*, 2010).

R. helvetica infection has been progressively becoming a clinical entity of its own. It was reported for the first time in 1999 in Sweden in two patients with fatal perimyocarditis, and it was recently reported in two Swedish patients with meningitis and septicaemia (Nilsson *et al.*, 1999). Serological associations with *R. helvetica* infections have been reported in patients in Europe (France, Italy, Switzerland) and Asia (Thailand), but additional evaluation and isolation of the bacterium from clinical samples are needed to confirm the pathogenicity of *R. helvetica* (Baumann *et al.*, 2003; Fournier *et al.* 2004; Ciceroni *et al.*, 2006; Nilsson *et al.* 2010). In 2002, an association of *R. helvetica* with sarcoidosis was proposed; however, the validity of this association has been questioned and later serological studies did not reveal the presence of anti-rickettsial antibodies in a group of Scandinavian sarcoidosis patients (Nilsson *et al.*, 2002; Planck *et al.*, 2004).

4.6.10 *Rickettsia monacensis*

R. monacensis was isolated and characterized from *I. ricinus* for the first time in Germany. Since then, most European countries have reported the presence of this agent, essentially based on molecular detection (Simser *et al.*, 2002). More recently, this *Rickettsia* species has also been isolated from *I. ricinus* in Portugal, and it was shown that *R. monacensis* was easily propagated and isolated in Vero cell lines at 28°C in conditions different from that for the isolate from Germany (Milhano *et al.*, 2010). *R. monacensis* was recently associated with febrile disease in humans in northern Spain (Jado *et al.*, 2007).

Table 1.25 in Appendix 1 elaborates on the geographical distribution of *Rickettsia* spp. and their reported diseases in Europe and in the Mediterranean basin. This table was generated independently of the main systematic literature review that is described in the introduction.

4.7 Lyme Borrelioses

Lyme borreliosis (LB) or Lyme disease is the most common tick-borne disease of humans in the northern hemisphere. It is a complex of several different zoonotic infections of which the aetiological agents are transmitted by hard ticks. At least 18 species or genospecies of spirochaetes in the *B. burgdorferi s.l.* complex have so far been described. Several are pathogenic to humans and domestic animals. They include *B. burgdorferi*, which is predominant in North America but also present in Eurasia, and *B. afzelii* and *B. garinii*, which are predominant in Eurasia. They are transmitted by tick species of the genus *Ixodes*, mainly *I. ricinus* in Europe, *I. persulcatus* in Eurasia, *I. pacificus* in the western USA and *I. scapularis* in the eastern USA. In Europe, at least another four *Borrelia* species, i.e. *B. bavariensis*, *B. valaisiana*, *B. spielmanii* and *B. lusitaniae* sometimes infect humans and may cause human LB (Richter *et al.*, 2004; Piesman and Gern, 2008; Rudenko *et al.*, 2011). In Europe, three tick species are considered to be vectors of LB spirochaetes, i.e. *Ixodes ricinus*, *I. hexagonus* and *I. uriae* (Piesman and Gern, 2008). Although the level of infection in the adult ticks in European populations of *I. ricinus* is higher (mean 17.4%; range 3–58%) than in the nymphs (mean 10.8%; range 2–43%), the nymphs are usually more important than the female adult ticks for transmission of the pathogens to humans (Hubálek and Halouzka, 1998). Larvae are rarely infected (mean 1.9%; range 0–11%; Hubálek and Halouzka, 1998). In *I. persulcatus*, however, the nymphs rarely feed on humans, so in this case it is the adult female ticks that are responsible for nearly all human infections with LB spirochaetes.

The enzootic cycle in general involves *Ixodes* spp. larvae and nymphs which become infected when feeding on infective wild bacteraemic mammals, particularly insectivores (shrews, hedgehogs), rodents (mice, voles, rats and squirrels) or hares. Certain bird species also serve as vertebrate reservoirs to the spirochaetes. Co-feeding transmission has been demonstrated to occur when sheep serve as a *Borrelia* reservoir (Ogden *et al.*, 1997).

It is important to distinguish between vertebrate hosts for the ticks and vertebrate reservoirs of the spirochaetes. Cervids appear to be refractory to the infection and usually do not serve as *Borrelia* reservoirs, but are extremely important hosts to *I. ricinus* females (Jaenson and Tälleklint, 1992). Many species of *Borrelia* may circulate in the same ecosystem, with the result that a single tick can be infected with two or more species of *Borrelia* – and with the TBE virus and other species of human-pathogenic bacteria. Throughout Europe, 13% of *Borrelia* infections in *I. ricinus* are mixed infections (Rauter and Hartung, 2005; Piesman and Gern, 2008). Multiple infection of a tick may occur because the host on which the tick was feeding had a multiple infection or because the tick had fed two or more times on hosts infected with different *Borrelia* spp. In Europe, *B. garinii* and *B. valaisiana* are predominant in the mixed infections, followed by mixed *B. garinii/B. afzelii* infection (Piesman and Gern, 2008). *B. afzelii* is mainly associated with rodents, while some serotypes of *B. garinii* and all serotypes of *B. valaisiana* are associated with birds. *B. lusitaniae* is associated with lizards in the Mediterranean countries, and often infects vector ticks more frequently than do the other genospecies in the complex (Richter and Matuschka, 2006).

LB is prevalent in most parts of Europe, but its prevalence is lower in southern Europe, such as in Portugal and Italy than in the former USSR to Japan, Mongolia and north-western China (Rauter and Hartung, 2005). The infection also occurs in some specific locations in North Africa. In North America, nearly all human LB cases are confined to the north-eastern USA, but the infection also occurs in other parts, including California, but with a lower prevalence than the north-eastern part of the country. LB is becoming increasingly prevalent in southern Canada and has also been reported from South America, including Mexico and Brazil. Climate change (not synonymous with global warming in the author's opinion, as climate change is a broad concept that includes all potential intentional and non-intentional changes in weather conditions) and an increasing abundance of deer could be associated with the spread of LB in Northern Europe (Gray *et al.*, 2009).

Human LB infections sometimes cause clinical disease, which can range from a relatively short influenza-like illness often accompanied by excruciating (nocturnal) pain, to a severe syndrome with neurological involvement, including meningitis, chronic severe arthritis and/or myocarditis. For instance, it has been estimated that in Sweden alone about 10,000 people annually contract the infection, resulting in clinical disease (Berglund, 2004). Among domesticated animals, clinical symptoms associated with LB have been reported in dogs, cattle and horses. Antibody titres against *B. burgdorferi s.l.* in dogs, cats and livestock can be high, but it is often difficult to establish a cause-and-effect relationship between exposure to the spirochaetes and clinical signs.

4.8 Recurrent (Relapsing) Fever

Relapsing fever is an infection caused by several spirochaetes of the genus *Borrelia* (Cutler, 2006). Relapsing fever borrelioses are characterized by recurrent febrile episodes and spirochaetaemia. Other than the louse-borne relapsing fever caused by *B. recurrentis* and transmitted by the body louse *Pediculus humanus*, endemic tick-borne relapsing fever is a zoonotic disease transmitted worldwide by soft tick species of the genus *Ornithodoros*. Within each region, specific relationships usually exist between the *Ornithodoros* vector species, *Borrelia* species and their distribution areas. Reservoir hosts are usually wild rodents. *Ornithodoros* ticks are included in the family Argasidae. They live close to their host, although the time spent on the host is relatively short. After each blood meal they are found in their habitats, typically in cracks and crevices of rodent burrows, but also in human shelters or just below the soil surface. Ticks become infected during a blood meal on a vertebrate with spirochaetaemia. Spirochaetes then invade all tissues of the tick, including ovaries (responsible for transmission between generations), salivary glands and excretory organs. Vertebrates and humans become infected during a blood meal through contamination of the feeding site by salivary and/or coxal secretions of the tick (Parola and Raoult, 2001).

B. hispanica is found in Spain, Portugal, Cyprus, Greece and North Africa. It has been isolated in *O. erraticus*, a tick commonly found in south-western Europe. This tick species usually lives in the burrows of wild rodents, its natural host. In Spain and Portugal; however, it has adapted to bite domestic pigs that are kept in continuous grazing and sometimes overnight in large burrows or inside old buildings, and this tick species has adapted to live in these habitats (Estrada-Peña and Jongejan, 1999). Humans may be bitten, and hence relapsing fever was sporadically reported in countries such as Spain during the 20th century, probably with an underestimated incidence (Sánchez-Yebra *et al.*, 1997). The disease caused by *B. hispanica* is one of the less severe in the relapsing fever group, and presents with neurological signs in less than 5% of cases (Cadavid and Barbour, 1998). In 1996, a new *Borrelia* species was isolated in southern Spain from three patients with relapsing fever and from *O. erraticus* ticks found in nearby areas (Anda *et al.*, 1996). The reservoir of this bacterium is still unknown and further records of the pathogen are unavailable. Although this new *Borrelia* species has not yet been cultured, molecular analyses have shown that it is closely related to *B. hispanica*, *B. duttoni* (an African species not present in Europe) and *B. crocidurae*.

On the borders of Europe, several other relapsing borrelioses are present. *B. persica*, the agent of Persian relapsing fever, is found in Israel, Syria, Egypt, Iran and Central Asia. It is transmitted by *O. tholozani* (Rodhain, 1998). This tick commonly lives in localities where livestock are housed, for example manmade shelters, caves and rocky overhangs (Estrada-Peña and Jongejan, 1999). The disease is sometimes severe (Cadavid and Barbour, 1998). *B. caucasica*, present in the Caucasus and Iraq, is transmitted by *O. verrucosus*, another argasid parasite of rodents. *B. latyschevii* is transmitted by *O. tartakovskyi* in Central Asia, the former USSR and Iran (Estrada-Peña and Jongejan, 1999; Rebaudet and Parola, 2006).

4.9 Piroplasmoses

4.9.1 Babesioses

Babesioses are caused by naturally tick-transmitted and generally host-specific intra-erythrocytic protozoan parasites of the genus *Babesia* (phylum Apicomplexa, order Piroplasmida). Babesiae, which are the second most common blood-borne parasites of mammals after the trypanosomes, are capable of infecting a wide variety of vertebrate mammalian species, including humans. More than 100 species have been identified, and these are traditionally divided on the basis of their morphology into the small and large babesiae (Telford *et al.*, 1993; Homer *et al.*, 2000). Molecular analysis suggests that the host range of many *Babesia* species is less restricted than previously believed and that still-unrecognized species may cause zoonotic infections in a variety of animals and humans (Gray and Weiss, 2008).

To date, only ixodid ticks have been identified as vectors for *Babesia* species. Some *Babesia* species can infect more than one genus of ticks; others can infect only ticks from the genus *Ixodes*. Several tick vectors can carry more than one *Babesia* species. The vectors become infected when ingesting the infected blood cells from a vertebrate reservoir that is competent in maintaining the *Babesia* organisms in an infectious state. Certain species of *Babesia*, such as *B. divergens* and *B. canis*, are transmitted transovarially to the next generation because they invade the female tick's ovaries. These species may persist in several generations of ticks, even without new infections. Some other *Babesia* species (e.g. *B. microti*) are only transmitted trans-stadially. All species of *Babesia* are naturally transmitted by the bite of infected female ticks. It has been demonstrated that male ticks may transmit *Babesia* species, although the epidemiological importance of male ticks in transmission has yet to be established.

Infection is initiated by inoculation of the sporozoites with the saliva of the vector tick into the bloodstream of the host. Transmission only occurs a few days after the tick has attached, because maturation of the sporozoites in the salivary glands of the vector is stimulated by feeding. *Babesia* species directly invade red blood cells, where their asexual multiplication most often results in two, sometimes four, daughter cells; these then leave the host cell and each enters another red cell (Homer *et al.*, 2000; Uilenberg, 2006). Animals and humans can also acquire the infection through the transfusion of contaminated blood products.

The two major factors involved in the pathogenesis of babesiosis are the release of pharmacologically active agents and intravascular haemolysis. The relative importance of each varies with the species of *Babesia*. The clinical features of babesiosis vary substantially from asymptomatic to life threatening, depending on the condition of the host and the parasite involved. During the acute babesial infection, the host may become severely ill as a result of host-mediated immunopathological mechanisms and erythrocyte lysis. Typically, the infected host can suffer high fever, anaemia and hyperbilirubinuria, possibly followed by alterations in the kidneys and other organs. All mammalian hosts examined have been able to develop immunity to *Babesia* species in which both humoral and cellular factors are involved. In endemic areas, all or almost all individuals of the host population are infected when they are young, with no or minimal clinical disease. The introduction of susceptible animals into endemic regions could lead to the recrudescence of babesiosis (Telford *et al.*, 1993; Homer *et al.*, 2000; Uilenberg, 2001; Hunfeld *et al.*, 2008).

Babesioses are well-recognized diseases of veterinary importance in Europe and North Africa. One of the most important and widespread *Babesia* species affecting cattle in temperate Europe is *B. divergens*, and this species probably occurs wherever the vector *I. ricinus* is present, which includes North Africa (Bouattour and Darghouth, 1996). *B. major*, another species in Europe, gives a much milder infection that is transmitted by *Ha. punctata* in western Europe. A low or non-pathogenic bovine *Babesia* species, *B. occultans*, was detected for the first time in unfed *Hy. marginatum* ticks collected in three bioclimatic regions of Tunisia. It was supposed that this species may have a wide distribution in the Mediterranean region, and not only in sub-Saharan Africa as previously described (Ros-García *et al.*, 2011). The taxonomy and even the geographical distribution of babesial parasites of small ruminants are not quite settled. The main species are *B. motasi* and *B. ovis*, which are transmitted by *Ha. punctata* and *Rh. bursa*, respectively. *B. motasi* in northwestern Europe produces a clinically milder disease than it does in the Mediterranean basin, and there are serological differences as well (Uilenberg, 2006). Babesiosis in the horse is mentioned with theileriosis (see Section 4.9.2).

On the basis of differences in vector specificity, geographical distribution, pathogenicity, antigenic properties and molecular investigations, the species *B. canis* of dogs has been subdivided into three subspecies, namely *B. c. canis*, *B. c. vogeli* and *B. c. rossi* (Kjemtrup *et al.*, 2000). These subspecies are currently considered to be separate species. *Babesia c. canis*, transmitted by *D. reticulatus* ticks, is the most common canine *Babesia* subspecies in temperate regions of Europe (Cacció *et al.*, 2002; Irwin, 2009). *Babesia c. vogeli*, transmitted by *Rh. sanguineus*, has also been reported in continental Europe (Cacció *et al.*, 2002; Duh *et al.*, 2004; Cardoso *et al.*, 2008). This subspecies is considered to be a mildly virulent subspecies, and commonly induces moderate clinical signs in dogs. Besides these large parasites, small babesiae such as *B. gibsoni*, transmitted by the brown dog tick, *Rh. sanguineus*, and the *B. microti*-like or 'Spanish isolate', with the proposed name of *Theileria annae*, have been reported from some European countries (Zahler *et al.*, 2000; Camacho, 2006; Beck *et al.*, 2009). Several studies have proved that *B. gibsoni* infection can be transmitted from dog to dog via bite wounds, saliva or ingested blood independently of the limitations of vector tick infestation (Jefferies *et al.*, 2007).

Babesioses are an emerging zoonotic problem caused by several species of protozoans in the genus *Babesia* (Homer *et al.*, 2000; Kjemtrup and Conrad, 2000; Herwaldt *et al.*, 2003; Hunfeld *et al.*, 2008).The first confirmed case of human babesiosis was diagnosed in a splenectomized Yugoslavian cattle farmer who died of a fatal *B. divergens* infection in 1956 (Skrabalo and Deanovic, 1957). To date, more than 60 cases of human babesiosis have been reported from Europe. At least 70% of the cases in Europe are associated with the cattle piroplasm, *B. divergens* (Genchi, 2007). In recent years, molecular studies have confirmed the responsibility of *B. divergens* in two cases of human babesiosis in the Canary Islands (Olmeda *et al.*, 1997) and in Portugal (Centeno-Lima *et al.*, 2003). A new European *B. divergens*-like organism (EU1), named *B. venatorum*, has been described from deer and from *I. ricinus* (Duh *et al.*, 2005; Gray, 2006; Bonnet *et al.*, 2007). This parasite species was involved in the first documented cases of human babesiosis in asplenic men in Italy, Austria and Germany (Herwaldt *et al.*, 2003; Häselbarth *et al.*, 2007). A human infection in a splenectomized patient, caused by a strain named *Babesia* EU3 that has high homology with *Babesia* EU1, was reported from Germany (Häselbarth *et al.*, 2007). *B. microti* infection has also been reported in Europe (Hildebrandt *et al.*, 2007). Sero-surveys suggest that a low percentage of Europeans (<3.4%) from several countries may be infected with *B. microti*. Most infected patients share splenectomy as a risk factor for acquiring the disease, but the rising number of HIV-positive individuals and the increasing population of immunocompromised patients may also serve to boost the number of human cases. The clinical course of human babesiosis varies according to the aetiological agent and ranges from subclinical infection to a severe disease with sudden onset. In many individuals, however, babesiosis is a mild, self-limited disease that requires only supportive therapy. Splenectomized or elderly patients infected with *B. microti* or *B. divergens* tend to develop severe and sometimes fatal illnesses (Homer *et al.* 2000; Gray, 2006).

4.9.2 Theilerioses

Theileria spp. (phylum Apicomplexa, order Piroplasmida) are tick-borne intracellular protozoan haemoparasites causing infection and often disease of veterinary and economic importance in livestock and wild animals in different regions of the world (Preston, 2001; Uilenberg, 2001). The genus *Theileria* differs from *Babesia* in that *Theileria* first penetrates lymphocytes or macrophages and develops there, and then enters red blood cells where the parasites multiply, forming tetrads often in the shape of a Maltese cross (Uilenberg, 2006).

The clinical signs of theileriosis in animals differ from babesioses in the absence of haemoglobinuria and the occurrence of a less severe anaemia. Ticks can only transmit these haemoparasites trans-stadially. There is no transovarial transmission because theileriae do not passage the ovaries and the eggs of the vectors. The newly hatched larvae of the ticks are never infected. Nymphs and adults become infective only if they were infected in the previous developmental stage. The transmission of parasites takes place by the injection of infected saliva of ticks, but it only occurs a few days after the tick has attached; the parasites have first to mature before they become infective (Mehlhorn and Schein, 1984; Preston, 2001).

A mild disease of domestic cattle is called tropical or Mediterranean theileriosis; this is caused by *T. annulata*, which is distributed in many areas of the world, extending from southern Europe to southern Asia (Brown, 1990). *T. annulata* is transmitted by a number of *Hyalomma* species, which are found in large numbers in the Mediterranean region, especially in semi-arid areas (Viseras and García-Fernández, 1999). *Hy. anatolicum anatolicum*, *Hy. detritum detritum* and *Hy. dromedarii* are considered as the main vectors in the field. Other species, such as *Hy. anatolicum excavatum* and *Hy. marginatum marginatum* may also play a role in the epidemiology of the disease in the field (Estrada-Peña *et al.*, 2004). *T. sergenti/buffeli/orientalis* causes a mild or asymptomatic disease in cattle known as bovine benign theileriosis (Uilenberg, 1981). *T. ovis* and *T. lestoquardi* (formerly *T. hirci*) are recognized as the species that can cause serious theileriosis in small ruminants, particularly sheep, where the disease occurs in the Old World (Schnittger *et al.*, 2000; Preston, 2001).

Equine piroplasmosis (EP) is a tick-borne intraerythrocytic protozoal disease of equids (horses, donkeys, mules and zebras) caused by *B. caballi* and *Theileria* (syn. *Babesia) equi*. *B. c. canis* of dogs has been reported in horses, but no clinical signs attributable to this parasitic species were described (Criado-Fornelio *et al.*, 2003; Hornok *et al.*, 2007a). This economically important protozoan disease of horses has been reported in many countries, thus making this disease a cause of great concern in the global horse industry. For this reason, testing of horses for EP is mandatory for the international movement of horses, either for participation in international events or for export. Only horses seronegative for both *T. equi* and *B. caballi* are qualified for importation to some countries, such as the USA, Canada, Australia and Japan (Bruning, 1996; Knowles, 1996). Within Europe, equine theileriosis is more prevalent in Portugal (Bashiruddin *et al.*, 1999), Spain (Camacho *et al.*, 2005) and Italy (Moretti *et al.*, 2009). The disease agent is mainly spread by competent ticks. To date, up to 12 species of ixodid ticks belonging to the genera *Hyalomma*, *Dermacentor* and *Rhipicephalus* have been identified as vectors of both *B. caballi* and *T. equi* (Bruning, 1996; Massaro *et al.*, 2003). Transmission of *T. equi* appears only to occur trans-stadially (de Waal, 1992). The parasites are also spread by the transfer of blood from infected to naive equids through shared needles, improperly shared equipment, and blood or serum transfusions (de Waal and Van Heerden, 2004). Transplacental transmission of *T. equi* from carrier mares to asymptomatic foals was recently confirmed (Allsopp *et al.*, 2007).

Clinical signs of infection with EP are not pathognomonic, especially in endemic areas, and vary from mild to severe. Acute and subacute cases are the most commonly observed. The mild form of the disease can cause equids to appear weak or show lack of appetite, while more severe cases may have fever, anaemia, jaundice, swollen abdomen, haemoglobinuria and bilirubinuria, and sometimes result in death (de Waal and Van Heerden, 2004). In the chronic phase, the horse can appear normal. In some cases of acute or chronic disease, mortality can reach up to 50% (de Waal, 1992). Infected animals that recover from acute or primary infection of *T. equi* remain lifelong carriers because anti-theilerial drugs suppress but do not eliminate the parasite.

The detection of apparently healthy carrier horses therefore remains a worldwide challenge for controlling the spread of the disease. Carrier mares may transmit the organism to their offspring and this may result in abortion or neonatal piroplasmosis. Some researchers suggest that foals may be born as carriers yet remain apparently healthy as colostral *T. equi* antibody may act to suppress parasitaemia in the newborn, reducing the incidence of clinical neonatal piroplasmosis (Allsopp *et al.*, 2007).

4.10 Hepatozoonosis

Hepatozoonosis is an arthropod-borne infection of both wild and domestic animals, including mammals, birds, reptiles and amphibians. It is caused by about 300 apicomplexan protozoal species from the genus *Hepatozoon* (Smith, 1996). The vertebrates are intermediate hosts, and several blood-sucking invertebrates, including hard ticks, are definitive hosts in which the sexual reproduction and sporogony of the protozoans occurs. Transmission of *H. canis* to the dog takes place by ingestion of a tick or parts of ticks containing *Hepatozoon* oocysts; this is because the protozoan is not disseminated within the tick but remains in the haemocoel. No salivary transfer of these parasites has been documented. In this respect, *Hepatozoon* differs from many other tick-borne protozoal and bacterial pathogens. The ticks become infected when they ingest infected neutrophils and gametocytes while feeding on the blood of a parasitaemic dog. Trans-stadial transmission of *H. canis* from the nymph to the adult stage in the tick vector has been recorded, whereas transovarial transmission could not be demonstrated (Baneth, 2006; Baneth *et al.*, 2007). Japanese scientists (Murata *et al.*, 1993) have reported that vertical transmission of *H. canis* occurred in puppies born from an infected dam and raised in a tick-free environment.

In the Old World, canine hepatozoonosis caused by *H. canis* is of veterinary importance. *H. canis* has been reported from the Mediterranean region of Europe (Spain, Portugal, Italy, Greece and France), Africa, the Middle East, the Far East and South America, where its main vector, the brown dog tick, *Rh. sanguineus*, is enzootic (Vincent-Johnson *et al.*, 1997; Baneth *et al.*, 2003). Recently, Italian scientists reported that *I. ricinus* might also be implicated in parasite transmission, thus explaining the occurrence of hepatozoonosis in areas considered *Rh. sanguineus* free (Gabrielli *et al.*, 2010). However, further studies are needed to confirm this hypothesis. Hepatozoonosis is of increasing importance in dogs in regions that have previously been considered free of the infection. *H. canis* is regularly introduced by dogs into north-west Europe after ingestion of infected *Rh. sanguineus* during visits to endemic regions (Holland, 2001). The first detection of *H. canis* in Slovakia (Majláthová *et al.*, 2007) and Italy (Gabrielli *et al.*, 2010) has been reported in naturally infected red fox (*Vulpes vulpes*). Prevalence data for dogs are still restricted to small areas and range from 0.9% in France (Criado-Fornelio *et al.*, 2009) to 71% in Croatia (Vojta *et al.*, 2009). According to a sero-survey of *H. canis* in Israel, this protozoal infection appears to be endemic in red fox populations and these may serve as reservoirs of the parasite for domestic dogs (Fishman *et al.*, 2004).

When the infected vector is ingested by the dog, *H. canis* mainly infects the haemolymphatic tissues and blood-forming organs, including the bone marrow, lymph nodes and spleen. The pathogenicity and clinical manifestations of hepatozoonosis vary according to the age of the host and the degree of infection. Canine hepatozoonosis is typically a mild clinical disease. However, clinical signs vary from an apparently subclinical infection with a low parasitaemia, to a life-threatening disease with lethargy, fever, cachexia and anaemia with a large number of circulating parasites. The grave and potentially fatal disease occurs mostly in young animals or in dogs suffering from a concurrent infection or immunosuppressive conditions. *H. canis* is commonly associated with co-infection with other diseases, in particular ehrlichiosis and leishmaniosis in endemic areas (Baneth, 2006).

4.11 Other Potentially Tick-borne Infections

The following infections are associated or suspected to be associated with ticks as one of the potential – but not the only – modes of transmission.

4.11.1 African horse sickness

African horse sickness (AHS) is caused by African horse sickness virus (AHSV), a member of the genus *Orbivirus*, subfamily *Sedoreovirinae* and the family *Reoviridae*. The virion is approximately 70 nm in diameter (Polson and Deeks, 1963), is non-enveloped and consists of a double-layered icosahedral capsid with 32 capsomeres (Pringle and Wickner, 2000). The double-capsid particles contain seven structural proteins (VP1–VP7), with the outer capsid made up of VP2 and VP5, and the inner capsid (the core) made up of VP3 and VP7 (as major proteins) and VP1, VP4 and VP6 (minor proteins) (Roy *et al.*, 1994). Among the capsid proteins, VP2 has been shown to be the most variable, and responsible for most of the antigenic variation of the virus (Iwata *et al.*, 1992; Mellor and Hamblin, 2004); hence VP2 is the major target of the host's neutralizing antibodies response (Burrage *et al.*, 1993, Roy *et al.*, 1994).

AHSV infects all equids, with horses being the most susceptible to clinical disease, which has a severe morbidity and high case fatality rate of 50–95%. Mules have a similar morbidity to horses, but with a lower case fatality rate of 50–70% (Coetzer and Guthrie, 2004). Donkeys and zebras are very resistant to the disease, and usually just develop subclinical infection (Theiler, 1921; Barnard, 1993; Coetzer and Guthrie, 2004). Donkeys in the Middle East can have a case fatality rate of up to 10%, and may be more susceptible to clinical disease than the South African donkey (Alexander, 1948; Hamblin *et al.*, 1998; Coetzer and Guthrie, 2004). Zebras only show mild fever when experimentally infected with the virus (Erasmus *et al.*, 1978; Coetzer and Guthrie, 2004); they maintain year-round infections and are considered a reservoir host for the virus in endemic regions (Davies and Otieno, 1977; Barnard, 1993). AHSV has also been isolated in blood samples 40 days post infection in zebras. In comparison, in experimentally challenged horses, viraemia usually lasts 4–8 days, but does not exceed 21 days (Barnard *et al.*, 1994; Coetzer and Guthrie, 2004). Donkeys were considered to be potential reservoirs, but the absence of viral antigens after 14–19 days post infection makes them unlikely long-term hosts, although they may play a small role in spread of the virus. Theiler demonstrated that AHSV can be transmitted to dogs and cause a similar pathology to that found in horses (Theiler, 1906). Dogs, however, are likely to be dead-end hosts for the virus (Braverman and Chizov-Ginzburg, 1996). The most common way for a dog to become infected with the virus is through the consumption of uncooked meat from an infected horse carcass, as was the case in Pretoria in 1980 (Van Rensburg *et al.*, 1981).

AHSV is considered to be endemic in north-eastern parts of South Africa, primarily Mpumalanga Province, but outbreaks regularly occur in other parts of South Africa (Lord *et al.*, 2002; Coetzer and Guthrie, 2004). In addition, major outbreaks of the virus have recently occurred in regions with close proximity to the Mediterranean Sea, such as in the Middle East (1959–1961), North Africa (1965, 1989, 1991), Spain (1966, 1987–1990) and Portugal (1989) (Rodríguez *et al.*, 1992; Coetzer and Guthrie, 2004; Mellor and Hamblin, 2004).

Clinical disease develops in susceptible animals after an incubation of about 5–7 days, depending on virulence and dose of virus (Theiler, 1921; Coetzer and Guthrie, 2004). While AHSV causes severe morbidity and mortality in the majority of infected horses, the pathogenesis and clinical disease that develops differs among cases. This difference is not completely understood, but is a function of both host factors (genetics and immune status) and virus factors (dose, route of infection in experimental inoculations and virulence phenotype) (Burrage and Laegreid, 1994).

The insect *Culicoides imicola* is the principal vector responsible for AHSV transmission, and its importance in AHSV transmission has been recognized for over 50 years. *C. imicola* has historically been found in Africa and South-east Asia (Meiswinkel *et al.*, 2004). Nevertheless tick species in general are also capable of transmitting the virus (Meiswinkel *et al.*, 2000; Mellor and Hamblin, 2004).

4.11.2 Bartonelloses

Bartonella infections are widespread in wild and domesticated mammals, and several new species have been described during the last few decades. These alphaproteobacteria infect erythrocytes and endothelial cells, leading to persistent infections of their mammalian hosts. As *Bartonella* species tend to infect the blood of their vertebrate hosts chronically, these microparasites can be ingested and potentially transmitted by blood-feeding arthropods. Confirmed vectors of *B. henselae* (the aetiological agent of cat-scratch disease), *B. bacilliformis*, *B. quintana*, *B. grahamii* and *B. taylorii* are *Ctenocephalides felis*, *Lutzomyia verrucarum*, *Pediculus humanus humanus* and *Ctenophthalmus nobilis*, respectively (Billeter *et al.*, 2008). *Bartonella* bacteria have been detected – based mainly on PCR – in several tick species, including *I. ricinus*, *I. scapularis*, *I. persulcatus*, *D. reticulatus*, *Rh. sanguineus* and *Carios kelleyi* (Billeter *et al.*, 2008). Some of the reasons why *Bartonella* species might be transmitted by ticks have been listed by Telford and Wormser (2010): other arthropods can transmit *Bartonella* species; the DNA of *Bartonella* species is often detected in ticks; human cases of bartonellosis preceded by tick bites are on record; and *Bartonella* species are commonly present in important hosts of *Ixodes* ticks, i.e. deer and rodents.

The mere detection of *Bartonella* DNA by PCR in blood-feeding arthropods is no evidence that these bacteria are viable and infective, or that these arthropods are competent vectors of *Bartonella* species. In fact, there is no conclusive evidence that any *Bartonella* species can, under natural circumstances, infect a vertebrate via tick bite, although a recent laboratory investigation using a membrane feeding technique suggested that *I. ricinus* is a competent vector for *B. henselae* (Cotté *et al.*, 2008). Trans-stadial transmission of the bacteria, their multiplication within the tick's salivary glands after a second blood meal, and transmission of viable and infective *B. henselae* from ticks to blood were recorded. However, this study may not be relevant to establishing the vector competence of ticks for bartonellae as certain of its parameters were unnatural: the ticks were fed continuously on blood containing exceedingly high numbers of bacteria; and the strain of *B. henselae* used is highly adapted to laboratory conditions and grows easily *in vitro*. As suggested by Telford and Wormser (2010), a more reliable proof of vector competence would be to feed an uninfected *Ixodes* sp. on a *B. henselae*-infected cat and then, after the tick has hatched, determine whether the nymph can transmit *B. henselae* by bite to an uninfected cat. However, additional (epidemiological) data would be necessary to conclusively prove that ticks are of importance as natural vectors of bartonellae.

4.11.3 Q fever

The aetiological agent of Q fever is the rickettsial parasite, *Coxiella burnetii*. This is an obligate intracellular bacterium, although it can survive for months or years outside host cells in such media as water, dried or frozen tissues and soil. The organism has been described as a possible biological weapon (Madariaga *et al.*, 2003). Ticks are one of a broad range of reservoirs for the organism. More than 40 species of ticks including *D. marginatus*, are naturally infected with *C. burnetii*. After feeding on bacteraemic hosts, nymphs or adults can transmit the pathogen trans-stadially, and females can pass it transovarially (Lang, 1990; Toledo *et al.*, 2009a).

Q fever is a worldwide zoonosis affecting mammals (including domestic animals), birds and arthropods in most areas in the world, including Europe (Aitken *et al.*, 1987; Lang, 1990). The most common reservoirs for human infections are cattle, sheep and goats. Dogs, cats, birds and reptiles are also susceptible to infection and may play a role in maintaining the infection in natural habitats. Although *C. burnetii* does not usually cause clinical disease in animals, it occasionally induces reproductive disorders such as stillbirths, abortion and metritis in pregnant goats and sheep, but is rarely documented in dairy cows (Aitken *et al.*, 1987; Lang, 1990). Infected animals develop high rickettsaemias and excrete large numbers of *C. burnetii* in their faeces for weeks, as well as shedding organisms via tissues and fluids.

The disease is epidemic, especially on farms or in farming communities when infected domestic animals are being handled, such as during wool shearing, lambing, calving and slaughtering. Therefore, farmers, abattoir workers, meat-packing workers, veterinarians and laboratory workers in contact with livestock are at high risk of infection. Humans can also be infected by inhalation of the organism or by the ingestion of infected milk and/or fresh dairy products. The organism may survive in contaminated milk and butter for up to 3 months. Close contact with infected animals is not required; infection may be from contaminated dust, straw or manure, as was reported for an outbreak of Q fever in an urban area in southern Wales, UK (Ayres *et al.*, 1996). After an incubation period of about 20 days, Q fever is characterized by mainly limited flu-like illness, pneumonia or hepatitis. It may become chronic and manifests with fatigue syndrome, chronic hepatitis, endocarditis and other endovascular infections. The fatality rate is less than 1% in acute cases but may rise to 30% in chronic cases. A scientific opinion on Q fever has been published by the EFSA Panel on Animal Health and Animal Welfare (2010b).

4.11.4 Tularaemia

Tularaemia is a zoonotic disease caused by one of the most infectious bacteria, *Francisella tularensis*, a Gram-negative obligate intracellular agent (Ellis *et al.*, 2002). This bacterial species causes great concern as a potential bioterrorism agent, and is listed among Class A biothreat agents (Dennis *et al.*, 2001; Oyston *et al.*, 2004). *F. tularensis* can be recovered from contaminated water, soil and vegetation. Four closely related subspecies of *F. tularensis* have been identified: *F. t. tularensis*, *F. t. holarctica*, *F. t. mediasiatica* and *F. t. novicida* (Forsman *et al.*, 1994). Subspecies *F. t. tularensis* and *F. t. holarctica* cause most human illness. *F. t. holarctica* is found throughout much of Europe except for the UK, Ireland and Iceland. Tularaemia is typically a disease of northern and central Europe and the countries of the former Soviet Union (Ellis *et al.*, 2002; Tarnvik *et al.*, 2004). Natural infections with *F. tularensis* have been reported in over 250 animal species and in a range of vertebrates, including mammals, birds, amphibians and fish, as well as in certain invertebrates (Morner and Addison, 2001; Oyston *et al.*, 2004; Santic *et al.*, 2006). However, tularaemia is primarily a disease of the orders Lagomorpha (rabbits and hares) and Rodentia (Friend, 2006). The European brown hare *(Lepus europaeus)* is a common host of *F. tularensis* in central Europe, where it causes a public health problem (Morner and Addison, 2001; Strauss and Pohlmeyer, 2001; Pikula *et al.*, 2004). In Scandinavia and Russia, tularaemia occurs frequently in mountain hares (*L. timidus*) (Morner and Addison, 2001).

Haematophagous arthropods such as deer and horse flies, ticks and mosquitoes are common vectors of *F. tularensis* (Petersen *et al.*, 2009). Ticks are believed to be the most important arthropods for *F. tularensis* as both mechanical and biological vectors (Hopla and Hopla, 1994). *D. reticulatus* plays an important role in the maintenance and transmission of *F. tularensis* among small and medium-sized mammals in central Europe (Guryčová *et al.*, 1995, 2001; Hubalek *et al.*, 1996). Other ticks, such as *I. ricinus*, *I. persulcatus*, *D. marginatus*, *Rh. rossicus* and *Ha. concinna*, have also been found to be naturally infected with *F. tularensis* in Europe (Hopla and Hopla, 1994; Keim *et al.*, 2007).

Tularaemia, in a range from subclinical infection to death, rarely occurs among domestic animals, and is most frequently observed in sheep during the lambing season, and in cats and dogs (Friend, 2006; O'Toole *et al.*, 2008). Humans are highly susceptible to *F. tularensis*. Infections in humans are typically sporadic, but outbreaks do occur (Matyas *et al.*, 2007). The disease is not contagious and is most often transmitted to humans by the bite of an infected arthropod vector, by direct contact with infected animals (e.g. voles, mice, water rats, squirrels, rabbits and hares), by contact with infected animal tissues or fluids, by ingestion of contaminated water or food, or by inhaling infected materials (Dennis *et al.*, 2001). Several emergences or re-emergences of tularaemia have recently been seen all over the world (Petersen and Schriefer, 2005): the highest incidences occurred in confined geographical areas of Finland and Sweden (Eliasson *et al.*, 2002), and the disease also appeared in Spain (Perez-Castrillon *et al.*, 2001), Kosovo (Reintjes *et al.*, 2002) and Germany (Kaysser *et al.*, 2008).

References

Aguirrebengoa, K., Portillo, A., Santibáñez, S., Marín, J.J., Montejo, M. and Oteo, J.A. (2008) Human *Rickettsia sibirica mongolitimonae* infection, Spain. *Emerging Infectious Diseases* 14, 528–529.

Aitken, I.D., Bogel, K., Cracea, E., Edlinger, E., Houwers, D., Krauss, H., Rady, M., Rehacek, J., Schiefer, H.G. and Schmeer, N. (1987) Q fever in Europe: current aspects of aetiology, epidemiology, human infection, diagnosis and therapy. *Reviews in Infection* 15, 323–327.

Aktas, M., Vatansever, Z., Altay, K., Aydin, M.F. and Dumanli, N. (2009) Molecular evidence for *Anaplasma phagocytophilum* in *Ixodes ricinus* from Turkey. *Transactions of the Royal Society of the Tropical Medicine and Hygiene* 104, 10–15.

Alberti, A. and Sparagano, O.A. (2006) Molecular diagnosis of granulocytic anaplasmosis and infectious cyclic thrombocytopenia by PCR-RFLP. *Annals of the New York Academy of Sciences* 1081, 371–378.

Alberti, A., Addis, M.F., Sparagano, O., Zobba, R., Chessa, B., Cubeddu, T., Parpaglia, M.L.P., Ardu, M. and Pittau, M. (2005) *Anaplasma phagocytophilum*, Sardinia, Italy. *Emerging Infectious Diseases* 11, 1322–1324.

Alekseev, A., Dubinina, H. and Schouls, H. (1998) First detection of *Ehrlichia* infected ticks among the primary vectors of the tick-borne encephalitis and borreliosis in the Russian Baltic region. *Bulletin of the Scandinavian Society of Parasitology* 8, 88–91.

Alexander, R.A. (1948) The 1944 epizootic of horsesickness in the Middle East. *The Onderstepoort Journal of Veterinary Science and Animal Industry* 23, 77–92.

Alexandre, N., Santos, A.S., Bacellar, F., Boinas, F.J., Núncio, M.S. and de Sousa, R. (2011) Detection of *Rickettsia conorii* strains in Portuguese dogs (*Canis familiaris*). *Ticks and Tick Borne Diseases* 2, 119–122.

Allsopp, M.T.E.P., Lewis, B.D. and Penzhorn, B.L. (2007) Molecular evidence for transplacental transmission of *Theileria equi* from carrier mares to their apparently healthy foals. *Veterinary Parasitology* 148, 130–136.

Amaro, M., Bacellar, F. and França, A. (2003) Report of eight cases of fatal and severe Mediterranean spotted fever in Portugal. *Annals of the New York Academy of Sciences* 990, 331–343.

Anda, P., Gebbia, J.A., Backenson, B., Coleman, J.L. and Benach, J.L. (1996) A glyceraldehyde-3-phosphate dehydrogenase homolog in *Borrelia burgdorferi* and *Borrelia hermsii*. *Infection and Immunology* 64, 262–268.

Anderson, J.F., Magnarelli, L.A., Burgdorfer, W., Casper, E.A. and Philip, R.N. (1981) Importation into the United States from Africa of *Rhipicephalus simus* on a boutonneuse fever patient. *American Journal of Tropical Medicine and Hygiene* 30, 897–899.

Andersson, S.G. and Kurland, C.G. (1998) Reductive evolution of resident genomes. *Trends in Microbiology* 6, 263–268.

Aradaib, I.E., Erickson, B.R., Mustafa, M.E., Khristova, M.L., Saeed, N.S., Elageb, R.M. and Nichol, S.T. (2010) Nosocomial outbreak of Crimean Congo haemorrhagic fever, Sudan. *Emerging Infectious Diseases* 16, 867–869.

Ayres, J.G., Smith, E.G. and Flint, N. (1996) Protracted fatigue and debility after acute Q fever. *The Lancet* 347, 978–979.

Azad, A.F. and Beard, C.B. (1998) Rickettsial pathogens and their arthropod vectors. *Emerging Infectious Diseases* 4, 179–186.

Bacellar, F. (1999) Ticks and rickettsiae in Portugal. In: Raoult, D. and Brouqui, P. (eds) *Rickettsiae and Rickettsial Diseases*. Elsevier, Paris, pp. 103–109.

Bacellar, F., Núncio, M.S., Alves, M.J. and Filipe, A.R. (1995) *Rickettsia slovaca*: an agent of the group of exanthematous fevers, in Portugal. *Enfermedades Infecciosas Microbiologia Clinica* 13, 218–223.

Bacellar, F., Beati, L., França, A., Poças, J., Regnery, R. and Filipe, A. (1999) Israeli spotted fever Rickettsia (*Rickettsia conorii* complex) associated with human diseases in Portugal. *Emerging Infectious Diseases* 5, 835–836.

Bakken, J.S. and Dumler, S. (2008) Human granulocytic anaplasmosis. *Infectious Disease Clinics of North America* 22, 433–448.

Baneth, G. (2006) *Hepatozoon canis* infection, In: Greene C.E. (ed.) *Infectious Diseases of the Dog and Cat*, 3rd edn. Saunders Elsevier, St Louis, Missouri, pp. 698–705.

Baneth, G., Mathew, J.S., Shkap, V., Macintire, D.K., Barta, R.R. and Ewing, S.A. (2003) Canine hepatozoonosis: two disease syndromes caused by separate *Hepatozoon* spp. *Trends in Parasitology* 19, 27–31.

Baneth, G., Samish, M. and Shkap, V. (2007) Life cycle of *Hepatozoon canis* (Apicomplexa: Adeleorina: Hepatozoidae) in the tick *Rhipicephalus sanguineus* and domestic dog (*Canis familiaris*). *Journal of Parasitology* 93, 283–299.

Barandika, J.F., Hurtado, A., García-Esteban, C., Gil, H., Escudero, R., Barral, M., Jado, I., Juste, R.A., Anda, P. and García-Pérez, A.L. (2007) Tick-borne zoonotic bacteria in wild and domestic small mammals in northern Spain. *Applied and Environmental Microbiology* 73, 6166–6171.

Barnard, B.J.H. (1993) Circulation of African horsesickness virus in zebra (*Equus burchelli*) in the Kruger National Park, South Africa, as measured by the prevalence of type specific antibodies. *The Onderstepoort Journal of Veterinary Research* 60, 111–117.

Barnard, B.J.H., Bengis, R., Keet, D. and Dekker, E.H. (1994) Epidemiology of African horsesickness: duration of viraemia in zebra (*Equus burchelli*). *The Onderstepoort Journal of Veterinary Research* 61, 391–393.

Bashiruddin, J.B., Cammà, C. and Rebêlo, E. (1999) Molecular detection of *Babesia equi* and *Babesia caballi* in horse blood by PCR amplification of part of the 16S rRNA gene. *Veterinary Parasitology* 84, 75–83.

Bastos, A.D.S., Penrith, M.L., Cruciere, C., Edrich, J.L., Hutchings, G., Roger, F., Couacy-Hymann, E.R. and Thomson, G. (2003) Genotyping field strains of African swine fever virus by partial *p72* gene characterization. *Archives of Virology* 148, 693–706.

Baumann, D., Pusterla, N., Péter, O., Grimm, F., Fournier, P.E., Schär, G., Bossart, W., Lutz, H. and Weber, R. (2003) Fever after a tick bite: clinical manifestations and diagnosis of acute tick bite-associated infections in northeastern Switzerland. *Deutsche Medizinische Wochenschrift* 128, 1042–1047.

Beati, L. and Raoult, D. (1993) *Rickettsia massiliae* sp. nov., a new spotted fever group rickettsia. *International Journal of Systematic Bacteriology* 43, 839–840.

Beati, L., Péter, O., Burgdorfer, W., Aeschlimann, A. and Raoult, D. (1993) Confirmation that *Rickettsia helvetica* sp. nov. is a distinct species of the spotted fever group of rickettsiae. *International Journal of Systematic Bacteriology* 4, 521–526.

Beati, L., Humair, P.F., Aeschlimann, A. and Raoult, D. (1994) Identification of spotted fever group rickettsiae isolated from *Dermacentor marginatus* and *Ixodes ricinus* ticks collected in Switzerland. *American Journal of Tropical Medicine and Hygiene* 51, 138–148.

Beati, L., Meskini, M., Thiers, B. and Raoult, D. (1997) *Rickettsia aeschlimannii* sp. nov., a new spotted fever group rickettsia associated with *Hyalomma marginatum* ticks. *International Journal of Systematic Bacteriology* 47, 548–554.

Beck, R., Vojta, L., Mrljak, V., Marinculi, A., Beck, A., Zivicnjak, T. and Cacció, S.M. (2009) Diversity of *Babesia* and *Theileria* species in symptomatic and asymptomatic dogs in Croatia. *International Journal for Parasitology* 39, 843–848.

Beninati, T., Lo, N., Noda, H., Esposito, F., Rizzoli, A., Favia, G.and Genchi, C. (2002) First detection of spotted fever group rickettsiae in *Ixodes ricinus* from Italy. *Emerging Infectious Diseases* 8, 983–986.

Bente, D.A., Alimonti, J.B., Shieh, W.J., Camus, G., Ströher, U., Zaki, S. and Jones, S.M. (2010) Pathogenesis and immune response of Crimean-Congo hemorrhagic fever virus in STAT-1 knockout mouse model. *Journal of Virology* 84, 11089–11100.

Bereczky, S., Lindegren, G., Karlberg, H., Akerström, S., Klingström, J. and Mirazimi, A. (2010) Crimean-Congo hemorrhagic fever virus infection is lethal for adult type I interferon receptor-knockout mice. *Journal of General Virology* 91, 1473–1477.

Berglund, J. (2004) Human borreliosis (Lyme borreliosis) was described already in 1909 in the *Journal of the Swedish Medical Association* [In Swedish]. *Läkartidningen* 101, 109–114.

Billeter, S.A., Levy, M.G., Chomel, B.B. and Breitschwerdt, E.B. (2008) Vector transmission of *Bartonella* species with emphasis on the potential for tick transmission. *Medical and Veterinary Entomology* 22, 1–15.

Bitam, I., Kernif, T., Harrat, Z., Parola, P. and Raoult, D. (2009) First detection of *Rickettsia aeschlimannii* in *Hyalomma aegyptium* from Algeria. *Clinical Microbiology Infection* 15 (Suppl. 2), 253–254.

Blaschitz, M., Narodoslavsky-Gföller, M., Kanzler, M., Walochnik, J. and Stanek, G. (2008) First detection of *Rickettsia helvetica* in *Ixodes ricinus* ticks in Austria. *Vector Borne Zoonotic Diseases* 8, 561–563.

Bodur, H., Akinci, E., Öngürü, P., Carhan, A., Uyar, Y., Tanrici, A., Cataloluk, O. and Kubar, A. (2010) Detection of Crimean-Congo hemorrhagic fever virus genome in saliva and urine. *International Journal of Infectious Diseases* 14, e247–e249.

Boillat, N., Genton, B., D'Acremont, V., Raoult, D. and Greub, G. (2008) Fatal case of Israeli spotted fever after Mediterranean cruise. *Emerging Infectious Diseases* 14, 1944–1946.

Bonnet, S., Jouglin, M., L'Hostis, M. and Chauvin, A. (2007) *Babesia* sp. EU1 from roe deer and transmission within *Ixodes ricinus*. *Emerging Infectious Diseases* 13, 1208–1210.

Bouattour, A. and Darghouth, M.A. (1996) First report of *Babesia divergens* in Tunisia. *Veterinary Parasitology* 63, 161–165.

Braverman, Y. and Chizov-Ginzburg, A. (1996). The role of dogs (*Canis domesticus*) as hosts for African horse sickness virus. *Veterinary Microbiology* 51, 19–25.

Brouqui, P., Bacellar, F., Baranton, G., Birtles, R.J., Bjoërsdorff, A., Blanco, J.R., Caruso, G., Cinco, M., Fournier, P.E., Francavilla, E., Jensenius, M., Kazar, J., Laferl, H., Lakos, A., Lotric-Furlan, S., Maurin, M., Oteo, J.A., Parola, P., Perez-Eid, C., Peter, O., Postic, D., Raoult, D., Tellez, A., Tselentis, Y. and Wilske, B. (2004) Guidelines for the diagnosis of tick-borne bacterial diseases in Europe. *Clinical Microbiology and Infection* 10, 1108–1132.

Brown, C.G.D. (1990) Control of tropical theileriosis (*Theileria annulata* infection) of cattle. *Parassitologia* 32, 23–31.

Bruning, A. (1996) Equine piroplasmosis: an update on diagnosis, treatment, and prevention. *British Veterinary Journal* 152, 139–151.

Burgdorfer, W. and Brinton, L.P. (1975) Mechanisms of transovarial infection of spotted fever rickettsiae in ticks. *Annals of the New York Academy of Sciences* 266, 61–72.

Burrage, T.G. and Laegreid, W.W. (1994) African horsesickness: pathogenesis and immunity. *Comparative Immunology Microbiology and Infectious Diseases* 17, 275–285.

Burrage, T.G., Trevejo, R., Stone-Marschat, M. and Laegreid W.W. (1993) Neutralizing epitopes of African horsesickness virus serotype 4 are located on VP2. *Virology* 196, 799–803.

Burt, F.J. and Swanepoel, R. (2005) Molecular epidemiology of African and Asian Crimean-Congo haemorrhagic fever isolates. *Epidemiology and Infection* 133, 659–666.

Cacció, S.M., Antunovic, B., Moretti, A., Mangili, V., Marinculic, A., Rafaj Baric, R., Slemenda, S.B. and Pieniazek, N.J. (2002) Molecular characterisation of *Babesia canis canis* and *Babesia canis vogeli* from naturally infected European dogs. *Veterinary Parasitology* 106, 285–292.

Cadavid, D. and Barbour, A.G. (1998) Neuroborreliosis during relapsing fever: review of the clinical manifestations, pathology, and treatment of infections in humans and experimental animals. *Clinical Infectious Diseases* 26, 151–164.

Camacho, A.T. (2006) Piroplasma infection in dogs in northern Spain. *Veterinary Parasitology* 138, 97–102.

Camacho, A.T., Guitian, F.J., Pallas, E., Gestal, J.J., Olmeda, A.S., Habela, M.A., Telford, S.R. and Spielman, A. (2005) *Theileria* (*Babesia*) *equi* and *B. caballi* infections in horses in Galicia, Spain. *Tropical Animal Health and Production* 37, 293–302.

Cardoso, L., Costa, A., Tuna, J., Vieira, L., Eyal, O., Yisaschar-Mekuzas, Y. and Baneth, G. (2008) *Babesia canis canis* and *Babesia canis vogeli* infections in dogs from northern Portugal. *Veterinary Parasitology* 156, 199–204.

Carelli, G., Decaro, N., Lorusso, E., Paradies, P., Elia, G., Martella, V., Buonavoglia, C. and Ceci, L. (2008) First report of bovine anaplasmosis caused by *Anaplasma centrale* in Europe. *Annals of the New York Academy of Sciences* 1149, 107–110.

Centeno-Lima, S., do Rosario, V., Parreira, R., Maia, A.J., Freudenthal, A.M., Nijhof, A.M. and Jongejan, F. (2003) A fatal case of human babesiosis in Portugal: molecular and phylogenetic analysis. *Tropical Medicine and International Health* 8, 760–764.

Chai, J.T., Eremeeva, M.E., Borland, C.D. and Karas, J.A. (2008) Fatal Israeli spotted fever in a UK traveler to South Portugal. *Journal of Travel Medicine* 15, 122–123.

Chamot, E., Chatelanat, P., Humair, L., Aeschilimann, A. and Bowessidjaou, J. (1987) Cinq cas de fièvre boutonneuse méditerranéenne en Suisse. *Annals Parasitology Humaine Comparée* 62, 371–379.

Chausov, E.V., Ternovoi, V.A., Protopova, E.V., Kononova, J.V., Konovalova, S.N., Pershikova, N.L., Romanenko, V.N., Ivanova, N.V., Bolshakova, N.P., Moskvitina, N.S. and Loktev, V.B. (2010) Variability of the tick-borne encephalitis virus genome in the 5′ noncoding region derived from ticks *Ixodes persulcatus* and *Ixodes pavlovskyi* in Western Siberia. *Vector-Borne Zoonotic Diseases* 10, 365–375.

Chmielewski, T., Podsiadly, E., Karbowiak, G. and Tylewska-Wierzbanowska, S. (2009) *Rickettsia* spp. in ticks, Poland. *Emerging Infectious Diseases* 15, 486–488.

Christova, I., Van De Pol, J., Yazar, S., Velo, E. and Schouls, L. (2003) Identification of *Borrelia burgdorferi sensu lato*, *Anaplasma* and *Ehrlichia* species, and spotted fever group rickettsiae in ticks from southeastern Europe. *European Journal of Clinical Microbiology and Infectious Diseases* 22, 535–542.

Ciceroni, L., Pinto, A., Ciarrocchi, S. and Ciervo, A. (2006) Current knowledge of rickettsial diseases in Italy. *Annals of the New York Academy of Sciences* 1078, 143–149.

Coetzer J.A.W. and Guthrie, A.J. (2004) African horse sickness. In: Coetzer J.A.W. and Tustin R.C. (eds) *Infectious Diseases of Livestock, Volume 2*, 2nd edn. Oxford University Press Southern Africa, Cape Town, South Africa, pp. 1231–1246.

Coetzer, J.A.W. and Tustin, R.C. (2004) *Infectious Diseases of Livestock*, 2nd edn. Oxford University Press Southern Africa, Cape Town, South Africa.

Corn, J.L., Barré, N., Thiebot, B., Creekmore, T.E., Garris, G.I. and Nettles, V.F. (1993) Potential role of cattle egrets, *Bubulcus ibis* (Ciconiformes: Ardeidae), in the dissemination of *Amblyomma variegatum* (Acari: Ixodidae) in the Eastern Caribbean. *Journal of Medical Entomology* 30, 1029–1037.

Cotté, V., Bonnet, S., Le Rhun, D., Le Naour, E., Chauvin, A., Boulouis, H-J., Lecuelle, B., Lilin, T. and Vayssier-Taussat, M. (2008) Transmission of *Bartonella henselae* by *Ixodes ricinus*. *Emerging Infectious Diseases* 14, 1074–1080.

Criado-Fornelio, A., Martinez-Marcos, A., Buling-Sarana, A. and Barbara-Carretero, J.C. (2003) Molecular studies on *Babesia*, *Theileria* and *Hepatozoon* in southern Europe: part I. Epizootiological aspects. *Veterinary Parasitology* 113, 189–201.

Criado-Fornelio, A., Buling, A., Pingret, J.L., Etievant, M., Boucraut-Baralon, C., Alongi, A., Agnone, A. and Torina, A. (2009) Hemoprotozoa of domestic animals in France: prevalence and molecular characterization. *Veterinary Parasitology* 159, 73–76.

Cutler, S.J. (2006) Possibilities for relapsing fever reemergence. *Emerging Infectious Diseases* 12, 369–374.

Danielova, V., Rudenko, N., Daniel, M., Holubova, J., Materna, J., Golovchenko, M. and Schawarzova, L. (2006) Extension of *Ixodes ricinus* ticks and agents of tick-borne diseases to mountain areas in the Czech Republic. *International Journal of Medical Microbiology* 296, 48–53.

Davies F.G. and Otieno, S. (1977) Elephants and zebras as possible reservoir hosts for African horse sickness virus. *Veterinary Record* 100, 291–292.

de la Fuente, J., García-García, J.C., Blouin, E.F., Saliki, J.T. and Kocan, K.M. (2002) Infection of tick cells and bovine erythrocytes with one genotype of the intracellular ehrlichia *Anaplasma marginale* excludes infection with other genotypes. *Clinical and Diagnostic Laboratory Immunology* (now *Clinical and Vaccine Immunology*) 9, 658–668.

de la Fuente, J., Naranjo, V., Ruiz-Fons, F., Höfle, U., Fernández de Mera, I.G., Villanúa, D., Almazán, C., Torina, A., Caracappa, S., Kocan, K.M. and Gortázar, C. (2005) Potential vertebrate reservoir hosts and invertebrate vectors of *Anaplasma marginale* and *A. phagocytophilum* in central Spain. *Vector-Borne and Zoonotic Diseases* 5, 390–401.

de la Fuente, J., Torina, A., Naranjo, V., Nicosia, S., Alongi, A., La Mantia, F. and Kocan, K.M. (2006) Molecular characterization of *Anaplasma platys* strains from dogs in Sicily, Italy. *BMC Veterinary Research* 2, 24, doi:10.1186/1746-6148-2-24.

de Sousa, R. and Bacellar, F. (2004) Morb-mortalidade por *R.conorii* em Portugal. *Revista Brasileira Parasitologia Veterinária* 13(Suppl. 1), 180–183.

de Sousa, R., Dória Nóbrega, S., Bacellar, F. and Torgal, J. (2003) Mediterranean spotted fever in Portugal: risk factors for fatal outcome in 105 hospitalized patients. *Annals of the New York Academy of Sciences* 990, 285–294.

de Sousa, R., Ismail, N., Dória Nóbrega, S., Costa, P., Abreu, T., França, A., Amaro, M., Proença, P., Brito, P., Poças, J., Ramos, T., Cristina, G., Pombo, G., Vitorino, L., Torgal, J., Bacellar, F. and Walker, D.H. (2005) The presence of eschars, but not greater severity, in Portuguese patients infected with Israeli spotted fever strain of *Rickettsia conorii*. *Annals of the New York Academy of Science* 1063, 197–202.

de Sousa, R., Parreira, P., Luz, T., Santos-Silva, M. and Bacellar, F. (2006a) Boutonneuse fever and climate variability. *Annals of the New York Academy of Sciences* 1078, 162–169.

de Sousa, R., Barata, C., Vitorino, L., Santos-Silva, M., Carrapato, C., Torgal, J., Walker, D. and Bacellar, F. (2006b) *Rickettsia sibirica* isolation from a patient and detection in ticks, Portugal. *Emerging Infectious Diseases* 12, 1103–1108.

de Sousa, R., Santos-Silva, M., Santos, A.S., Barros, S., Torgal, J., Walker, D.H. and Bacellar, F. (2007) *Rickettsia conorii* Israeli tick typhus strain isolated from *Rhipicephalus sanguineus* ticks in Portugal. *Vector-Borne and Zoonotic Diseases* 7, 444–447.

de Sousa, R., Duque, L., Anes, M., Poças, J., Torgal, J., Bacellar, F., Olano, J.P. and Walker, D.H. (2008a) Lymphangitis in a Portuguese patient infected with *Rickettsia sibirica*. *Emerging Infectious Diseases* 14, 529–530.

de Sousa, R., França, A., Dória Nóbrega, S., Belo, A., Amaro, M., Abreu, T., Poças, J., Proença, P., Vaz, J., Torgal, J., Bacellar, F., Ismail, N. and Walker, D.H. (2008b) Host- and microbe-related risk factors for and pathophysiology of fatal *Rickettsia conorii* infection in Portuguese patients. *Journal of Infectious Diseases* 198, 576–585.

De Tray, D.E. (1963) African swine fever. *Advanced Veterinary Science* 8, 299–333.

de Waal, D.T. (1992) Equine piroplasmosis: a review. *British Veterinary Journal* 148, 6–14.

de Waal, D.T. and Van Heerden, J. (2004) Equine piroplasmosis. In: Coetzer, J.A.W. and Tustin, R.C. (eds) *Infectious Diseases of Livestock, Volume 1*, 2nd edn. Oxford University Press Southern Africa, Cape Town, South Africa, pp. 425–433.

Dennis, D.T., Inglesby, T.V., Henderson, D.A., Bartlett, J.G., Ascher, M.S., Eitzen, E., Fine, A.D., Friedlander, A.M., Hauer, J., Layton, M., Lillibridge, S.R., McDade, J.E., Osterholm, M.T., O'Toole, T., Parker, G., Perl, T.M., Russell, P.K. and Tonat, K. (2001) Tularemia as a biological weapon: medical and public health management. *Journal of the American Medical Association* 285, 2763–2773.

Deyde, V.M., Khristova, M.L., Rollin, P.E., Ksiazek, T.G. and Nichol, S.T. (2006) Crimean-Congo hemorrhagic fever virus genomics and global diversity. *Journal of Virology* 80, 8834–8842.

Dixon, L.K. and Wilkinson, P.J. (1988) Genetic diversity of African swine fever virus isolates from soft ticks (*Ornithodoros moubata*) inhabiting warthog burrows in Zambia. *Journal of General Virology* 69, 2981–2993.

Donatein, A. and Lestoquard, F. (1935) Existence on allergie d'une rickettsia du chien. *Bulletin de la Societé de Pathologie Exotique* 28, 418–419.

Donoso Mantke, O., Schädler, R. and Niedrig, M. (2008) A survey on cases of tick-borne encephalitis in European countries. *Eurosurveillance* 13, 1–9.

Duh, D., Tozon, N., Petrovec, M., Strasek, K. and Avsic-Zupanc, T. (2004) Canine babesiosis in Slovenia: molecular evidence of *Babesia canis canis* and *Babesia canis vogeli*. *Veterinary Research* 35, 363–368.

Duh, D., Petrovec, M. and Avsic-Zupanc, T. (2005) Molecular characterization of human pathogen *Babesia* EU1 in *Ixodes ricinus* ticks from Slovenia. *Journal of Parasitology* 91, 463–465.

Dumler, J.S., Barbet, A.F., Bekker, C.P.J., Dasch, G.A., Palmer, G.H., Ray, S.C., Rikihisa, Y. and Rurangirwa, F.R. (2001) Reorganization of genera in the families Rickettsiaceae and Anaplasmataceae in the order Rickettsiales: unification of some species of *Ehrlichia* with *Anaplasma*, *Cowdria* with *Ehrlichia* and *Ehrlichia* with *Neorickettsia*, descriptions of six new species combinations and designation of *Ehrlichia equi* and 'HGE agent' as subjective synonyms of *Ehrlichia phagocytophila*. *International Journal of Systematic and Evolutionary Microbiology* 51, 2145–2165.

EFSA Panel on Animal Health and Welfare (AHAW) (2010a) Scientific opinion on African swine fever. *EFSA Journal* 8(3), 1556.

EFSA Panel on Animal Health and Animal Welfare (AHAW) (2010b) Scientific opinion on Q fever. *EFSA Journal* 8(5), 1595.

EFSA Panel on Animal Health and Welfare (AHAW) (2010c) Scientific opinion on the role of tick vectors in the epidemiology of Crimean Congo hemorrhagic fever and African swine fever in Eurasia. *EFSA Journal* 8(8), 1703.

Elata, A.T., Karsany, M.S., Elageb, R.M., Hussain, M.A., Eltom, K.H., Elbashir, M.I. and Aradaib, I.E. (2011) A nosocomial transmission of Crimean-Congo hemorrhagic fever to an attending physician in north Kordufan, Sudan. *Virology Journal* 8, 303, doi:10.1186/1743-422X-8-303.

Eliasson, H., Lindback, J., Nuorti, J.P., Arneborn, M., Giesecke, J. and Tegnell, A. (2002) The 2000 tularemia outbreak: a case-control study of risk factors in disease-endemic and emergent areas, Sweden. *Emerging Infectious Diseases* 8, 956–960.

Ellis, J., Oyston, P.C., Green, M. and Titball, R.W. (2002) Tularemia. *Clinical Microbiology Reviews* 15, 631–646.

Erasmus, B.J., Young, E., Pietrse, L.M. and Boshoff, S.T. (1978) The susceptibility of zebra and elephants to African horsesickness virus. In: Bryans, J.T. and Gerber, H. (eds). *Equine Infectious Diseases. Proceedings of the Fourth International Conference on Equine Infectious Diseases*. Veterinary Publications, Princeton, New Jersey, pp. 410–413.

Eremeeva, M.E., Beati, L., Makarova, V., Fetisova, N.F., Tarasevich, I.V., Balayeva, N.M. and Raoult, D. (1994) Astrakhan fever rickettsiae: antigenic and genotypic analysis of isolates obtained from human and *Rhipicephalus pumilio* ticks. *American Journal of Tropical Medicine and Hygiene* 51, 697–706.

Estrada-Peña, A. and Jongejan, F. (1999) Ticks feeding on humans: a review of records on human-biting Ixodoidea with special reference to pathogen transmission. *Experimental and Applied Acarology* 23, 685–715.

Estrada-Peña, A., Bouattour, A., Camicas, J. and Walker, L.A.R. (2004) *Ticks of Domestic Animals in the Mediterranean Region. A Guide to Identification of Species*. Produced for the European Union Project Integrated Consortium on Ticks and Tick-borne Diseases (ICTTD-2). Available on request from F. Jongejan@uu.nl.

Ewing, S.A. (1981) Transmission of *Anaplasma marginale* by arthropods. In: *Proceedings of the 7th National Anaplasmosis Conference, Mississippi State. College of Veterinary Medicine, Stockville, Mississippi*, pp. 395–423.

Fernández-Soto, P., Encinas-Grandes, A. and Pérez-Sánchez, R. (2003) *Rickettsia aeschlimannii* in Spain: molecular evidence in *Hyalomma marginatum* and five other tick species that feed on humans. *Emerging Infectious Diseases* 9, 889–890.

Fernández-Soto, P., Pérez-Sánchez, R., Encinas-Grandes, A. and Sanz R.A. (2004) Detection and identification of *Rickettsia helvetica* and *Rickettsia* sp. IRS3/IRS4 in *Ixodes ricinus* ticks found on humans in Spain. *European Journal of Clinical Microbiology of Infectious Diseases* 23, 648–649.

Fisgin, N.T., Doganci, L., Tanyel, E. and Tulek, N. (2010) Initial high rate of misdiagnosis in Crimean-Congo haemorrhagic fever patients in an endemic region of Turkey. *Epidemiology and Infection* 138, 139–144.

Fishman, Z., Gonen, L., Harrus, S., Strauss-Ayali, D., King, D. and Baneth, G. (2004) A serosurvey of *Hepatozoon canis* and *Ehrlichia canis* antibodies in wild red foxes (*Vulpes vulpes*) from Israel. *Veterinary Parasitology* 119, 21–26.

Forsman, M., Sandström, G. and Sjöstedt, A. (1994) Analysis of 16S ribosomal DNA sequences of *Francisella* strains and utilization for determination of the phylogeny of the genus and for identification of strains by PCR. *International Journal of Systematic Bacteriology* 44, 38–46.

Fournier, P.E., Roux, V. and Raoult, D. (1998) Phylogenetic analysis of spotted fever group rickettsiae by study of the outer surface protein rOmpA. *International Journal of Bacteriology* 48, 839–849.

Fournier, P.E., Xeridat, B. and Raoult, D. (2003a) Isolation of a rickettsia related to Astrakhan fever rickettsia from a patient in Chad. *Annals of the New York Academy of Sciences* 990, 152–157.

Fournier, P.E., Durand, J.P., Rolain, J.M., Camicas, J.L., Tolou, H. and Raoult, D. (2003b) Detection of Astrakhan fever rickettsia from ticks in Kosovo. *Annals of the New York Academy of Sciences* 990, 158–161.

Fournier, P.E., Allombert, C., Supputamongkol, Y., Caruso, G., Brouqui, P. and Raoult, D. (2004) An eruptive fever associated with antibodies to *Rickettsia helvetica* in Europe and Thailand. *Journal of Clinical Microbiology* 42, 816–818.

Fournier, P.E., Gouriet, F., Brouqui, P., Lucht, F. and Raoult, D. (2005) Lymphangitis-associated rickettsiosis, a new rickettsiosis caused by *Rickettsia sibirica mongolotimonae*: seven new cases and review of the literature. *Clinical Infectious Diseases* 40, 1435–1444.

Friedhoff, K.T. (1997) Tick-borne diseases of sheep and goats caused by *Babesia*, *Theileria* or *Anaplasma* spp. *Parassitologia* 39, 99–109.

Friend, M. (2006) Tularemia, st ed. Circular 1297, US Geological Survey, Reston, Virginia.

Gabrielli, S., Kumlien, S., Calderini, P., Brozzi, A., Iori, A. and Cancrini, G. (2010) The first report of *Hepatozoon canis* identified in *Vulpes vulpes* and ticks from Italy. *Vector-Borne and Zoonotic Diseases* 10, 855–859.

García-García, J.C., Portillo, A., Núñez, M.J., Santibáñez, S., Castro, B. and Oteo, J.A. (2010) A patient from Argentina infected with *Rickettsia massiliae*. *American Journal of Tropical Medicine and Hygiene* 82, 691–692.

Genchi, C. (2007) Human babesiosis, an emerging zoonosis. *Parassitologia* 49(Suppl. 1), 22–31.

Georges, K., Loria, G.R., Riili, S., Greco, A., Caracappa, S., Jongejan, F. and Sparagano, O. (2001) Detection of haemoparasites in cattle by reverse line blot hybridisation with a note on the distribution of ticks in Sicily. *Veterinary Parasitology* 99, 273–286.

Giammanco, G.M., Mansueto, S., Ammatuna, P. and Vitale, G. (2003) Israeli spotted fever *Rickettsia* in Sicilian *Rhipicephalus sanguineus* ticks. *Emerging Infectious Diseases* 9, 892–893.

Giammanco, G.M., Vitale, G., Mansueto, S., Capra, G., Caleca, M.P. and Ammatuna, P. (2005) Presence of *Rickettsia conorii* subsp. *israelensis*, the causative agent of Israeli spotted fever. in Sicily, Italy, ascertained in a retrospective study. *Journal of Clinical Microbiology* 43, 6027–6031.

Gilbert, L., Jones, L.D., Laurenson, M.K., Gould, E.A., Reid, H.W. and Hudson, P.J. (2004) Ticks need not bite their red grouse hosts to infect them with louping ill virus. *Proceedings of the Royal Society of London Series B, Biological Sciences* 271, 202–205.

Gilot, B. (1984) Biologie et écologie de *Rhipicephalus sanguineus* (Latreille, 1806) (Acariens, Ixodoidea) dans le sud-est de la France. Intéret dans l'épidémiologie de la fièvre boutonneuse (première partie). *Bulletin de la Société Scientifique de Médicine Vétérinaire* 86, 25–33.

Goethert, H.K. and Telford, S.R. (2003). Enzootic transmission of *Anaplasma bovis* in Nantucket cottontail rabbits. *Journal of Clinical Microbiology* 41, 3744–3747.

Goldwasser, R.A., Steiman, Y., Klingberg, W., Swartz, T.A. and Klingberg, M.A. (1974) The isolation of strains of rickettsiae of the spotted fever group in Israel and their differentiation from other members of the group by immunofluorescence methods. *Scandinavian Journal of Infectious Diseases* 6, 53–62.

Golovljova, I., Vene, S., Brus-Sjölander, K., Vasilenko, V., Plyusnin, A. and Lundkvist, Å. (2004) Characterization of tick-borne encephalitis virus from Estonia. *Journal of Medical Virology* 74, 580–588.

Gouriet, F., Fenollar, F., Patrice, J.Y., Drancourt, M. and Raoult, D. (2005) Use of shell-vial cell culture assay for isolation of bacteria from clinical specimens: 13 years of experience. *Journal of Clinical Microbiology* 43, 4993–5002.

Grard, G., Moureau, G., Charrel, R.N., Lemasson, J.J., Gonzalez, J.P., Gallian, P., Gritsun, T.S., Holmes, E.C., Gould, E.A. and de Lamballerie, X. (2007) Genetic characterization of tick-borne flaviviruses: new insights into evolution, pathogenetic determinants and taxonomy. *Virology* 361, 80–92.

Gray, J.S. (2006) Identity of the causal agents of human babesiosis in Europe: Mini-Review. *International Journal of Medical Microbiology* 296(S1), 131–136.

Gray, J.S. and Weiss, L.M. (2008) *Babesia microti* In: Khan, N. (ed.) *Emerging Protozoan Pathogens*. Taylor and Francis, Abingdon, UK, pp. 303–349.

Gray, J.S., Dautel, H., Estrada-Peña, A., Kahl, A. and Lindgren, E. (2009) Effects of climate change on ticks and tick-borne diseases in Europe. *Interdisciplinary Perspectives on Infectious Diseases* 2009, Article ID 593232, doi: 10.1155/2009/593232.

Gritsun, T.S., Lashkevich, V.A. and Gould, E.A. (2003) Tick-borne encephalitis. *Antiviral Research* 57, 129–146.

Gross, E.M. and Yagupsky, P. (1987) Israeli rickettsial spotted fever in children. A review of 54 cases. *Acta Tropica* 44, 91–96.

Groves, M.G., Dennis, G.L., Amyx, H.L. and Huxsoll, D.L. (1975) Transmission of *Ehrlichia canis* to dogs by ticks (*Rhipicephalus sanguineus*). *American Journal of Veterinary Research* 36, 937–940.

Guryčová, D., Kociánová, E., Výrostekovl, V. and Řeháček, J. (1995) Prevalence of ticks infected with *Francisella tularensis* in natural foci of tularemia in western Slovakia. *European Journal of Epidemiology* 11, 469–474.

Guryčová, D., Výrosteková, V., Khanakah, G., Kocianová, E. and Stanek G. (2001) Importance of surveillance of tularemia natural foci in the known endemic area of Central Europe, 1991–1997. *Wiener Klinische Wochenschrift* 113, 433–438.

Hamblin, C., Salt, J.S, Mellor, P.S., Graham, S.D., Smith, P.R. and Wohlsein, P. (1998) Donkeys as reservoirs of African horse sickness virus. *Archives of Virology* 14(Suppl.), 37–47.

Harrus, S., Perlman-Avrahami, A., Mumcuoglu, K.Y., Morick, D., Eyal, O. and Baneth, G. (2011) Molecular detection of *Ehrlichia canis, Anaplasma bovis, Anaplasma platys, Candidatus midichloria mitochondrii* and *Babesia canis vogeli* in ticks from Israel. *Clinical Microbiology and Infection* 17, 459-463.

Harvey, J.W., Simpson, C.F. and Gaskin, J.M. (1978) Cyclic thrombocytopenia induced by a *Rickettsia*-like agent in dogs. *Journal of Infectious Diseases* 137, 182–188.

Häselbarth, K., Tenter, A.M., Brade, V., Krieger, G. and Hunfeld, K.P. (2007) First case of human babesiosis in Germany – clinical presentation and molecular characterization of the pathogen. *International Journal of Medical Microbiology* 297, 197–204.

Herndon, D.R., Palmer, G.H., Shkap, V., Knowles, D.P. Jr and Brayton, K.A. (2010) Complete genome sequence of *Anaplasma marginale* subsp. *centrale*. *Journal of Bacteriology* 192, 379–380.

Herwaldt, B.L., Caccio, S., Gherlinzoni, F., Aspöck, H., Slemenda, S.B., Piccaluga, P., Martinelli, G., Edelhofer, R., Hollenstein, U., Poletti, G., Pampiglione, S., Loschenberger, K., Tura, S. and Pieniazek, N.J. (2003) Molecular characterization of a non-*Babesia divergens* organism causing zoonotic babesiosis in Europe. *Emerging Infectious Diseases* 9, 942–948.

Heyman, P., Duh, D., Van Der Kuylen, B., Cochez, C., Van Esbroeck, M., Vandenvelde, C. and Avsic-Zupanc, T. (2007) Molecular and serological evidence for *Anaplasma platys* and *Babesia* sp. infection in a dog, imported in Belgium, from southern Spain. *Journal of Veterinary Medicine, Series A, Physiology Pathology Clinical Medicine* 54, 276–279.

Hildebrandt, A., Hunfeld, K.P., Baier, M., Krumbholz, A., Sachse, S., Lorenzen, T., Kiehntopf, M., Fricke, H.J. and Straube, E. (2007) First confirmed autochthonous case of human *Babesia microti* infection in Europe. *European Journal of Clinical Microbiology and Infectious Diseases* 26, 595–601.

Hinnebusch, J. and Barbour, A.G. (1991) Linear plasmids of *Borrelia burgdorferi* have a telomeric structure and sequence similar to those of a eukaryotic virus. *Journal of Bacteriology* 173, 7233–7239.

Hirsch, M. and Pantchev, N. (2008) Occurrence of the travel diseases leishmaniasis, ehrlichiosis, babesiosis and dirofiloriosis in dogs living in Germany. *Kleintierpraxis* 53, 154–165.

Holland, M. (2001) Emerging diseases in northern Europe. *Journal of Small Animal Practice* 42, 205–206.

Homer, M.J., Aguilar-Delfin, I., Teleford, S.R., Krause, P.J. and Persing, D.H. (2000) Babesiosis. *Clinical Microbiology Reviews* 13, 451–469.

Hopla, C.E. and Hopla, A.K. (1994) Tularemia. In: Beran, G.W. (ed.) *CRC Handbook of Zoonoses, Section A, Bacterial, Rickettsial and Mycotic Diseases*, 2nd edn. CRC Press, Boca Raton, Florida, pp. 113–125.

Horak, I.G., Swanepoel, R. and Gummow, B. (2001) The distribution of *Hyalomma* spp. and human cases of Crimean-Congo haemorrhagic fever in South Africa. In: Proceedings of 10th Conference of the Association of Institutions for Tropical Veterinary Medicine, Copenhagen, Denmark, 20–23 August 2001, pp. 501–509.

Hornok, S., Edelhofer, R., Földvári, G., Joachim, A. and Farkas, R. (2007a) Serological evidence for *Babesia canis* infection of horses and an endemic focus of *B. caballi* in Hungary. *Acta Veterinaria Hungarica* 55, 491–500.

Hornok, S., Elek, V., de la Fuente, J., Naranjo, V., Farkas, R., Majoros, G. and Foldvari, G. (2007b) First serological and molecular evidence on the endemicity of *Anaplasma ovis* and *A. marginale* in Hungary. *Veterinary Microbiology* 122, 316–322.

Hornok, S., Foldvari, G., Elek, V., Naranjo, V., Farkas, R. and de la Fuente, J. (2008) Molecular identification of *Anaplasma marginale* and rickettsial endosymbionts in blood-sucking flies (Diptera: Tabanidae, Muscidae) and hard ticks (Acari: Ixodidae). *Veterinary Parasitology* 154, 354–359.

Hornok, S., Meli, M.L., Perreten, A., Farkas, R., Willi, B., Beugnet, F., Lutz, H. and Hofmann-Lehmann, R. (2010) Molecular investigation of hard ticks (Acari: Ixodidae) and fleas (Siphonaptera: Pulicidae) as potential vectors of rickettsial and mycoplasmal agents. *Veterinary Microbiology* 140, 98–104.

Hubálek, Z. and Halouzka, J. (1998) Prevalence rates of *Borrelia burgdorferi sensu lato* in host-seeking *Ixodes ricinus* ticks in Europe. *Parasitology Research* 84, 167–172.

Hubálek, Z., Treml, F., Halouzka, J., Juricova, Z., Huňady, M. and Janík, V. (1996) Frequent isolation of *Francisella tularensis* from *Dermacentor reticulatus* ticks in an enzootic focus of tularaemia. *Medical and Veterinary Entomology* 10, 241–246.

Hunfeld, K.P., Hildebrandt, A. and Gray, J.S. (2008) Babesiosis: recent insights into an ancient disease. *International Journal for Parasitology* 38, 1219–1237.

Ibarra, V., Oteo, J.A. and Portillo, A. (2006) *Rickettsia slovaca* infection: DEBONEL/TIBOLA. *Annals of the New York Academy of Sciences* 1078, 206–214.

Inokuma, H., Raoult, D. and Brouqui, P. (2000) Detection of *Ehrlichia platys* DNA from brown dog ticks (*Rhipicephalus sanguineous*) in Okinawa Island, Japan. *Journal of Clinical Microbiology* 38, 4219–4221.

Irwin, P.I. (2009) Canine babesiosis: from molecular taxonomy to control. Review. *Parasites and Vectors* 2(Suppl. 1), 1–9.

Iwata, H., Yamagawa, M. and Roy, P. (1992) Evolutionary relationships among the gnat-transmitted orbiviruses that cause African horse sickness, bluetongue, and epizootic hemorrhagic disease as evidenced by their capsid protein sequences. *Virology* 191, 251–261.

Jado, I., Oteo, J.A., Aldámiz, M., Gil, H., Escudero, R., Ibarra, V., Portu, J., Portillo, A., Lezaun, M.J., García-Amil, C., Rodríguez-Moreno, I. and Anda, P. (2007) *Rickettsia monacensis* and human disease, Spain. *Emerging Infectious Diseases* 13, 1405–1407.

Jaenson, T.G.T. and Tälleklint, L. (1992) Incompetence of roe deer as reservoirs of the Lyme borreliosis spirochete. *Journal of Medical Entomology* 29, 813–817.

Jefferies, R., Ryan, U.M., Jardine, J., Broughton, D.K., Robertson, I.D. and Irwin, P.J. (2007) Blood, bull terriers and babesiosis: further evidence for direct transmission of *Babesia gibsoni* in dogs. *Australian Veterinary Journal* 85, 459–463.

Jones, L.D., Gaunt, M., Hails, R.S., Laurenson, K., Hudson, P.J., Reid, H., Henbest, P. and Gould, E.A. (1997) Transmission of louping ill virus between infected and uninfected ticks co-feeding on mountain hares. *Medical and Veterinary Entomology* 11, 172–176.

Jori, F. and Bastos, A.D.S. (2009) Role of wild suids in the epidemiology of Africa swine fever. *EcoHealth* 6, 296–310.

Karti, S., Odabasi, Z., Korten, V., Yilmaz, M., Sonmez, M., Caylan, R., Akdogan, E., Eren, N., Koksal, I., Ovali, E., Erickson, B.R., Vincent, M.J., Nichol, S.T., Comer, J.A., Rollin, P.E. and Ksiazek, T.G. (2004) Crimean–Congo hemorrhagic fever in Turkey. *Emerging Infectious Diseases* 10, 1379–1384.

Kaysser, P., Seibold, E., Matz-Rensing, K., Pfeffer, M., Essbauer, S. and Splettstoesser, W.D. (2008) Re-emergence of tularemia in Germany: presence of *Francisella tularensis* in different rodent species in endemic areas. *BMC Infectious Diseases* 8, 157, doi:10.1186/1471-2334-8-157.

Keim, P.S., Johansson, A. and Wagner, D.M. (2007) Molecular epidemiology, evolution and ecology of *Francisella. Annals of the New York Academy of Sciences* 1105, 30–66.

Keshtkar-Jahromi, M., Kuhn, J.H., Christova, I., Bradfute, S.B., Jahrling, P.B. and Bavari, S. (2011) Crimean-Congo hemorrhagic fever: current and future prospects of vaccines and therapies. *Antiviral Research* 90, 85–92.

Kjemtrup, A.M. and Conrad, P.A. (2000) Human babesiosis: an emerging tick-borne disease. *International Journal for Parasitology* 30, 1323–1337.

Kjemtrup, A.M., Kocan, A.A., Whitworth, L., Meinkoth, J., Birkenheuer, A.J., Cummings, J., Boudreaux, M.K., Stockham, S.L., Irizarry-Rovira, A. and Conrad, P.A. (2000) There are at least three genetically distinct small piroplasms from dogs. *International Journal for Parasitology* 30, 1501–1505.

Kleiboeker, S.B., Scoles, G.A., Burrage, T.G. and Sur, J.H. (1999) African swine fever virus replication in the midgut epithelium is required for infection of *Ornithodoros* ticks. *Journal of Virology* 73, 8587–8598.

Knowles, D. (1996) Equine babesiosis (piroplasmosis): a problem in the international movement of horses. *British Veterinary Journal* 152, 123–126.

Kocan, K.M., de la Fuente, J., Blouin, E.F. and García-García, J.C. (2004) *Anaplasma marginale* (Rickettsiales: Anaplasmataceae): recent advances in defining host–pathogen adaptations of a tick-borne rickettsia. *Parasitology* 129, 285–300.

Kocan, K.M., de la Fuente, J., Blouin, E.F., Coetzee, J.F. and Ewing, S.A. (2010) The natural history of *Anaplasma marginale*. *Veterinary Parasitology* 167, 95–107.

Kuloglu, F., Rolain, J.M., Fournier, P.E., Akata, F., Tugrul, M. and Raoult, D. (2004) First isolation of *Rickettsia conorii* from humans in the Trakya (European) region of Turkey. *European Journal of Clinical Microbiology and Infectious Diseases* 23, 609–614.

Labuda, M. and Nuttall, P.A. (2008) Viruses transmitted by ticks. In: Bowman, A.S. and Nuttall, P.A. (eds) *Ticks: Biology, Disease and Control*. Cambridge University Press, Cambridge, UK, pp. 253–280.

Lakos, A. (1997) Tick-borne lymphadenopathy – a new rickettsial disease? *The Lancet* 350(9083), 1006.

Lakos, A. (2002) Tick-borne lymphadenopathy (TIBOLA). *Wiener Klinische Wochenschrift* 114, 648–654.

Lakos, A. and Raoult, D. (1999) Tick-borne lymphadenopathy (TIBOLA) a *Rickettsia slovaca* infection? In: Raoult, D. and Brouqui, D.R. (eds) *Rickettsiae and Rickettsial Diseases at the Turn of the Third Millennium*. Elsevier. Paris, pp. 258–261.

Lambert, M.T., Dugernier, G., Bigaignon, J., Rahier, V. and Piot, P. (1984) Mediterranean spotted fever in Belgium. *The Lancet* 2(8410),1038.

Lang, G.H. (1990) Coxiellosis (Q fever) in animals. In: Marrie, T.J. (ed.) *Q Fever, Volume 1. The Disease*. CRC Press, Boca Raton, Florida, pp. 23–48.

Lindquist, L. and Vapalahti, O. (2008) Tick-borne encephalitis. *The Lancet* 371(9627), 1861–1871.

Lobetti, R. (2007) Viral diseases of the dog. Presented to: 32nd World Small Animal Veterinary Association Congress, Sydney Convention Centre, Darling Harbour, Australia, 19–23 August 2007.

Lord, C.C., Venter, G.J., Mellor, P.S., Paweska, J.T. and Woolhouse, M.E. (2002). Transmission patterns of African horse sickness and equine encephalosis viruses in South African donkeys. *Epidemiology and Infection* 128, 265–275.

Lubisi, B.A., Bastos, A.D.S., Dwarka, R.M. and Vosloo, W. (2005) Molecular epidemiology of African swine fever in East Africa. *Archives of Virology* 150, 2439–2452.

MacLeod, J. (1936) Studies on tick-borne fever in sheep. II. Experiments on transmission and distribution of the disease. *Parasitology* 28, 320–329.

Madariaga, M.G., Rezai, K., Trenholme, G.M. and Weinstein, R.A. (2003) Q fever: a biological weapon in your backyard. *The Lancet Infectious Diseases* 3, 709–712.

Maeda, K., Markowitz, N., Hawley, R.C., Ristic, M., Cox, D. and McDade, J.E. (1987) Human infection with *Ehrlichia canis*, a leukocytic rickettsia. *New England Journal of Medicine* 316, 853–856.

Majláthová, V., Hurníková, Z., Majláth, I. and Petko, B. (2007) *Hepatozoon canis* infection in Slovakia: imported or autochthonous? *Vector-Borne and Zoonotic Diseases* 7, 199–202.

Massaro, W., Palmer, G.H., Kappmeyer, L.S., Scoles, G.A. and Knowles, D.P. (2003) Expression of *B. equi* merozoite antigen 2 during development of *Babesia equi* in the midgut and salivary gland of the vector tick *Boophilus microplus*. *Journal of Clinical Microbiology* 41, 5803–5809.

Matsumoto, K., Parola, P., Brouqui, P. and Raoult, D. (2004) *Rickettsia aeschlimannii* in *Hyalomma* ticks from Corsica. *European Journal of Clinical Microbiology and Infectious Diseases* 23, 732–734.

Matyas, B.T., Nieder, H.S. and Telford, S.R. (2007) Pneumonic tularemia on Martha's Vineyard: clinical, epidemiologic, and ecological characteristics. *Annals of the New York Academy of Sciences* 1105, 351–377.

Maxey, E.E. (1899) Some observations of the so-called spotted fever of Idaho. *Medical Sentinel* 10, 433–438.

McDonald, J., MacLean, J.D. and McDade, J.E. (1988) Imported rickettsial disease: clinical and epidemiological features. *American Journal of Medicine* 185, 799–805.

Mehlhorn, H. and Schein, E. (1984) The piroplasms: life cycle and sexual stages. *Advances in Parasitology* 23, 37–103.

Meiswinkel, R., Baylis, M. and Labuschagne, K. (2000) Stabling and the protection of horses from *Culicoides bolitinos* (Diptera: Ceratopogonidae), a recently identified vector of African horse sickness. *Bulletin of Entomological Research* 90, 2509–2515.

Meiswinkel, R., Venter, G.J. and Nevill, E.M. (2004) Vectors, *Culicoides* spp. In: Coetzer J.A.W. and Tustin, R.C. (eds) *Infectious Diseases of Livestock, Volume 1*, 2nd edn. Oxford University Press Southern Africa, Cape Town, South Africa, pp. 93–136.

Mellor, P.S. and Hamblin, C. (2004) African horse sickness. *Veterinary Research* 35, 445–466.

Menn, B., Lorentz, S. and Naucke, T.J. (2010) Imported and travelling dogs as carriers of canine vector-borne pathogens in Germany. *Parasites and Vectors* 3, 34, doi:10.1186/1756-3305-3-34.

Michaud, V.P., Gil, P., Kwiatek, S., Prome, S., Dixon, L., Romero, L., Le Potier, M.F., Arias, M., Couacy-Hymann, E., Roger, F., Libeau, G. and Albina, E. (2007) Long-term storage at tropical temperature of dried-blood filter papers for detection and genotyping of RNA and DNA viruses by direct PCR. *Journal of Virological Methods* 146, 257–265.

Milhano, N., De Carvalho, I., Alves, A.S., Arroube, S., Soares, J., Rodriguez, P., Carolino, M., Núncio, M.S., Piesman, J. and de Sousa, R. (2010) Coinfections of *Rickettsia slovaca* and *Rickettsia helvetica* with *Borrelia lusitaniae* in ticks collected in a Safari Park, Portugal. *Ticks and Tickborne Diseases* 1, 172–177.

Mittermayer, T., Brezina, R. and Urvölgyi, J. (1980) First report of an infection with *Rickettsia slovaca*. *Folia Parasitologica* (Praha) 27, 373–376.

Mokrani, N., Parola, P., Tebbal, S., Dalichaouche, M., Aouati, A. and Raoult, D. (2008) *Rickettsia aeschlimannii* infection, Algeria. *Emerging Infectious Diseases* 14, 1814–1815.

Moretti, A., Mangili, A., Salvatori, R., Maresca, C., Scoccia, E., Torina, A., Moretta, I., Gabrielli, S., Tampieri, M.P. and Pietrobelli, M. (2009) Prevalence and diagnosis of *Babesia* and *Theileria* infections in horses in Italy: a preliminary study. *Veterinary Journal* 184, 346–350.

Morner, T. and Addison, E. (2001) Tularemia. In: Williams, E.S. and Barker, I.K. (eds) *Infectious Diseases of Wild Mammals*, 3rd edn. Iowa State University Press, Ames, Iowa, pp. 303–312.

Movila, A., Rolain, J.M., Podavalenko, A., Toderas, I., Tkachenco, L., Naglov, V. and Raoult, D. (2009) Detection of spotted fever group rickettsiae and family Anaplasmataceae in *Ixodes ricinus* ticks from Republic of Moldova and Eastern Ukraine. *Clinical Microbiology and Infection* 15(Suppl. 2), 32–33.

Movila, A., Reye, A.L., Dubinina, H.V., Tolstenkov, O.O., Toderas, I., Hübschen, J.M., Muller, C.P. and Alekseev, A. (2011) Detection of *Babesia* sp. EU1 and members of spotted fever group rickettsiae in ticks collected from migratory birds at Curonian Spit, North-western Russia. *Vector-Borne Zoonotic Diseases* 11, 89–91.

Murata, T., Inoue, M., Tateyama, S., Taura, Y. and Nakama, S. (1993) Vertical transmission of *Hepatozoon canis* in dogs. *Journal of Veterinary Medical Science* 55, 867–868.

Mylonakis, M.E., Koutinas, A.F., Breitschwerdt, E.B., Hegarty, B.C., Billinis, C.D., Leontides, L.S. and Kontos, V.S. (2004) Chronic canine ehrlichiosis (*Ehrlichia canis*): a retrospective study of 19 natural cases. *Journal of the American Animal Hospital Association* 40, 174–184.

Naderi, H.R., Sarvghad, M.R., Bojdy, A., Hadizadeh, M.R., Sadeghi, R. and Sheybari, F. (2011) Nosocomial outbreak of Crimean-Congo haemorrhagic fever. *Epidemiology and Infection* 139, 862–866.

Naranjo, V., Ruiz-Fons, F., Höfle, U., Fernández de Mera, I.G., Villanúa, D., Almazán, C., Torina, A., Caracappa, S., Kocan, K.M., Gortázar, C. and de la Fuente, J. (2006) Molecular epidemiology of human and bovine anaplasmosis in southern Europe. *Annals of the New York Academy of Sciences* 1078, 95–99.

Ngwamidiba, M., Blanc, G., Ogata, H., Raoult, D. and Fournier, P.E. (2005) Phylogenetic study of *Rickettsia* species using sequences of the autotransporter protein-encoding gene *sca2*. *Annals of the New York Academy of Sciences* 1063, 94–99.

Ngwamidiba, M., Blanc, G., Raoult, D. and Fournier, P.E. (2006) *Sca1*, a previously undescribed paralog from autotransporter protein-encoding genes in *Rickettsia* species. *BMC Microbiology* 6, 12, doi:10.1186/1471-2180-6-12.

Nijhof, A.M., Bodaan, C., Postigo, M., Nieuwenhuijs, H., Opsteegh, M., Franssen, L., Jebbink, F. and Jongejan, F. (2007) Ticks and associated pathogens collected from domestic animals in the Netherlands. *Vector-Borne Zoonotic Diseases* 7, 585–595.

Nilsson, K., Lindquist, O. and Påhlson, C. (1999) Association of *R. helvetica* with chronic perimyocarditis in sudden cardiac death. *The Lancet* 354, 1169–1173.

Nilsson, K., Påhlson, C., Lukinius, A., Eriksson, L., Nilsson, L. and Lindquist, O. (2002) Presence of *Rickettsia helvetica* in granulomatous tissue from patients with sarcoidosis. *Journal of Infectious Diseases* 185, 1128–1138.

Nilsson, K., Elfving, K. and Påhlson, C. (2010) *Rickettsia helvetica* in a patient with meningitis, Sweden, 2006. *Emerging Infectious Diseases* 16, 490–492.

Ogden, N.H., Nuttall, P.A. and Randolph, S.E. (1997) Natural Lyme disease cycles maintained via sheep co-feeding ticks. *Parasitology* 115, 591–599.

Ogden, N.H., Bown, K., Horrocks, B.K., Woldehiwet, Z. and Bennett, M. (1998) Granulocytic *Ehrlichia* infection in ixodid ticks and mammals in woodlands and uplands of the U.K. *Medical and Veterinary Entomology* 12, 423–429.

Olmeda, A.S., Armstrong, P.M., Rosenthal, B.M., Valladares, B., del Castillo, A., de Armas, F., Miguelez, M., Gonzalez, A., Rodriguez, J.A., Spielman, A. and Telford, S.R. (1997) A subtropical case of human babesiosis, *Acta Tropica* 67, 229–234.

Oteo, J.A., Portillo, A., Santibáñez, S., Pérez-Martínez, L., Blanco, J.R., Jiménez, S., Ibarra, V., Pérez-Palacios, A. and Sanz, M. (2006) Prevalence of spotted fever group *Rickettsia* species detected in ticks in La Rioja, Spain. *Annals of the New York Academy of Sciences* 1078, 320–323.

Otero, R., Fenoll, A., Casal, J., Gross, E.M., Yagupsky, P., Torok, V. and Goldwasser, R.A. (1982) Resurgence of Mediterranean spotted fever. *The Lancet* 320(8307) [formerly 2(8307)], 1107.

O'Toole, D., Williams, E.S., Woods, L.W., Mills, K., Boerger-Fields, A., Montgomery, D.L., Jaeger, P., Edwards, W.H., Christensen, D. and Marlatt, W. (2008) Tularemia in range sheep: an overlooked syndrome. *Journal of Veterinary Diagnostic Investigation* 20, 508–513.

Oyston, P.C., Sjostedt, A. and Titball, R.W. (2004) Tularaemia: bioterrorism defence renews interest in *Francisella tularensis*. *Nature Reviews Microbiology* 2, 967–978.

Papadopoulos, B., Morel, P.C. and Aeschlimann, A. (1996) Ticks of domestic animals in the Macedonia region of Greece. *Veterinary Parasitology* 63, 25–40.

Parola, P. and Raoult, D. (2001) Ticks and tick-borne bacterial human diseases, an emerging infectious threat. *Clinical Infectious Diseases* 32, 897–928.

Parola, P., Beati, L., Cambon, M. and Raoult, D. (1998) First isolation of *Rickettsia helvetica* from *Ixodes ricinus* ticks in France. *European Journal of Clinical Microbiology and Infectious Diseases* 17, 95–100.

Parola, P., Rovery, C., Rolain, J.M., Brouqui, P., Davoust, B. and Raoult, D. (2009) *Rickettsia slovaca* and *R. raoultii* in tick-borne rickettsioses. *Emerging Infectious Diseases* 15, 1105–1108.

Pascucci, I., Conte, A. and Scacchia, M. (2007) Use of geographic information systems to identify areas at risk of introducing *Amblyomma variegatum* and *A. hebraeum* to Italy. *Veterinaria Italiana* 43, 655–661.

Paweska, J. (2007) Epidemiology and laboratory diagnosed cases of CCHF in Southern Africa. Presentation No. 21 006 at International Meeting on Emerging Infectious Diseases, Vienna, 23–25 February 2007.

Perez, M., Bodor, M., Zhang, C., Xiong, Q. and Rikihisa, Y. (2006) Human infection with *Ehrlichia canis* accompanied by clinical signs in Venezuela. *Annals of the New York Academy of Sciences*, 1078, 110–117.

Perez-Castrillon, J.L., Bachiller-Luque, P., Martin-Luquero, M., Mena-Martin, F.J. and Herreros, V. (2001) Tularemia epidemic in northwestern Spain: clinical description and therapeutic response. *Clinical Infectious Diseases* 33, 573–576.

Petersen, J.M., Mead, P.S. and Schriefer, M.E. (2009) *Francisella tularensis*: an arthropod-borne pathogen. *Veterinary Research* 40(2), 7, doi: 10.1051/vetres:2008045.

Petersen, J.M. and Schriefer, M.E. (2005) Tularemia: emergence/re-emergence. *Veterinary Research* 36, 455–467.

Piesman, J. and Gern, L. (2008) Lyme borreliosis in Europe and North America. In: Bowman, A.S. and Nuttall, P.A. (eds) *Ticks, Biology, Disease and Control*. Cambridge University Press, Cambridge, UK, pp. 220–252.

Pikula, J., Beklova, M., Holesovoska, Z. and Treml, F. (2004) Ecology of European brown hare and distribution of natural foci of tularemia in the Czech Republic. *Acta Veterinaria Brno* 73, 267–273.

Planck, A., Eklund, A., Grunewald, J. and Vene, S. (2004) No serological evidence of *Rickettsia helvetica* infection in Scandinavian sarcoidosis patients. *European Respiratory Journal* 24, 811–813.

Plowright, W. (1977) Vector transmission of African swine fever virus. In: *Seminar on Hog Cholera/Classical Swine Fever and African Swine Fever*. Commission of the European Communities, Brussels, Publication No. Eur. 5904 En, pp. 575–587.

Pluta, S., Hartelt, K., Oehme, R., Mackenstedt, U. and Kimmig, P. (2010) Prevalence of *Coxiella burnetii* and *Rickettsia* spp. in ticks and rodents in southern Germany. *Ticks and Tick-Borne Diseases* 1, 145–147.

Polson, A. and Deeks, D. (1963) Electron microscopy of neurotropic African horse-s ckness virus. *Journal of Hygiene* 61, 149–153.

Potgieter, F.T. and Stoltsz, W.H. (2004) Bovine anaplasmosis. In: Coetzer, J.A.W. and Tustin, R.C. (eds) *Infectious Diseases of Livestock, Volume 1*, 2nd edn. Oxford University Press Southern Africa, Cape Town, South Africa, pp. 594–616.

Preston, P.M. (2001) Theilerioses. In: Service, M.W. (ed.) *Encyclopaedia of Arthropod-transmitted Infections of Man and Domesticated Animals*. CAB International, Wallingford, UK, pp. 487–502.

Pretorius, A.M. and Birtles, R.J. (2002) *Rickettsia aeschlimannii*: a new pathogenic spotted fever group rickettsia, South Africa. *Emerging Infectious Diseases* 8(8), 874, doi: 10.3201/eid0808.020199.

Pringle, C.R. and Wickner, R.B. (2000) Part three. The viruses. Genus: Orbivirus. In: van Regenmortel, M.H.V., Fauquet, C.M., Bishop, D.H.L., Carstens, E.B., Estes, M.K., Lemon, S.M., Maniloff, J., Mayo, M.A., McGeoch, D.J., Pringle, C.R. and Wickner, R.B. (eds) *Virus Taxonomy: Classification and Nomenclature of Viruses. Seventh Report of the International Committee of Taxonomy of Viruses*. Academic Press, San Diego, California, pp. 408–421.

Prosenc, K., Petrovec, M., Trilar, T., Duh, D. and Avsic-Zupanc, T. (2003) Detection of rickettsiae in *Ixodes ricinus* ticks in Slovenia. *Annals of the New York Academy of Sciences* 990, 201–204.

Psaroulaki, A., Germanakis, A., Gikas, A., Scoulica, E. and Tselentis, Y. (2005a) First isolation and genotypic identification of *Rickettsia conorii* Malish 7 from a patient in Greece. *European Journal of Clinical Microbiology and Infectious Diseases* 24, 297–298.

Psaroulaki, A., Germanakis, A., Gikas, A., Scoulica, E. and Tselentis, Y. (2005b) Simultaneous detection of *Rickettsia mongolotimonae* in a patient and a tick in Greece. *Journal of Clinical Microbiology* 43, 3558–3559.

Psaroulaki, A., Ragiadakou, D., Kouris, G., Papadopoulos, B., Chaniotis, B. and Tselentis, Y. (2006) Ticks, tick-borne rickettsiae, and *Coxiella burnetii* in the Greek Island of Cephalonia. *Annals of the New York Academy of Sciences* 1078, 389–399.

Punda-Polic, V., Petrovec, M., Trilar, T., Duh, D., Bradaric, N., Klismanic, Z. and Avsic-Zupanc, T. (2002) Detection and identification of spotted fever group rickettsiae in ticks collected in southern Croatia. *Experimental and Applied Acarology* 28, 169–176.

Raoult, D. and Parola, P. (eds) (2007) *Rickettsial Diseases*. Informa Healthcare, New York.

Raoult, D. and Roux, V. (1997) Rickettsioses as paradigms of new or emerging infectious diseases. *Clinical Microbiology Reviews* 10, 694–719.

Raoult, D., Nicolas, D., De Micco, P., Gallais, H. and Casanova, P. (1985) Epidemiologic aspects of Mediterranean boutonneuse fever in the south of Corsica. *Bulletin de la Société de Pathologie Exotique et de ses Filiales* 78, 446–451.

Raoult, D., Brouqui, P. and Roux, V. (1996) A new spotted-fever-group rickettsiosis. *The Lancet* 348, 412.

Raoult, D., Berbis, P., Roux, V., Xu, W. and Maurin, M. (1997) A new tick-transmitted disease due to *Rickettsia slovaca*. *The Lancet* 350, 112–113.

Raoult, D., Lakos, A., Fenollar, F., Beytout, J., Brouqui, P. and Fournier, P.E. (2002) Spotless rickettsiosis caused by *Rickettsia slovaca* and associated with *Dermacentor* ticks. *Clinical Infectious Diseases* 34, 1331–1336.

Rauter, C. and Hartung, T. (2005) Prevalence of *Borrelia burgdorferi sensu lato* genospecies in *Ixodes ricinus* ticks in Europe: a metaanalysis. *Applied and Environmental Microbiology* 71, 7203–7216.

Rebaudet, S. and Parola, P. (2006) Epidemiology of relapsing fever borreliosis in Europe. *FEMS Immunology and Medical Microbiology* 48, 11–15.

Regev-Yochay, G., Segal, E. and Rubinstein, E. (2000) Glucose-6-phosphate dehydrogenase deficiency: possible determinant for a fulminant course of Israeli spotted fever. *Israeli Medical Association Journal* 2, 781–782.

Rehacek, J. (1984) *Rickettsia slovaca*, the organism and its ecology. *Acta Scientiarum Naturalium Academiae Scientiarum Bohemoslovacae*, Brno 18, 1–50.

Rehacek, J. and Tarasevich, I.V. (1988) *Acari-Borne Rickettsiae and Rickettsioses in Euroasia*. Veda Publishing House, Bratislava, Slovakia.

Reid, H.W. (1988) Louping ill. In: Monath, T.P. (ed.) *The Arboviruses: Epidemiology and Ecology, Volume III.* CRC Press, Boca Raton, Florida, 117–135.

Reintjes, R., Dedushaj, I., Gjini, A., Jorgensen, T.R., Cotter, B., Lieftucht, A., D'Ancona, F., Dennis, D.T., Kosoy, M.A., Mulliqi-Osmani, G., Grunow, R., Kalaveshi, A., Gashi, L. and Humolli, I. (2002) Tularemia outbreak investigation in Kosovo: case control and environmental studies. *Emerging Infectious Diseases* 8, 69–73.

Richter, D. and Matuschka, F.R. (2006) Perpetuation of the Lyme disease spirochete *B. lusitaniae* by lizards. *Applied Environmental Microbiology* 702, 4627–4632.

Richter, D., Schlee, D.B., Allgöwer, R. and Matuschka, F.R. (2004) Relationships of a novel Lyme disease spirochete, *Borrelia spielmani* sp. nov. with its hosts in Central Europe. *Applied Environmental Microbiology* 70, 6414–6419.

Rikihisa, Y. (2006) New findings on members of the family Anaplasmataceae of veterinary importance. *Annals of the New York Academy of Sciences*, 1078, 438–445.

Rodhain, F. (1998) Joseph Desire Tholozan and the Persian relapsing fever. *History of Science and Medicine* 32, 309–313.

Rodriguez, M., Hooghuis, H. and Castaño, M. (1992) African horse sickness in Spain. *Veterinary Microbiology* 33, 129–142.

Ros-García, A., M'Ghirbi, Y., Bouarrour, A. and Hurtado, A. (2011) First detection of *Babesia occultans* in *Hyalomma* ticks from Tunisia. *Parasitology* 138, 578–582.

Roux, V. and Raoult, D. (1995) Phylogenetic analysis of genus *Rickettsia* by 16S rDNA sequencing. *Research Microbiology* 146, 385–396.

Roux, V. and Raoult, D. (2000) Phylogenetic analysis of members of the genus *Rickettsia* using the gene encoding the outer-membrane protein rOmpB (*ompB*). *International Journal of Systematic Evolution Microbiology* 50, 1449–1455.

Roux, V., Rydkina, E., Eremeeva, M. and Raoult, D. (1997) Citrate synthase gene comparison, a new tool for phylogenetic analysis, and its application for the rickettsiae. *International Journal of Systematic Bacteriology* 47, 252–261.

Roy, P., Mertens, P.P.C. and Casal, I. (1994) African horse sickness virus structure. *Comparative Immunology, Microbiology and Infectious Diseases* 17, 243–273.

Rudenko, N., Golovchenko, M., Grubhoffer, L. and Oliver, J.H. Jr. (2011) Updates on *Borrelia burgdorferi sensu lato* complex with respect to public health. *Ticks and Tick-borne Diseases* 2, 123–128.

Ruiz-Beltran, R., Herrero-Herrero, J.I., Martin, A.M., Vicent, G.V., Sanz, O.F., Mateos, S.A., Querol, P.R. and Portugal, A.J. (1985) Formas graves de fiebre exantemática mediterránea. Análisis prospectivo de 71 enfermos. *Anales de Medicina Interna* 2, 365–368.

Rydkina, E., Roux, V., Rudakov, N., Gafarova, M., Tarasevich, I. and Raoult, D. (1999) New rickettsiae in ticks collected in territories of the former Soviet Union. *Emerging Infectious Diseases* 5, 811–814.

Rymaszewska, A. and Grenda, S. (2008) Bacteria of the genus *Anaplasma* – characteristics of *Anaplasma* and their vectors: a review. *Veterinarni Medicina* 53, 573–584.

Sainz, A., Amusategui, I. and Tesouro, M.A. (1999) *Ehrlichia platys* infection and disease in dogs in Spain. *Journal of Veterinary Diagnostic Investigation* 11, 382–384.

Sambri, V., Marangoni, A., Storni, E., Cavrini, F., Moroni, A., Sparacino, M. and Cevenini, R. (2004) Tick borne zoonosis: selected clinical and diagnostic aspect. *Parassitologia* 46, 109–113.

Sánchez-Vizcaíno J.M. (2006) African swine fever. In: Straw, B., D'Allaire, S., Mengeling, W. and Taylor, D. (ed.) *Diseases of Swine*, 9th edn. Iowa State University, Ames, Iowa, pp. 291–298.

Sánchez-Yebra, W., Diaz, Y., Molina, P., Giner, P., Vitutia, M.M. and Anda, P. (1997) Tick-borne recurrent fever. Description of 5 cases. *Enfermedades Infecciosas y Microbiologia Clinica* 15, 77–81.

Sanogo, Y.O., Parola, P., Shpynov, S., Camicas, J.L., Brouqui, P., Caruso, G. and Raoult, D. (2003) Genetic diversity of bacterial agents detected in ticks removed from asymptomatic patients in northeastern Italy. *Annals of the New York Academy of Sciences* 990, 182–190.

Santic, M., Molmeret, M., Klose, K.E. and Abu Kwaik, Y. (2006) *Francisella tularensis* travels a novel, twisted road within macrophages. *Trends in Microbiology* 14, 37–44.

Santos-Silva, M., Sousa, R., Santos, A., Melo, P., Encarnação, V. and Bacellar, F. (2006) Ticks parasitizing wild birds in Portugal: detection of *Rickettsia aeschlimannii, R. helvetica* and *R. massiliae*. *Experimental and Applied Acarology* 39, 331–338.

Sardelic, S., Fournier, P.E., Polic, V.P., Bradaric, N., Grgic, D., Ivic, I., Ledina, D., Luksic, B., Milas, I. and Raoult, D. (2003) First isolation of *Rickettsia conorii* from human blood in Croatia. *Croatian Medical Journal* 44, 630–634.

Schnittger, L., Yin, H., Jianxun, L., Lugwing, W., Shayan, P., Rahbari, S., Voss-Holtmann, A. and Ahmed, J.S. (2000) Ribosomal small subunit RNA gene-sequence analysis of *Theileria lestoquardi* and a *Theileria* species highly pathogenic for small ruminants in China. *Parasitology Research* 85, 352–358.

Scolovschi, C., Barbarot, S., Lebebvre, M., Parola, P. and Raoult, D. (2010) *Rickettsia sibirica mongolitimonae* in a traveler from Egypt [Letter]. *Emerging Infectious Diseases* 16, 1495–1496.

Sekeyova, Z., Roux, V. and Raoult, D. (2001) Phylogeny of *Rickettsia* spp. inferred by comparing sequences of 'gene D', which encodes an intracytoplasmic protein. *International Journal of Systematic and Evolutionary Microbiology* 51, 1353–1360.

Selmi, M., Bertolotti, L., Tomassone, L. and Mannelli, A. (2008) *Rickettsia slovaca* in *Dermacentor marginatus* and tick-borne lymphadenopathy, Tuscany, Italy. *Emerging Infectious Diseases* 14, 817–820.

Shpynov, S., Rudakov, N., Tohkov, Y., Matushchenko, A., Tarasevich, I. and Raoult, D. (2009) Detection of *Rickettsia aeschlimannii* in *Hyalomma marginatum* ticks in western Russia. *Clinical Microbiology and Infection* 15(Suppl. 2), S315–S316.

Silaghi, C., Hamel, D., Thiel, C., Pfister, K. and Pfeffer, M. (2011) Spotted fever group rickettsiae in ticks, Germany. *Emerging Infectious Diseases* 17, 890–892.

Simpson, R.M., Gaunt, S.D., Hair, J.A., Kocan, K.M., Henk, W.G. and Casey, H.W. (1991) Evaluation of *Rhipicephalus sanguineus* as a potential biologic vector of *Ehrlichia platys*. *American Journal of Veterinary Research* 52, 1537–1541.

Simser, J.A., Palmer, A.T., Fingerle, V., Wilske, B., Kurtti, T.J. and Munderloh, U.G. (2002) *Rickettsia monacensis* sp. nov., a spotted fever group *Rickettsia*, from ticks (*Ixodes ricinus*) collected in a European city park. *Applied Environmental Microbiology* 68, 4559–4566.

Skarphédinsson, S., Lyholm, B.F., Ljungberg, M., Søgaard, P., Kolmos, H.J. and Nielsen, L.P. (2007) Detection and identification of *Anaplasma phagocytophilum*, *Borrelia burgdorferi*, and *Rickettsia helvetica* in Danish *Ixodes ricinus* ticks. *Acta Pathologica, Microbiologica et Immunologica Scandinavica* 115, 225–230.

Skotarczak, B. (2003) *Canine ehrlichiosis*. *Annals of Agricultural and Environmental Medicine* 10, 137–141.

Skrabalo, Z. and Deanovic, Z. (1957) Piroplasmosis in man; report of a case. *Documenta de Medicina Geographica et Tropica* 9, 11–16.

Smith, T.G. (1996) The genus *Hepatozoon* (Apicomplexa: Adeleina). *Journal of Parasitology* 82, 565–585.

Sparagano, O.A., de Vos, A.P., Paoletti, B., Camma, C., de Santis, P., Otranto, D. and Giangaspero, A. (2003) Molecular detection of *Anaplasma platys* in dogs using polymerase chain reaction and reverse line blot hybridization. *Journal of Veterinary Diagnostic Investigation* 15, 527–534.

Sréter-Lancz, Z., Sréter, T., Széll, Z. and Egyed, L. (2005) Molecular evidence of *Rickettsia helvetica* and *R. monacensis* infections in *Ixodes ricinus* from Hungary. *Annals of Tropical Medicine and Parasitology* 99, 325–330.

Strauss, E. and Pohlmeyer, K. (2001) Population density of European hares (*Lepus europaeus* Pallas, 1778) and hunting activity in Lower Saxony. *Zeitschrift für Jagdwissenschaft* 47, 43–62.

Strle, F. (2004) Human granulocytic ehrlichiosis in Europe. *International Journal of Medical Microbiology* 293, 27–35.

Stuen, S. (2007) *Anaplasma phagocytophilum* – the most widespread tick-borne infection in animals in Europe. *Veterinary Research Communications* 31, 79–84.

Swanepoel, R. (1998) Crimean-Congo haemorrhagic fever. In: *Zoonoses*. Palmer, S.R. Soulsby, E.J.L. and Simpson, D.I.H. (eds) Oxford University Press, Oxford, UK, pp. 461–470.

Swanepoel, R. (2006) *Crimean Congo Haemorrhagic Fever – Epidemiology*. Available at: http://www.mzcp-zoonoses.gr/pdf/Session5_1_SwanepoelTickEpidemiology.pdf (accessed 24 January 2008).

Swanepoel, R., Shepherd, A.J., Leman, P.A., Shepherd S.P., McGillivray, G.M., Erasmus, M.J., Searle, L.A. and Gill, D.E. (1987) Epidemiologic and clinical features of Crimean-Congo hemorrhagic fever in southern Africa. *American Journal of Tropical Medicine and Hygiene* 36, 120–132.

Tamura, A., Ohashi, N., Urakami, H. and Miyamura, S. (1995) Classification of *Rickettsia tsutsugamushi* in a new genus, *Orientia* gen. nov., as *Orientia tsutsugamushi* comb. nov. *International Journal Systematic Bacteriology* 45, 589–591.

Tarasevich, I.V. and Mediannikov, O.Y. (2006) Rickettsial diseases in Russia. *Annals of the New York Academy of Sciences* 1078, 48–59.

Tarasevich, I.V., Makarova, V.A., Fetisova, N.F., Stepanov, A.V., Miskarova, E.D. and Raoult, D. (1991) Studies of a "new" rickettsiosis "Astrakhan" spotted fever. *European Journal of Epidemiology* 7, 294–298.

Tarnvik, A., Priebe, H.S. and Grunow, R. (2004) Tularemia in Europe: an epidemiological overview. *Scandinavian Journal of Infectious Diseases* 36, 350–355.

Telford, S.R. III and Goethhert, H.K. (2008) Emerging and emergent tick-borne infections. In: Bowman, A.S. and Nuttall, P.A. (eds) *Ticks. Biology, Disease and Control*. Cambridge University Press, Cambridge, UK, pp. 344–376.

Telford, S.R. III and Wormser, G.P. (2010) *Bartonella* spp. transmission by ticks not established. *Emerging Infectious Diseases* 16, 379–384.

Telford, S.R., Gorenflot, A., Brasseur, P. and Spielman, A. (1993) Babesial infection in human and wildlife. In: Kreier, J.P. (ed.) *Parasitic Protozoa, Volume 5*. Academic Press, San Diego, California, pp. 1–47.

Teysseire, N. and Raoult, D. (1992) Comparison of Western immunoblotting and microimmunofluorescence for diagnosis of Mediterranean spotted fever. *Journal of Clinical Microbiology* 30, 455–460.

Theiler, A. (1906) Transmission of horse sickness into dogs. *Report of the Government Veterinary Bacteriologist1906–1906*.Transvaal Department of Agriculture, Pretoria, South Africa, pp. 160–162.

Theiler, A. (1921) African horse sickness (*Pestis equorum*). *South African Department of Agricultural Science Bulletin* 19, 1–30.

Tijsse-Klasen, E., Jameson, L.J., Fonville, M., Leach, S., Sprong, H. and Medlock, J.M. (2011) First detection of spotted fever group rickettsiae in *Ixodes ricinus* and *Dermacentor reticulatus* ticks in the UK. *Epidemiology and Infection* 139, 524–529.

Toledo, A., Jado, I., Olmeda, A.S., Casado-Nistal, M.A., Gil, H., Escudero, R. and Anda, P. (2009a) Detection of *Coxiella burnetii* in ticks collected from Central Spain. *Vector-Borne and Zoonotic Diseases* 9, 465–468.

Toledo, A., Olmeda, A.S., Escudero, R., Jado, I., Valcárcel, F., Casado-Nistal, M.A., Rodríguez-Vargas, M., Gil, H. and Anda, P. (2009b) Tick-borne zoonotic bacteria in ticks collected from central Spain. *American Journal of Tropical Medicine and Hygiene* 81, 67–74.

Torina, A., Vicente, J., Alongi, A., Scimeca, S., Turla, R., Nicosia, S., Di Marco, V., Caracappa, S. and de la Fuente, J. (2007) Observed prevalence of tick-borne pathogens in domestic animals in Sicily, Italy during 2003–2005. *Zoonoses Public Health* 54, 8–15.

Torina, A., Alongi, A., Naranjo, V., Scimeca, S., Nicosia, S., Di Marco, V., Caracappa, S., Kocan, K.M. and de la Fuente, J. (2008) Characterization of anaplasma infections in Sicily, Italy. *Annals of the New York Academy of Sciences* 1149, 90–93.

Uilenberg, G. (1981) Theilerial species of domestic livestock. In: Irvin, A.D., Cunningham, M.P. and Young, A.S. (eds), *Advances in the Control of Theileriosis*. Martinus Nijhoff, The Hague, The Netherlands, pp. 4–37.

Uilenberg, G. (1997) General review of tick-borne diseases of sheep and goats world-wide. *Parassitologia* 39, 161–165.

Uilenberg, G. (2001) Babesiosis. In: Service, M.W. (ed.) *Encyclopaedia of Arthropod-transmitted Infections of Man and Domesticated Animals*. CAB International, Wallingford, UK, pp. 53–60.

Uilenberg, G. (2006) *Babesia* – a historical review. *Veterinary Parasitology* 138, 3–10.

Valbuena, G. and Walker, D.H. (2009) Infection of the endothelium by members of the order Rickettsiales. *Thrombosis and Haemostasis* 102, 1071–1079.

Van Rensburg, I.B., De Clerk, J., Groenewald, H.B. and Botha, W.S. (1981) An outbreak of African horse-sickness in dogs. *Journal of the South African Veterinary Association/Tydskrif van die Suid-Afrikaanse Veterinere Vereniging* 52, 323–325.

Vene, S.L. (1989) Diagnosis of rickettsial infections in Swedish patients: detection of specific IgM. *European Journal of Epidemiology* 5, 436–437.

Vincent-Johnson, N., Macintire, D.K.E. and Baneth, G. (1997) Canine hepatozoonosis: pathophysiology, diagnosis, and treatment. *Small Animals* 19, 51–62.

Viseras, J. and García-Fernández, P. (1999) Studies on theileriosis in Southern Spain. *Parassitologia* 41(Suppl. 1), 111–115.

Vitale, G., Mansuelo, S., Rolain, J.M. and Raoult, D. (2006) *Rickettsia massiliae* human isolation. *Emerging Infectious Diseases* 12, 174–175.

Vojta, L., Mrljak, V., Čurković, S., Živičnjak, T., Marinculić, A. and Beck, R. (2009) Molecular epizootiology of canine hepatozoonosis in Croatia. *International Journal of Parasitology* 39, 1129–1136.

Walker, A.R., Alberdi, M.P., Urquhart, K.A. and Rose, H. (2001) Risk factors in habitats of the tick *Ixodes ricinus* influencing human exposure to *Ehrlichia phagocytophila* bacteria. *Medical and Veterinary Entomology* 15, 40–49.

Walker, D.H. (1990) The role of host factors in the severity of spotted fever and typhus rickettsiosis. *Annals of the New York Academy of Sciences* 590, 10–19.

Walker, D.H. (2003) Rickettsial diseases in travelers. *Travel Medicine and Infectious Diseases* 1, 35–40.

Walker, D.H. (2006) Targeting rickettsia. *New England Journal of Medicine* 354, 1418–1420.

Walker, D.H. (2007) Rickettsiae and rickettsial infections: the current state of knowledge. *Clinical Infectious Diseases* 45(Suppl. 1), S39–S44.

Walker, D.H. and Fishbein, D.B. (1991) Epidemiology of rickettsial diseases. *European Journal of Epidemiology* 7, 237–245.

Walker, D.H., Herrero-Herrero, J.I., Ruiz-Beltrán, R., Bullón-Sopelana, A. and Ramos-Hidalgo, A. (1987) The pathology of fatal Mediterranean spotted fever. *American Journal of Clinical Pathology* 87, 669–672.

Wieland, B., D'Hollander, S., Salman, M. and Koenen, F. (2011) Qualitative risk assessment in a data-scarce environment: a model to assess the impact of control measures on spread of African swine fever. *Preventive Veterinary Medicine* 99, 4–14.

Wolach, B., Franco, S., Bogger-Goren, S., Drucker, M., Goldwasser, R.A., Sadan N. and Bernheim, J. (1989) Clinical and laboratory findings of spotted fever in Israeli children. *Pediatric Infectious Diseases Journal* 8, 152–155.

Woldehiwet, Z. (2006) *Anaplasma phagocytophilum* in ruminants in Europe. *Annals of the New York Academy of Sciences* 1078, 446–460.

Yabsley, M.J., McKibben, J., Macpherson, C.N., Cattan, P.F., Cherry, N.A., Hegarty, B.C., Breitschwerdt, E.B., O'Connor, T., Chandrashekar, R., Paterson, T., Perea, M.L., Ball, G., Friesen, S., Goedde, J., Henderson, B. and Sylvester, W. (2008) Prevalence of *Ehrlichia canis, Anaplasma platys, Babesia canis vogeli, Hepatozoon canis, Bartonella vinsonii berkhoffii,* and *Rickettsia* spp. in dogs from Grenada. *Veterinary Parasitology* 151, 279–285.

Yoshikawa, H., Kimura, M., Ogawa, M., Rolain, J.M. and Raoult, D. (2005) Laboratory-confirmed Mediterranean spotted fever in a Japanese traveler to Kenya. *American Journal of Tropical Medicine and Hygiene* 73, 1086–1089.

Yu, X., Fan, M., Xu, G., Liu, Q. and Raoult, D. (1993) Genotypic and antigenic identification of two new strains of spotted fever group rickettsiae isolated from China. *Journal of Clinical Microbiology* 31, 83–88.

Zahler, M., Rinder, H., Schein, E. and Gothe, R. (2000) Detection of a new pathogenic *Babesia microti*-like species in dogs. *Veterinary Parasitology* 89, 241–248.

Znazen, A., Hammami, B., Lahiani, D., Ben Jemaa, M., Hammami, A. (2011) Israeli spotted fever, Tunisia. *Emerging Infectious Diseases* 17, 1328–1330.

5 Maps of Reported Occurrence of Tick-borne Pathogens*

Agustín Estrada-Peña,[1] Frank Koenen,[2] Ilaria Pascucci,[3] Thomas G.T. Jaenson,[4] Maxime Madder,[5] Rita de Sousa[6] and Róbert Farkas[7]

[1]*University of Zaragoza, Spain;* [2]*Veterinary and Agrochemical Research Centre, CODA-CERVA, Brussels, Belgium;* [3]*Istituto Zooprofilattico Sperimentale dell'Abruzzo e del Molise "G. Caporale", Teramo, Italy;* [4]*Uppsala University, Sweden;* [5]*Institute of Tropical Medicine, Antwerp, Belgium and University of Pretoria, South Africa;* [6]*National Institute of Health Doutor Ricardo Jorge, IP (INSA), Águas de Moura, Portugal; and* [7]*Szent István University, Budapest, Hungary*

The maps that are presented in this chapter on the geographical occurrence of tick-borne pathogens have been produced using three sources of data: (i) a systematic literature review based on scientific papers retrieved from the databases integrated in the ISI Web of Knowledge and PubMed, published from January 2000 to March 2010 (see Introduction); (ii) a pool of scientific papers considered relevant by the experts of the Working Group on Ticks and Tick-borne Pathogens of the Animal Health and Animal Welfare (AHAW) Unit of the European Food Safety Authority (EFSA), and derived from their private collections regardless of the time frame concerned; and (iii) published historical data (for the period of approximately 1970–2000) of the Integrated Consortium on Ticks and Tick-borne Diseases (the Sixth Framework Programme funded ICTTD-3 European project). Consequently, the data presented in the maps were classified according to their chronology into those published in the years 2000 to 2010, or as historical data (published before the year 2000).

The data included in the maps were classified according to the reported geographic location using either coordinates (latitude/longitude) or, in the cases where coordinates were not indicated or could not be found because the name of the location was not specified in the corresponding paper, according to the smallest administrative region in which the findings were made. In this last case, the symbols used in the maps are placed either in the middle of the smallest administrative region described in the scientific paper (for the countries that do not have official nomenclature of territorial units for statistics – NUTS), or in the middle of the NUTS containing the specified location. Countries or areas not showing cases of the tick-borne pathogen under investigation are not

* Frank Koenen wrote Section 5.1; Ilaria Pascucci, Section 5.2; Thomas G.T. Jaenson, Sections 5.3, 5.6; Maxime Madder, Section 5.4; Rita de Sousa, Section 5.5; Róbert Farkas, Sections 5.7, 5.8, 5.9, 5.11, 5.12; Agustín Estrada-Peña produced the maps with the advice of the authors and Thomas G.T. Jaenson, Section 5.10.

necessarily free of that pathogen. Rather, data on that species may not be presented in the literature available for this report. The presence of the pathogens represented in the maps is based on confirmed detection of the antigen for that pathogen. Serological evidence was not included, so these maps may underestimate the real distribution of the pathogens.

5.1 African Swine Fever

The few existing records for African swine fever virus (ASFV) have been obtained from active surveillance of the infection in domestic pigs, wild boars and soft ticks associated with foci of outbreaks. The exception to this is the island of Sardinia, which is considered to be endemic for ASFV. The *Ornithodoros* tick is not present on this island, and no records exist of wild boar infested with ticks. Therefore, ticks do not play a role in the maintenance of ASFV on that island. The endemicity is mainly related to the traditional extensive production system. The distribution recorded on the map (Fig. 5.1) displays the records for ASFV that were found in the available literature of the last 10 years (solid black circles) and data from before 2000 (solid black diamonds). Both symbols indicate records in which the coordinates were not provided, just the name of the location (see the introduction to this chapter). The distribution is expected to be biased because the culinary habits in wide regions of northern Africa and parts of the eastern Mediterranean erase the domestic pig as the main host of the virus. The most recent outbreaks are in the Caucasus. ASFV in Portugal and Spain was eradicated in 1999 and 1995, respectively (Sánchez-Vizcaíno, 2006).

5.2 Crimean-Congo Haemorrhagic Fever

Clinical cases of Crimean-Congo haemorrhagic fever (CCHF) in humans, as well as ticks infected with the CCHF virus (CCHFV) under natural (field) conditions have been recorded only in an area covering southern Russia, Turkey, Albania, Kosovo, Greece,

Fig. 5.1. Reported occurrence of African swine fever virus (ASFV). • Smallest administrative region or territorial unit for statistics (NUTS), data from 2000–2010; ♦ Smallest administrative region or territorial unit for statistics (NUTS), historical data (before 2000).

Romania and Bulgaria (Fig. 5.2). The distribution map displays the records for CCHFV that were found in the available literature of the last 10 years, and are indicated by solid black circles. These symbols represent records in which the coordinates were not provided, just the name of the location (see the introduction to this chapter). The disease is widespread in Turkey, following the areas where the main tick vector, *Hyalomma marginatum*, exists. No published reports exist for the distribution and incidence of the disease in the Balkans and Russian areas other than reports collated from ProMED Mail (http://www.promedmail.org, from the International Society for Infectious Diseases). It is likely that the virus circulates in the Balkans (as suggested by the recent first report of CCHF in humans in Greece; Papa *et al.*, 2008). This is probably the most exposed area for spread of the virus to other parts of Europe. Both the virus and signs of clinical infection in humans have been partly surveyed in Europe, but with negative results. This is a cause of concern, because we then ignore the reasons why the disease (or the virus) is absent in places where the tick vectors are widespread (see the comments on *Hy. marginatum* and other hard ticks in Chapter 2, Section 2.2, and Chapter 6, Section 6.2).

5.3 Tick-borne Encephalitis Group Viruses

The map shown in Fig. 5.3 displays the records of the tick-borne encephalitis (TBE) group viruses that were found in the available literature of the last 10 years. The solid black circles indicate the coordinates (latitude/longitude). The stars indicate records in which the coordinates were not provided, just the name of the location (see the introduction to this chapter).

Tick-borne encephalitis virus (TBEV) includes three viral subtypes, i.e. Western TBEV, Siberian TBEV and Far Eastern TBEV. *Ixodes ricinus* is the main vector of the Western subtype and the Siberian and Far Eastern subtypes are vectored mainly by *I. persulcatus*. Although the main vector of the Western

Fig. 5.2. Reported occurrence of Crimean-Congo haemorrhagic fever virus (CCHFV) (2000–2010).
• Smallest administrative region or territorial unit for statistics (NUTS).

Fig. 5.3. Reported occurrence of tick-borne encephalitis group viruses (2000–2010). ★ Smallest administrative region or territorial unit for statistics (NUTS); • Coordinate (latitude/longitude).

subtype is *I. ricinus,* several other species, including *I. hexagonus, I. arboricola, Haemaphysalis* (*Ha.*) *concinna, Ha. inermis* and *Ha. punctata* are competent but secondary vectors (Labuda and Nuttall, 2008). Records of TBE group viruses published during the last 10 years and shown on the map originate from the following countries: Czech Republic, Denmark, Estonia, Finland (including the Åland Islands), Germany, Greece, Hungary, Italy, Latvia, Liechtenstein, Lithuania, Norway, Poland, Russia, Sweden, Switzerland and Turkey. The map (Fig. 5.3) does not include records published before 2000. 'Countries with increased risk of TBE include Austria, Slovakia, Hungary, the Czech Republic, the Baltic States, southern Germany and southern and eastern Sweden' (ECDC, 2010). TBE is prevalent in Russia all the way from the western border with Europe to its eastern border. The TBE virus subtypes in Russia are in general vectored by *I. persulcatus* and usually cause more severe disease than the central European strain.

The map does not show the presence of the louping ill virus (LIV). This virus is endemic in sheep farming areas of Northern Ireland, the UK (Scotland, Cornwall and Wales) and Norway.

The geographical distribution of TBE-infected ticks within regions and countries is such that the risk of infection can change dramatically from one area to another within short distances. TBE incidence fluctuates from year to year, but the number of reported human TBE cases has increased in Europe in the last two decades. 'The mean number of TBE infections in Europe, excluding Russia, between 1990 and 2007 was 2805 cases per year, up 193% from a mean of 1452 cases per year (1976–1989)' (ECDC 2010). An increase in TBE incidence has been noted in some northern countries and new TBE foci have appeared, especially in the last decade, in many areas, e.g. Denmark, Norway, Sweden and Finland (Jääskeläinen *et al.*, 2006; Formsgaard *et al.*, 2009). The main determinants for such changes in TBE incidence may vary from one time to another and from one region to another region. Also, the determining factors are often many and they interact – often in complex manners. Among the main determinants should be mentioned: the changing climate, which may affect the vector directly as well as indirectly via the plant and host communities; sociopolitical changes, which may increase or decrease human activities in TBEV foci; and technological factors, e.g. better diagnostic methods and increased awareness (Donoso Mantke *et al.*, 2008; Telford and Goethert, 2008; Randolph, 2010).

5.4 Anaplasmoses and Ehrlichioses

The map shown in Fig. 5.4 displays the records of *Anaplasma* and *Ehrlichia* species that were found in the available literature of the last 10 years. The solid black circles indicate the coordinates (latitude/longitude). The stars indicate records in which the coordinates were not provided, just the name of the location (see the introduction to this chapter).

The distribution of *E. canis* has expanded with the distribution of its vector, *Rhipicephalus* (*Rh.*) *sanguineus*. For *E. ruminantium*, no records have yet been found in Europe and its vector ticks seem to be absent from this continent, except for some imports of immature stages on birds.

For anaplasmoses, *A. phagocytophilum* has been recorded in most European countries. *A. marginale*, normally recorded from tropical and subtropical regions, has a more limited distribution in Europe with records from Sicily, Hungary and Spain. *A. ovis* has so far been found in the same countries as *A. marginale*. In Italy, *A. centrale*, a similar parasite to *A. marginale*, was first recorded in 2008 (Carelli *et al.*, 2008). Similarly, *A. bovis* has also been recorded in Italy although it is most commonly present in other continents, including South America, Africa and the Indian subcontinent.

A. platys has been detected worldwide, and in Europe cases of infection have been reported from Spain, France, Greece and Italy (de la Fuente *et al.*, 2006).

5.5 Rickettsioses

The map shown in Fig. 5.5 displays the records of *Rickettsia* species that were found

Fig. 5.4. Reported occurrence of *Anaplasma* spp. and *Ehrlichia* spp. (2000–2010). ★ Smallest administrative region or territorial unit for statistics (NUTS); • Coordinate (latitude/longitude).

Fig. 5.5. Reported occurrence of *Rickettsia* spp. (2000–2010). ★ Smallest administrative region or territorial unit for statistics (NUTS).

in the available literature of the last 10 years. The stars indicate records in which the coordinates were not provided, just the name of the location (see the introduction to this chapter).

The map represents records of all the tick-transmitted rickettsiae recorded in Europe, irrespective of the species or strains. Each of the species has a specific area of distribution, *R. conorii* Malish, which causes Mediterranean spotted fever (MSF) is mainly present in the Mediterranean, where its vector, *Rh. sanguineus*, is present. The *R. conorii* Israeli spotted fever strain was first reported from Israel but was later also detected in Portugal, Sicily, Tunisia and Libya. As most of the *Rickettsia* species are transmitted by different tick species, of which the distribution is not always overlapping, these rickettsial pathogens are found in areas determined by their vector.

Although no detailed information is available about the presence and absence of the different rickettsiae, it must be said that more and more recent studies, especially molecular studies, reveal the presence of the *Rickettsia* species in new member states of the European Union (EU), and not necessarily in adjacent countries from which the parasite was originally described.

5.6 Borrelioses

The map shown in Fig. 5.6 displays the records of *Borrelia* species that were found in the available literature of the last 10 years. The solid black circles indicate the coordinates (latitude/longitude). The stars indicate records in which the coordinates were not provided, and only the name of the location was included in the original report (see the introduction to this chapter).

As described above, this map shows the more recent records of *B. afzelii*, *B. burgdorferi s.s.*, *B. garinii*, *B. lusitaniae*, *B. spielmanii*, *B. valaisiana* and *B. burgdorferi s.l. B. burgdorferi s.l.* is considered to consist of at least 15 genospecies, eight of which occur in Europe, namely: *B. burgdorferi s.s.*, present in Europe and the USA, but rare in Russia and apparently absent from Asia; and *B. afzelii*,

Fig. 5.6. Reported occurrence of *Borrelia* spp. (2000–2010). ★ Smallest administrative region or territorial unit for statistics (NUTS); • Coordinate (latitude/longitude).

B. bavariensis (formerly *B. garinii* OspA serotype 4), *B. garinii*, *B. valaisiana*, *B. spielmanii*, *B. lusitaniae* and *B. bissettii* in Eurasia. *B. garinii* and *B. afzelii* are the most commonly encountered and most widely distributed of the 'European' Lyme disease genospecies. In Portugal, Tunisia and Morocco, *B. lusitaniae* is common and predominates over the other genospecies in *I. ricinus* populations (Piesman and Gern, 2008).

Although *I. ricinus* and *I. persulcatus* are apparently the only tick species responsible for infecting humans with the Lyme disease spirochaetes in Europe and Asia, *B. burgdorferi s.l.* has been found in several other tick species (EUCALB, 2010). These 'carrier species' are not considered to be competent natural vectors of *B. burgdorferi s.l.*

There are indications from northern Europe about an increasing incidence of Lyme borrelioses and TBE as a result of increased abundance of the vector, *I. ricinus*, and the expansion of its range, probably due to an extended vegetation period caused by climate change (Tälleklint and Jaenson, 1998; Lindgren *et al.*, 2000; Gray *et al.*, 2009). Thus, the ranges of at least some of the Lyme borreliosis genospecies are likely to change significantly in the near future.

Note that, as mentioned above, the present map is based solely on recent records (published after 1 January 2000). Therefore, many earlier records of *B. burgdorferi s.l.* and some genospecies, especially *B. afzelii*, *B. garinii* and *B. burgdorferi s.s.*, from large areas of Europe are not shown on the map. Thus, while not revealed by this map, it should be noted that *B. burgdorferi s.l.* is common in most or all *I. ricinus* populations in southern and central Sweden, southern and central Finland, Estonia, Latvia, Russia, Byelorussia, Kirghizia, Moldavia, Ukraine and Ireland.

5.7 Babesioses

The map shown in Fig. 5.7 displays the records of *Babesia* species that were found in the literature of the last 10 years. The empty circles indicate the coordinates (latitude/longitude). The solid black circles indicate records in which the coordinates were not provided, just the name of the location (see the introduction to this chapter).

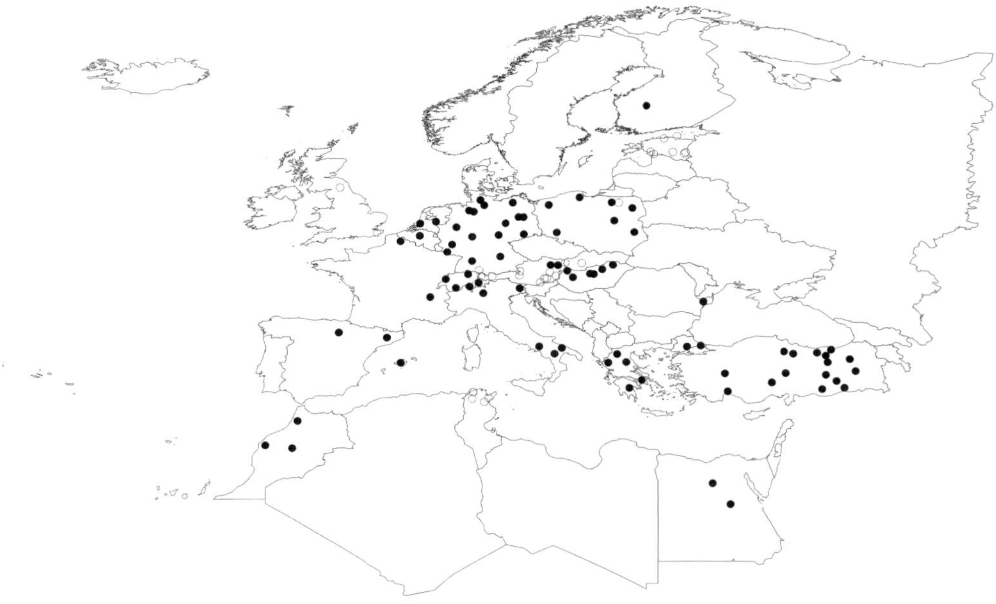

Fig. 5.7. Reported occurrence of *Babesia* spp. (2000–2010). • Smallest administrative region or territorial unit for statistics (NUTS); ○ Coordinate (latitude/longitude).

The occurrence of *Babesia* species has been reported from many European countries. Several tick vectors can carry more than one species of *Babesia*. Some *Babesia* species can infect more than one genus of ticks; others can only infect ticks from the genus *Ixodes*. The most common species is *B. canis canis* which exists in several foci of wide areas in Europe, from France to Poland, where its known vector species, *Dermacentor reticulatus*, is present. Recently a new zoonotic *Babesia* species (Babesia EU-1; proposed name *B. venatorum*) has been detected in roe deer and in humans, with *I. ricinus* identified as its vector. In endemic areas, all or almost all individuals of the host population are infected when they are young, with no or minimal clinical disease. The introduction of susceptible animals into endemic regions could lead to a recrudescence of babesiosis.

5.8 Theilerioses

The map shown in Fig. 5.8 displays the records of *Theileria* species that were found in the available literature of the last 10 years. The empty circles indicate the coordinates (latitude/longitude). The solid black circles indicate records in which the coordinates were not provided, just the name of the location (see the introduction to this chapter).

There have been very few reports on theileriosis of domesticated ruminants from the European countries. Mediterranean theileriosis, caused by *T. annulata* is known to occur in southern Europe, North Africa and Turkey. However, no published reports exist for its incidence and distribution in the Balkans and Iberian Peninsula where ixodid ticks such as *Hy. anatolicum*, *Hy. marginatum* and *Hy. excavatum*, which are known to transmit *T. annulata*, are found in large numbers, especially in semi-arid areas. Most papers published in the last 10 years reported the occurrence of horse theileriosis caused by *T.* (*Babesia*) *equi* in Morocco, Turkey, Italy, Croatia and Slovenia (see Fig. 5.9, equine piroplasmoses). Based on these reports, the exact geographical distribution and the incidence of theileriosis in domesticated and wild animals cannot be mapped properly in Europe.

Fig. 5.8. Reported occurrence of *Theileria* spp. (2000–2010). • Smallest administrative region or territorial unit for statistics (NUTS); ○ Coordinate (latitude/longitude).

Fig. 5.9. Reported occurrence of equine piroplasmoses (2000–2010). ■ Smallest administrative region or territorial unit for statistics (NUTS).

5.9 Equine Piroplasmoses

The map shown in Fig. 5.9 displays the records of equine piroplasmoses that were found in the available literature of the last 10 years. The solid black squares indicate records in which the coordinates were not provided, just the name of the location (see the introduction to this chapter).

The distribution and incidence of equine piroplasmoses caused by two species of apicomplexan protozoa, *Theileria* (syn. *Babesia) equi* and *Babesia caballi*, have been reported from a few European countries. Piroplasmosis seems to be more prevalent in France, Italy and Turkey than in other countries. However, the disease may occur in other countries because several species of ixodid ticks belonging to the genera *Hyalomma*, *Dermacentor* and *Rhipicephalus* have been identified as vectors of both *B. caballi* and *T. equi* in a wide area in Europe. Besides infected tick vectors, both parasite species can be spread by infected horses without clinical signs, by the transfer of blood from infected to naive equids through shared needles, improperly shared equipment, and blood or serum transfusions.

5.10 Bartonelloses

The map shown in Fig. 5.10 displays the records of *Bartonella* species that were found in the available literature of the last 10 years. The empty circles indicate the coordinates (latitude/longitude). The solid black circles indicate records in which the coordinates were not provided, just the name of the location (see the introduction to this chapter).

This map shows the geographical locations only of recent (2000–2010) records of *B. henselae*, *B. quintana*, *B. vinsonii* and unidentified *Bartonella* species. The genus at present comprises 28 species and subspecies (DSMZ, 2010). Thus, the map does not give a complete picture of the geographical distribution in Europe of the genus *Bartonella*.

Because *Bartonella* species usually infect their vertebrate hosts chronically, these microparasites can be ingested and potentially transmitted by blood-feeding arthropods. Confirmed vectors of *B. henselae* (the aetiological agent of cat-scratch disease), *B. bacilliformis*, *B. quintana*, *B. grahamii* and *B. taylorii* are *Ctenocephalides felis*, *Lutzomyia verrucarum*, *Pediculus humanus humanus* and *Ctenophthalmus nobilis*, respectively (Billeter *et al.*, 2008). *Bartonella* bacteria have been detected – based mainly on PCR – in several tick species, including *I. ricinus*, *I. scapularis*, *I. persulcatus*, *D. reticulatus*, *Rh. sanguineus* and *Carios kelleyi* (Billeter *et al.*, 2008). Some of the reasons that *Bartonella* species might be transmitted by ticks have been listed by Telford and Wormser (2010): other arthropods can transmit *Bartonella* species; the DNA of

Fig. 5.10. Reported occurrence of *Bartonella* spp. (2000–2010). • Smallest administrative region or territorial unit for statistics (NUTS); ○ Coordinate (latitude/longitude).

Bartonella species is often detected in ticks; human cases of bartonellosis preceded by tick bites are on record; and *Bartonella* species are commonly present in important hosts of *Ixodes* ticks, i.e. deer and rodents.

Although many researchers have indicated that ticks might be vectors of bartoneloses, it should be clear that there is, so far, no evidence that any tick species is a natural, competent vector of any *Bartonella* species (Telford and Wormser, 2010).

5.11 Tularaemia

The map shown in Fig. 5.11 displays the records of *Francisella tularensis* infection in humans or animals that were found in the available literature of the last 10 years. The solid black squares indicate records in which the coordinates were not provided, just the name of the location (see the introduction to this chapter).

Not all of these cases are necessarily related to tick transmission. Although vector-borne transmission of tularaemia to mammalian hosts has an important role in the pathogenesis of the disease (Petersen *et al.*, 2009), there are several vectors in addition to ticks that are related to the transmission and maintenance of the pathogen. The map indicates the sporadic reported cases by locations; the sparsity of the data is mainly due to the low percentage of compliance with reporting this disease and the similarity of clinical signs to those of several other diseases.

5.12 Q Fever

The map shown in Fig. 5.12 displays the records of *Coxiella burnetii* infection in animals and Q fever disease in humans or animals that were reported in the available literature of the last 10 years. The solid black squares indicate records in which the coordinates were not provided, just the name of the location (see the introduction to this chapter).

In domestic animals, *C. burnetii* is endemic in most European member states,

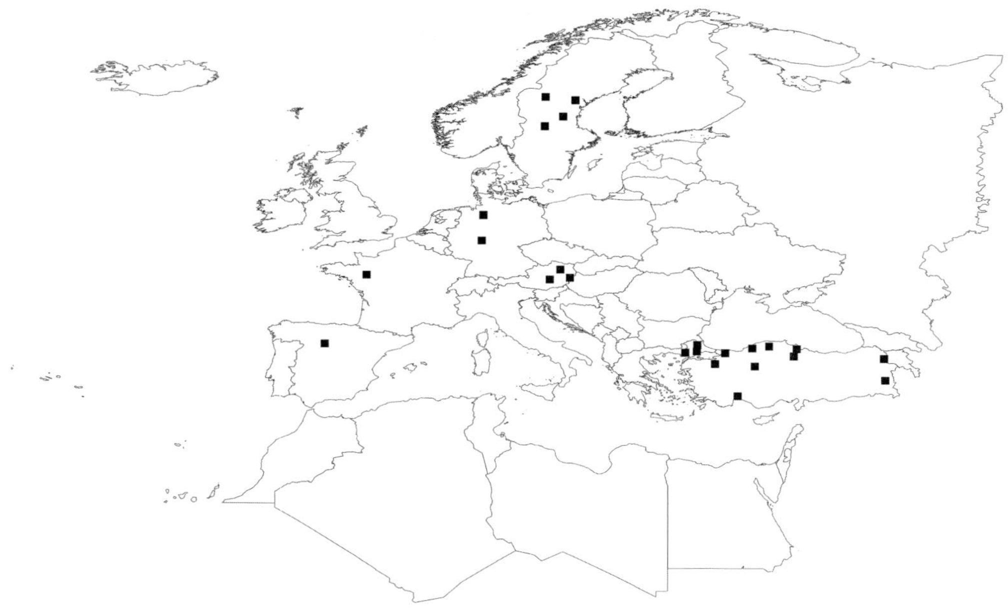

Fig. 5.11. Reported occurrence of *Francisella tularensis* (2000–2010). ■ Smallest administrative region or territorial unit for statistics (NUTS).

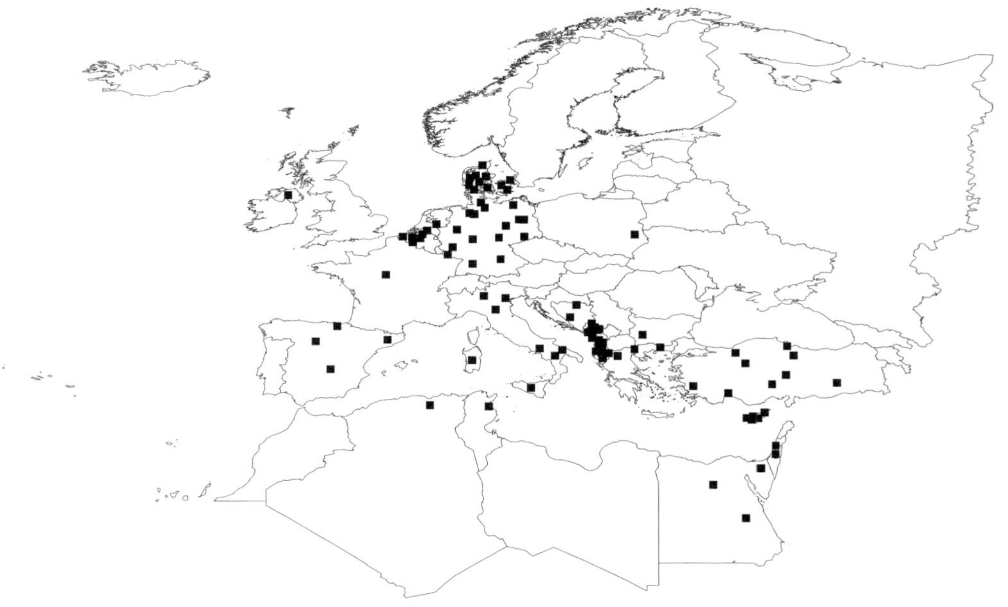

Fig. 5.12. Reported occurrence of *Coxiella burnetii* (2000–2010). ■ Smallest administrative region or territorial unit for statistics (NUTS).

but has a limited impact on human and animal health. Most countries have a long history of infection in domestic animals and sporadic human cases. In the Netherlands, a Q fever outbreak in humans first emerged in 2007 and is considered to have been the largest community outbreak ever recorded (EFSA, 2010).

References

Billeter, S.A., Levy, M.G., Chomel, B.B. and Breitschwerdt, E.B. (2008) Vector transmission of *Bartonella* species with emphasis on the potential for tick transmission. *Medical and Veterinary Entomology* 22, 1–15.

Carelli, G., Decaro, N., Lorusso, E., Paradies, P., Elia, G., Martella, V., Buonavoglia C. and Ceci, L. (2008) First report of bovine anaplasmosis caused by *Anaplasma centrale* in Europe. *Annals of the New York Academy of Sciences* 1149, 107–110.

de la Fuente, J., Torina, A., Naranjo, V., Nicosia, S., Alongi, A., La Mantia, F. and Kocan, K.M. (2006) Molecular characterization of *Anaplasma platys* strains from dogs in Sicily, Italy. *BMC Veterinary Research* 2, 24, doi:10.1186/1746-6148-2-24.

Donoso Mantke, O., Schädler, R. and Niedrig, M. (2008) A survey on cases of tick-borne encephalitis in European countries. *Eurosurveillance* 13, 1–9.

DSMZ (Deutsche Sammlung von Mikroorganismen und Zellkulturen) (2010) Bacterial Nomenclature Up-to-date. Available at: http://www.dsmz.de/bacterial-diversity/bacterial-nomenclature-up-to-date.html (accessed 2 July 2012).

ECDC (European Centre for Disease Prevention and Control) (2010) TBE epidemiology In Europe; TBE surveillance in Europe; TBE situation in individual European countries. Available at: http://www.ecdc.europa.eu/en/healthtopics/spotlight/spotlight_tickborne/Pages/TBE_epidemiology_surveillance.aspx (accessed 2 July 2012).

EFSA Panel on Animal Health and Welfare (AHAW) (2010) Scientific opinion on Q Fever. *EFSA Journal* 8(5), 1595.

EUCALB (European Union Concerted Action on Lyme Borreliosis) (2010) Biology: Vector Competence. Available at: http://meduni09.edis.at/eucalb/cms_15/index.php?view=article&catid=57%3Abiology-cat1&id=56%3Atick-vector-competence&format=pdf&option=com_content&Itemid=90 (accessed 2 July 2012).

Formsgaard, A., Christiansen, C.B. and Bødker, R. (2009) First identification of tick-borne encephalitis in Denmark outside of Bornholm, August 2009. *Eurosurveillance* 14(36):pii:19325.

Gray, J.S., Dautel, H., Estrada-Peña, A., Kahl, A. and Lindgren, E. (2009) Effects of climate change on ticks and tick-borne diseases in Europe. *Interdisciplinary Perspectives on Infectious Diseases* 2009, Article ID 593232, doi: 10.1155/2009/593232.

Jääskeläinen, A.E., Tikkakoski, T., Uzcátegui, N.Y., Alekseev, A.N., Vaheri, A. and Vapalahti, O. (2006) Siberian subtype tick-borne encephalitis virus, Finland. *Emerging Infectious Diseases* 12(10), 1568–71.

Labuda, M. and Nuttall, P.A. (2008) Viruses transmitted by ticks. In: Bowman, A.S. and Nuttall, P.A. (eds) *Ticks. Biology, Disease and Control.* Cambridge University Press, Cambridge, UK, pp. 253–280.

Lindgren, E., Tälleklint, L. and Polfeldt, T. (2000) Impact of climatic change on the northern latitude limit and population density of the disease-transmitting European tick *Ixodes ricinus*. *Environmental Health Perspectives* 108, 119–123.

Papa, A., Maltezou, H.C., Tsiodras, S., Dalla, V.G., Papadimitriou, T., Pierroutsakos, I., Kartalis, G.N. and Antoniadis, A. (2008) A case of Crimean-Congo haemorrhagic fever in Greece, June 2008. *Eurosurveillance* 13(33):piii=18952.

Petersen, J.M., Mead, P.S. and Schriefer, M.E. (2009) *Francisella tularensis*: an arthropod-borne pathogen. *Veterinary Research* 40(2), 7, doi: 10.1051/vetres:2008045.

Piesman, J. and Gern, L. (2008) Lyme borreliosis in Europe and North America. In: Bowman, A.S. and Nuttall, P.A. (eds) *Ticks. Biology, Disease and Control.* Cambridge University Press, Cambridge, UK, pp. 220–252.

Randolph, S.E. (2010) Human activities predominate in determining changing incidence of tick-borne encephalitis in Europe. *Eurosurveillance* 15(27):pii=19606.

Sánchez-Vizcaíno, J.M. (2006) African swine fever. In: Straw, B.E., Zimmerman, J.J., D'Allaire, S. and Taylor, D.J. (eds) *Diseases of Swine*, 9th edn. Blackwell Publishing, Ames, Iowa, pp. 291–298.

Tälleklint, L. and Jaenson, T.G. (1998) Increasing geographical distribution and density of *Ixodes ricinus* (Acari: Ixodidae) in central and northern Sweden. *Journal of Medical Entomology* 35, 521–526.

Telford, S.R. III and Goethhert, H.K. (2008) Emerging and emergent tick-borne infections. In: Bowman, A.S. and Nuttall, P.A. (eds) *Ticks. Biology, Disease and Control.* Cambridge University Press, Cambridge, UK, pp. 344–376.

Telford, S.R. III and Wormser, G.P. (2010) *Bartonella* spp. transmission by ticks not established. *Emerging Infectious Diseases* 16, 379–384.

6 Maps of Reported Occurrence of Ticks*

Agustín Estrada-Peña,[1] Thomas G.T. Jaenson,[2] Róbert Farkas[3] and Ilaria Pascucci[4]

[1]University of Zaragoza, Spain; [2]Uppsala University, Sweden; [3]Szent István University, Budapest, Hungary; and [4]Istituto Zooprofilattico Sperimentale dell'Abruzzo e del Molise "G. Caporale", Teramo, Italy

The maps that are presented in this chapter on the occurrence of ticks have been drawn up using three sources of data: (i) a systematic literature review previously published by EFSA (2010) based on scientific papers published from January 2000 to March 2010 and retrieved from the databases integrated in the ISI Web of Knowledge and PubMed; (ii) a pool of scientific papers considered relevant by the experts of the Working Group on Ticks and Tick-borne Pathogens of the Animal Health and Animal Welfare (AHAW) Unit of the European Food Safety Authority (EFSA), and derived from their private collections regardless of the time frame concerned; and (iii) published historical data (for the period of approximately years 1970–2000) of the Integrated Consortium on Ticks and Tick-Borne Diseases (the Sixth Framework Programme funded ICTTD-3 European project).

The data included in the maps were classified according to their geographic accuracy using either coordinates (latitude/longitude) or, in the cases where coordinates were not indicated or could not be found because the name of the location was not specified in the corresponding paper, according to the smallest administrative region in which the findings were made. In the latter case, the symbol representing the data was placed either in the middle of that region (for the countries that do not have official nomenclature of territorial units for statistics – NUTS), or in the middle of the NUTS containing the specified location. Countries or areas not showing the presence of a tick genus or species are not necessarily free of it. Rather, that genus or species may not be represented in the literature that was available for this monograph.

Owing to the lack of reported data in the area of concern, no maps were produced for the genera *Argas* and *Amblyomma*.

6.1 Soft Ticks (Argasidae), Genus *Ornithodoros*

The map shown in Fig. 6.1 covers the known distribution of the species in the genus *Ornithodoros* in the Palaearctic region, and shows historical data (before 2000). The solid black circles indicate the coordinates (latitude/longitude). The stars indicate records in which the coordinates were not provided, just

* Agustín Estrada-Peña produced the maps with the advice of the authors and wrote Sections 6.1, 6.2.4; Thomas G.T. Jaenson, Section 6.2.1; Róbert Farkas, Sections 6.2.2, 6.2.5; Ilaria Pascucci, Section 6.2.3.

Fig. 6.1. Reported occurrence of *Ornithodoros* spp. ∗ Smallest administrative region or territorial unit for statistics (NUTS), historical data (before 2000); • Coordinate (latitude/longitude), historical data (before 2000).

the name of the location (see the introduction to this chapter). As mentioned previously (Section 2.1.2), the lack of adequate morphological features renders the separation of possible species a difficult task. Hence, the map includes only the accurate records of the species in the group of *O. erraticus*. The reported distribution of these ticks is poor, as they seem to be restricted to parts of the western Mediterranean and portions of Georgia, Armenia and Turkey. Species in the genus are also known to be present in parts of Morocco, Tunisia and Egypt. Currently, there is fragmentary knowledge about the distribution of this species, although it has been hypothesized that *O. erraticus* would be present in the Iberian Peninsula and parts of Morocco, while *O. sonrai* would be the species found in Egypt and southern Morocco. The mechanisms operating in the separation of these species in Morocco are unknown. It is interesting to note, however, that there are no recent records (years 2000–2010) of the ticks in Europe, so, as noted above, the mapping has been carried out with historical records (years 1950–2000). This is because the tick has received some attention owing to the occurrence of the epidemics of African swine fever (ASF) in Europe. All of the species in the *O. erraticus* group are xerophilous ticks, and therefore live in open and dry habitats commonly associated with rodent burrows.

The apparent lack of records of the tick in the centre of the target area seems to be the result both of a lack of interest in the soft ticks and the lack of knowledge on the morphology of these species. Further sampling efforts should be done in wide areas of the European Mediterranean region, with special emphasis on nests of rodents, which are the preferred hosts of this group of species.

6.2 Hard Ticks (Ixodidae)

6.2.1 Genus *Ixodes*: *Ixodes ricinus*

The map shown in Plate 20 displays the records of *I. ricinus* ticks that were found in the available literature of the last 10 years (solid red circles and black stars) and those that are from historical data older than 2000 (solid green circles). The circles indicate the coordinates (latitude/longitude). The stars indicate records in which the coordinates were not provided, just the name of the location (see the introduction to this chapter).

I. ricinus is the 'common tick of Europe' and is widely distributed in north-western Europe through much of the Palaearctic, from Iceland and Ireland through central and southern Europe eastwards to central Asia (Iran) and southwards to North Africa. It is present in relatively dry Mediterranean habitats in northern Africa and the Iberian Peninsula, in the damp sheep pastures of Ireland, Scotland, Wales and England, and in relatively humid, mixed coniferous/deciduous woodland biotopes throughout most of Europe, including Scandinavia, Finland and western Russia. In Europe, changes in climate during the last few decades appear to have caused an expansion in the tick's range to higher altitudes (Materna *et al.*, 2005) and latitudes (Tälleklint and Jaenson, 1998; Lindgren *et al.*, 2000; Lindgren and Jaenson, 2006; Jaenson and Lindgren, 2011; Jaenson *et al.*, 2012), as well as affecting its population density. Moreover, with the changing range of *I. ricinus*, its role as an important vector of infections of humans and/or domesticated mammals, e.g. tick-borne encephalitis (TBE), including louping ill of sheep, Lyme borrelioses due to *Borrelia burgdorferi* s.l., anaplasmosis (*Anaplasma phagocytophilum*), babesiosis (*Babesia divergens*), rickettsiosis (*Rickettsia helvetica*), and tularaemia (*Francisella tularensis*), needs to be monitored in the future.

6.2.2 Genus *Haemaphysalis*

Haemaphysalis punctata

The map shown in Fig. 6.2 displays the records of *Haemaphysalis* (*Ha.*) *punctata* ticks that were found in the available literature of the last 10 years (stars), and in historical data published before 2000 (solid black circles). The circles indicate the coordinates (latitude/longitude). The stars show records in which the coordinates were not provided, just the

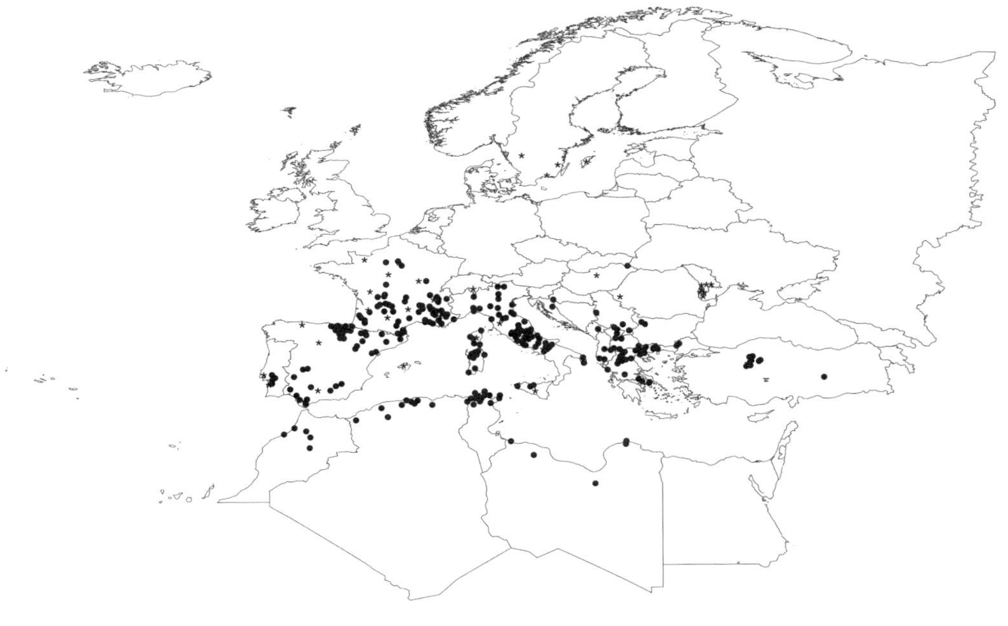

Fig. 6.2. Reported occurrence of *Haemaphysalis punctata*. ★ Smallest administrative region or territorial unit for statistics (NUTS), data for 2000–2010; • Coordinate (latitude/longitude), historical data (before 2000).

name of the location (see the introduction to this chapter).

Ha. punctata is widely present in Europe and in the whole Mediterranean area. It occurs mainly in a dense forest environment. Adults feed on domestic and wild ungulates; immature stages, besides feeding on small mammals, can also feed on birds, including migratory birds. Migratory birds are carriers of immature ticks and could potentially introduce them into free areas. Nevertheless, most reports show this tick species to be concentrated in southern-central Europe while northern Africa appears to be the southern limit of its distribution.

Haemaphysalis concinna

The map shown in Fig. 6.3 displays the records of *Ha. concinna* ticks that were found in the available literature of the last 10 years. The stars indicate records in which the coordinates were not provided, just the name of the location (see the introduction to this chapter).

Ha. concinna is a palaearctic species scattered in the Eurasian region from the Atlantic coast to Japan. It is mainly present in temperate climates and is well adapted to different biotopes. Adults feed on domestic and wild ungulates; immature stages, besides feeding on small mammals, can also feed on birds, including migratory birds.

Haemaphysalis inermis

The map shown in Fig. 6.4 displays the records of *Ha. inermis* ticks that were found in the available literature of the last 10 years. The stars indicate records in which the coordinates were not provided, just the name of the location (see the introduction to this chapter).

Ha. inermis is present in the southern-central Eurasian region. It is an ancient species and its geographic origin has been identified in the Caucasus. *Ha. inermis* is well adapted to different biotopes. Adults feed on domestic and wild ungulates, but also on canids and lagomorphs. Immature stages, besides feeding on small mammals, can also feed on lizards, on ground-feeding birds, and migratory birds.

Fig. 6.3. Reported occurrence of *Haemaphysalis concinna*. ∗ Smallest administrative region or territorial unit for statistics (NUTS), data for 2000–2010.

Fig. 6.4. Reported occurrence of *Haemaphysalis inermis*. ✳ Smallest administrative region or territorial unit for statistics (NUTS), data for 2000–2010.

6.2.3 Genus *Hyalomma*

Hyalomma marginatum

Hyalomma marginatum is well known from the Mediterranean region, where it can be found populating areas in northern Africa and southern Europe. The map shown in Plate 21 displays the records of *Hy. marginatum* ticks that were found in the available literature of the last 10 years (solid red circles and black stars) and those that are from historical data older than 2000 (solid green circles). The circles indicate the coordinates (latitude/longitude). The stars indicate records in which the coordinates were not provided, just the name of the location (see the introduction to this chapter).

The tick is restricted to the oro-Mediterranean (temperate) zone; it appears that a too-cold or too-dry climate may restrict the colonization of potentially available areas. Accurate records exist for the tick in wide areas from Turkey to Portugal. The presence of the tick has been recorded from southern Russia and neighbouring Republics (but has not been documented with any geographical information). The tick is also known from other sites in northern Africa, but those records lack adequate referencing. Scattered records of the tick exist in northern Europe, but these records came from migratory birds, and they do not represent established populations. Migratory birds may bring immature ticks into free areas in spring, but the current climate in the Scandinavian countries jeopardizes their establishment. Concerns exist about the probable spread of the tick into northern latitudes where the tick is currently absent. Permanent populations of the tick are currently established at around 46°N, and these seem to be restricted in movement because of the Alps and the Balkan mountain ranges. In any case, considerable interest exists about the possible spread of the tick, and more field studies are necessary to understand the basic phenological processes behind the observed distribution.

6.2.4 Genus *Rhipicephalus*

There are serious concerns about the identity of the ticks of the genus *Rhipicephalus* in the

Palearctic region. Currently, the presence has been reported of *Rhipicephalus* (*Rh.*) *annulatus* (formerly in the genus *Boophilus*) and of three more species, namely *Rh. bursa*, *Rh. sanguineus* and *Rh. turanicus*. There is no problem in differentiating specimens of *Rh. bursa*, but there are serious troubles in the differentiation of both *Rh. sanguineus* and *Rh. turanicus*. Therefore, we prefer to report on specimens of both *Rh. sanguineus* and *Rh. turanicus* as the *Rh. sanguineus* complex (group).

Rhipicephalus sanguineus *complex* (Rh. sanguineus *and* Rh. turanicus)

The map shown in Fig. 6.5 displays the records of ticks of the *Rh. sanguineus* complex that were found in the available literature of the last 10 years (stars and solid black circles) and those that are from historical data older than 2000 (open circles). The open circles indicate the coordinates (latitude/longitude). The stars and solid black circles indicate records in which the coordinates were not provided, just the name of the location (see the introduction to this chapter).

Species in the *Rh. sanguineus* complex have a wide Mediterranean distribution (see Fig. 6.5) linked to an endophilous (nidicolous) habitat associated mainly with domestic carnivores. Other than this anthropophilous habitat, the ticks of this complex are common in open areas with different degrees of moisture content, but are always associated with warm areas. It is difficult to ascertain the degree to which *Rh. sanguineus* ticks can colonize northern latitudes. While the climate is much too cold for the tick at such latitudes, the association of the tick with endophilous sites may protect its populations from the extremely cold temperatures. Therefore, it could be expected that the tick may be found at more northern latitudes than the currently reported range.

Rhipicephalus bursa

The map shown in Plate 22 displays the records of *Rh. bursa* ticks that were found in

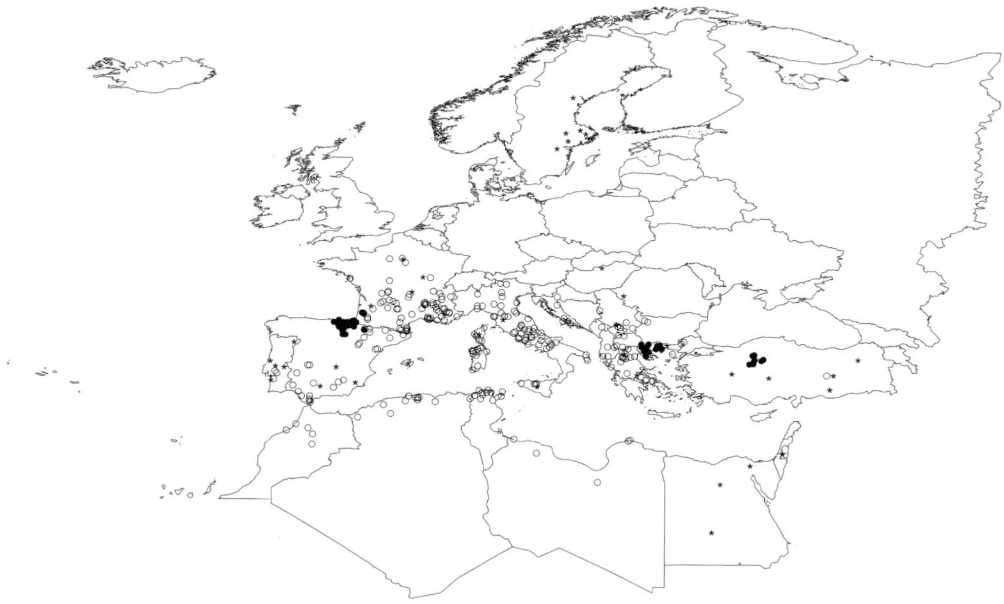

Fig. 6.5. Reported occurrence of *Rhipicephalus sanguineus* group (*Rh. sanguineus* and *Rh. turanicus*). ★ Smallest administrative region or territorial unit for statistics (NUTS), data for 2000–2010; • Smallest administrative region or territorial unit for statistics (NUTS), data for 2000–2010; ○ Coordinate (latitude/longitude), historical data (before 2000).

the available literature of the last 10 years (red stars and solid red circles) and those that are from historical data older than 2000 (green stars and solid green circles). The circles indicate the coordinates (latitude/longitude). The stars indicate records in which the coordinates were not provided, just the name of the location (see the introduction to this chapter).

The distribution of *Rh. bursa* overlaps with that of *Hy. marginatum* and *Dermacentor marginatus* in parts of its Mediterranean range. Such a distribution extends to approximately 45°N latitude. However, *Rh. bursa* prefers drier areas while *Hy. marginatum* and *D. marginatus* colonize colder sites. It is interesting to note the different distribution areas of *D. marginatus* and *Rh. bursa* in France, where similar tick field collection efforts on both species have been carried out in the last 30–40 years. *Rh. bursa* is more abundant in the oro-Mediterranean zone than *D. marginatus*, while *D. marginatus* can colonize colder and more humid sites. The tick seems to be absent from many parts of southern Russia and is probably replaced by other tick species in this region, but this extreme has not been reported.

Rhipicephalus annulatus (*formerly* Boophilus)

The map shown in Fig. 6.6 displays the records of *Rh. annulatus* ticks that were found in the available literature of the last 10 years (stars), and in historical data older than 2000 (solid black circles). The circles indicate the coordinates (latitude/longitude). The stars indicate records in which the coordinates were not provided, just the name of the location (see the introduction to this chapter).

The data gathered about the distribution of *Rh. annulatus* in the Mediterranean region seems to be discontinuous in nature. The tick is well represented in many areas of the Mediterranean region (see Fig. 6.6). It seems that the apparent lack of *Rh. annulatus* at sites that should support populations of the species may be a result of the control of ticks on cattle, which has resulted in an apparent absence of the species. In any case, *Rh. annulatus*

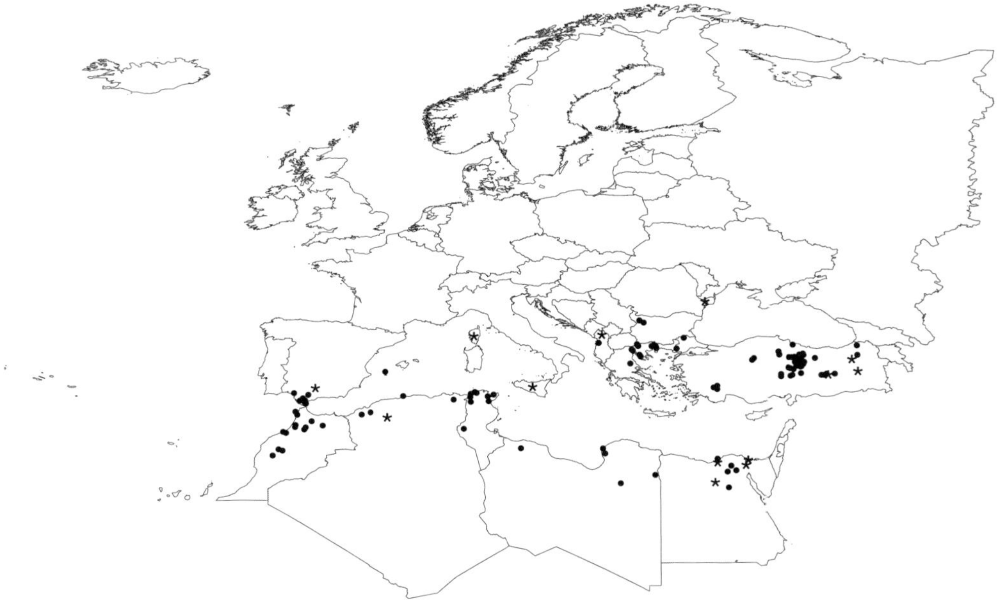

Fig. 6.6. Reported occurrence of *Rhipicephalus* (*Boophilus*) *annulatus*. ★ Smallest administrative region or territorial unit for statistics (NUTS), data for 2000–2010; • Coordinate (latitude/longitude), historical data (before 2000).

has a patchy area of distribution because it is restricted to sites with adequate moisture content. In many parts of northern Africa, the tick is found only in areas where a permanent or temporal flow of water is available.

6.2.5 Genus *Dermacentor*

Dermacentor marginatus

The map shown in Plate 23 displays the records of *D. marginatus* ticks that were found in the available literature of the last 10 years (solid red circles and black stars) and those that are from historical data published before 2000 (solid green circles). The circles indicate the coordinates (latitude/longitude). The stars show records in which the coordinates were not provided, just the name of the location (see the introduction to this chapter).

There are no conclusive studies about the involvement of *D. marginatus* in the transmission of Crimean-Congo Haemorrhagic Fever (CCHF) virus among animals and humans. However, the area of distribution of this tick species overlaps in many parts with that of *Hy. marginatum* (the proven vector) and *Rh. bursa* (a suspected vector of some CCHF viral strains). *D. marginatus* is well known from the Mediterranean region (see Plate 23), and it seems that this species can spread further than its previously known distribution area. It is present in areas north of the Balkans, being recorded from as far north as Hungary and southern Germany. The tick *D. reticulatus*, which has a northern distribution in Europe, has been wrongly reported as *D. marginatus* in parts of Germany. Therefore, the actual northern limits of *D. marginatus* are not well known, although the species does seem to be widespread in France in areas of a climate similar to that found in southern Germany. *D. marginatus* has not been recorded from Belgium or the Netherlands. Established populations are well known in southern Russia, but there are no geographically accurate reports about this.

Dermacentor reticulatus

The map shown in Fig. 6.7 displays the records of *D. reticulatus* ticks that were found in the available literature of the last 10 years

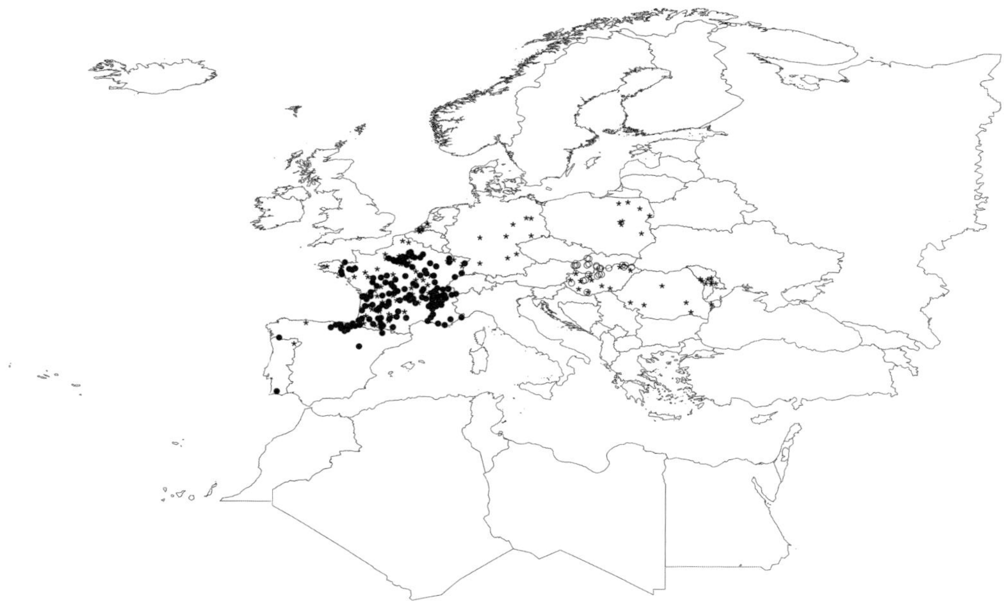

Fig. 6.7. Reported occurrence of *Dermacentor reticulatus*. ✱ Smallest administrative region or territorial unit for statistics (NUTS), data for 2000–2010; ○ Coordinate (latitude/longitude), data for 2000–2010; • Coordinate (latitude/longitude), historical data (before 2000).

(stars and open circles) and those that are from historical data older than 2000 (solid black circles). The circles indicate the coordinates (latitude/longitude). The stars show records in which the coordinates were not provided, just the name of the location (see the introduction to this chapter).

Accurate records of *D. reticulatus* (syn. *D. pictus*) also known as the 'ornate dog tick' exist in wide areas in Europe, from France to Moldavia. The area of distribution of this tick species (Fig. 6.7) overlaps in many parts of Europe with that of *D. marginatus* and it has been wrongly reported as the latter species. *D. reticulatus* seems to be widespread in France, Germany, Hungary and the eastern part of Poland. It prefers cold regions with an adequate amount of air relative humidity, therefore it is absent in the Mediterranean region where the relative humidity is not conducive to survival. It is absent from mountain regions, but very abundant in low-altitude hills. The actual geographical distribution and northern limits of this species are not well known. Concerns exist about the probable spread of *D. reticulatus* into other areas of countries that are already infested (e.g. Germany) and into new countries where the tick is currently absent.

References

EFSA Panel on Animal and Welfare (AHAW) (2010) Scientific opinion on the role of tick vectors in the epidemiology of Crimean Congo hemorrhagic fever and African swine fever in Eurasia. *EFSA Journal* 8(8), 1703.

Jaenson, T.G.T. and Lindgren, E. (2011) The range of *Ixodes ricinus* and the risk of contracting Lyme borreliosis will increase northwards when the vegetation period becomes longer. *Ticks and Tick-Borne Diseases* 2, 44–49.

Jaenson, T.G.T., Jaenson, D.G.E., Eisen, L., Petersson, E. and Lindgren, E. (2012) Changes in the geographical distribution and abundance of the tick *Ixodes ricinus* during the past 30 years in Sweden. *Parasites and Vectors* 5, 8, doi:10.1186/1756-3305-5-8.

Lindgren, E. and Jaenson, T.G.T. (2006) Lyme borreliosis in Europe: influences of climate and climate change, epidemiology, ecology and adaptation measures. Report No. EUR/04/5046250, Regional Office for Europe, World Health Organization, Copenhagen.

Lindgren, E., Tälleklint, L. and Polfeldt, T. (2000) Impact of climatic change on the northern latitude limit and population density of the disease-transmitting European tick *Ixodes ricinus*. *Environmental Health Perspectives* 108, 119–123.

Materna, J., Daniel, M. and Danielová, V. (2005) Altitudinal distribution limit of the tick *Ixodes ricinus* shifted considerably towards higher altitudes in central Europe: results of three years monitoring in the Krkonose Mts. (Czech Republic). *Central European Journal of Public Health* 13, 24–28.

Tälleklint, L. and Jaenson, T.G. (1998) Increasing geographical distribution and density of *Ixodes ricinus* (Acari: Ixodidae) in central and northern Sweden. *Journal of Medical Entomology* 35, 521–526.

7 Surveillance and Control Measures*

Róbert Farkas,[1] Mo Salman[2] and Agustín Estrada-Peña[3]
[1]Szent István University, Budapest, Hungary; [2]Colorado State University, USA; and [3]University of Zaragoza, Spain

Tick surveys often are conducted to determine whether tick control is needed or not. The species that are collected can provide an indication of the potential risks of tick-borne diseases in a given area. The most common method for sampling ticks is the use of a flag or drag cloth pulled or dragged through the vegetation. Ticks collected on the cloth are counted as the number of a given species per unit of distance dragged (e.g. 100 m). Absolute measures of tick population densities cannot be obtained with this method, although the relative abundance of ticks in the sampling area can be determined. An alternative to dragging/flagging is the use of carbon dioxide traps. The gas is an attractant for crawling ticks, which then adhere to the trap. When more reliable estimates of tick abundance are required, a mark-and-recapture method can be used. Ticks can also be sampled from wild or domestic hosts, or from people (Lindenmayer *et al.*, 1991; Nicholson and Mather, 1996). For tick-infested cattle, horses, mules and other livestock, a time-tested method is the scratching technique, whereby livestock inspectors pass their hands over different regions of the animal's body to detect attached ticks. A similar technique is used in combination with visual inspection to examine wild animals. For example, investigators can be assigned to deer-check stations during the hunting season to count all ticks on hunter-killed animals. Another technique for sampling ticks is to trap small and medium-sized wild animals and hold them over trays filled with water or alcohol to catch fed ticks as they detach. These methods provide more accurate information, but are expensive and labour intensive. Passive surveillance utilizes information collected from tick laboratories or hospitals to assess tick or disease distribution (White, 1993); this passive approach is less expensive and requires less effort, and it can provide useful information of appropriate types, although the value of the results is sometimes limited by unidentifiable biases in data collection (Johnson *et al.*, 2004).

7.1 Introduction to Surveillance and its Aim

7.1.1 What are surveillance and monitoring?

A surveillance system is defined as the systematic, ongoing or repeated, measurement,

* Mo Salman wrote Section 7.1; Róbert Farkas, Sections 7.2, 7.3.1; Agustín Estrada-Peña, Section 7.3.2.

collection, collation, analysis, interpretation and timely dissemination of health-related data that is essential for describing health hazard occurrence and for the planning, implementation and evaluation of risk mitigation measures (OIE, 2010). The term monitoring is defined as the systematic, ongoing or repeated, measurement, collection, collation, analysis, interpretation and timely dissemination information of health-related data without an associated predefined plan of action. Thus, the difference between these two systems is related to the anticipated action plan for the expected outcome from the corresponding system. It should be noticed that monitoring the presence, species, density and geographical distribution of ticks is an essential component for an early warning system for specific tick-borne diseases (TBDs) in selected regions of the world. This type of monitoring is considered as part of a surveillance system for TBDs in order to increase the likelihood that there will be timely detection of undefined (emerging) or unexpected (exotic or re-emerging) threats of TBDs.

A surveillance system for TBDs may require a combination of measurements related to the host as well as the ticks that are involved in the transmission of the pathogens. The collection of tick specimens can support at least the essential components of the surveillance: (i) tick species identification; and (ii) presence or absence of the pathogens in the collected specimens. Tick collection, however, should follow specific field protocol in order to maximize the available specimens to be collected.

The key feature of surveillance is that it has an associated corrective action when the presence of disease exceeds a predetermined threshold. Because of the requirement of action at certain thresholds, surveillance is a critical component of control programmes (Salman, 2003). The surveillance process can be broken down into three distinct components based on the fact that it is an action-oriented programme. There is a data gathering and analysis component via a defined monitoring system, a dissemination of the collected data or analysis findings to stakeholders and a predefined response or action component based on when the findings of the data analysis exceed a predefined disease level threshold (Salman, 2003; Thrusfield, 2007a). Further objectives of surveillance include identifying hypotheses for research, evaluating the effectiveness of intervention programmes, disease forecasting, and developing policy and public education campaigns (Akhtar and White, 2003).

This requirement that surveillance has an action associated with a predefined threshold is its key difference from disease monitoring. Disease monitoring is the continuous adaptable efforts used to assess the disease or health status of a population and the risk factors associated with disease (Salman, 2003). This can be a continuous effort to gather data about general health or production metrics, or it can be the monitoring of a specific health or production metric. The data gathered under general health monitoring are commonly known as vital statistics (Thrusfield, 2007a). The use of monitoring data is in its ability to provide information on the prevalence of a particular disease or changes in the rate or direction of the spread of that disease. Typically, the same approaches used in surveillance are used in monitoring, and the terms have been used interchangeably, but the key difference is in the objectives of the system (Salman, 2003). For example, the US Department of Agriculture conducts monitoring programmes through its National Animal Health Monitoring System in which it investigates and collects data over time on disease states, animal characteristics and production characteristics. This body of data can be used to recognize a need for further research or the need for a surveillance programme, but in and of itself is only informative. In monitoring, the objectives are purely an observational undertaking.

7.1.2 Necessary components and considerations of monitoring and surveillance systems

A monitoring and surveillance system (MoSS) is built on three basic principles. First is the identification of the disease(s) to be assessed based on a risk assessment, the epidemiology of the selected diseases, the disease pathophysiology and the inherent agent characteristics (Zepeda and Salman, 2003). This surveillance can be general, focusing on a broad range of diseases or syndromes, or specific to a particular disease of interest (OIE, 2010). Secondly, there is the identification of the goals and objectives of the surveillance system. These can be monitoring for trends, disease control or disease eradication; the latter two tasks use surveillance systems specifically because an action is implied. Therefore, thresholds need to be identified that will result in action should a disease-positive animal be found or should population levels of disease exceed a predefined prevalence or incidence threshold (Christensen, 2003; Salman, 2003). The identification of the sampling scheme must also be decided upon, which also has an impact on the type of data collection done: either passive or active. Finally, the limitations of diagnostic tests and other disease measurement methods need to be incorporated into the system in order to understand its overall validity (Stark, 2003; Jebara, 2004).

A risk assessment can be used to identify the diseases of concern for an area. The protocol consists of the identification of disease hazards, the characterization of those hazards and the likelihood of exposure to the ranges of hazards or outcomes that can be caused by a disease. Ideally, the assessment quantifies the probability of experiencing the full range of hazards associated with a disease threat. The hazards that are identified can then be prioritized based on their public health, production and international trade implications. The risk management of the prioritized hazards uses a monitoring and surveillance system (Dennis *et al.*, 2002). Risk communication is the role of the OIE (World Organisation for Animal Health), public health systems and appropriate national government organizations. The role of the OIE in risk communications is through its requirement that OIE-listed diseases and significant animal health concerns be reported to the organization on a weekly, monthly and annual basis via the required reports for all member states (OIE, 2010).

Monitoring and surveillance is demanding of resources. It does not just include finding cases, but also requires the following components: public education campaigns, indemnities, producer awareness, response resources such as vaccines, and the appropriate training of personnel involved in sample collection and disease determination (Christensen, 2003; Salman, 2003; Jebara, 2004; Thrusfield, 2007a).

Proper training of personnel is even more important when attempting to exclude or prevent diseases with a public health and socio-economic impact. This is because few people have experience with the disease in their day-to-day clinical life and they are unfamiliar with its clinical manifestations.

The design of a surveillance system has many parts that are necessary to define before initiating a surveillance programme. These are: an established case definition; standardized methods of collecting data; a defined sampling method and sample size; an understanding of how the data will be analysed and used by decision makers; methods of communication of the results from the surveillance system; and an evaluation system to allow for regular critiquing and improvement of the programme.

Case definitions increase the specificity of a surveillance programme, especially of a disease-specific programme. Typically, a case is defined by the clinical manifestation of disease and laboratory results. Definitions that include the serological testing authorized to determine suspect and confirmed cases are most specific when the tests used are defined and the order of laboratory tests is explicitly stated (Akhtar and White, 2003; OIE, 2010). Screening tests are used to identify suspect cases. By nature, screening tests are typically of higher sensitivity so as to reduce false negatives, but are also known for having lower specificity than other tests and thus have a higher rate of false positives. Therefore, it is necessary to confirm suspect cases with tests of higher specificity or diagnostic tests (Dohoo *et al.*, 2003b). Those developing the surveillance system should also consider the reproducibility of test sensitivity and specificity between different laboratories, as well as the laboratory burden and capacity to test for the disease(s) of interest. Another issue that needs to be considered for TBDs is the species identification and the detection of the pathogens among the collected ticks associated with specific TBDs.

Data collection methods and, more importantly, storage and tracking methods must also be standardized. Using the same collection forms across all areas of the surveillance system will prevent differences in reporting from different sources and will increase consistency. This will also ensure that all necessary information is captured and reported by the reporting agencies. Lastly, it will reduce errors when data are recorded into a database by narrowing the need to interpret different questions and answers as to where they fit into the database. When a standardized form is used, the data entry can be standardized too (Thrusfield, 2007b). The need for this standardization comes from the need of the surveillance system and its results to be reliable. Decision makers will be more likely to consider the MoSS information when they are confident that what it is reporting has internal validity and, when necessary, external validity as well.

External validity, or the extrapolation of the surveillance results to the greater population of interest, is determined by the sampling method (Dohoo *et al.*, 2003c). There are different populations that can be considered for surveillance based on whether probability or non-probability sampling is used. Validity requires probability sampling and, more specifically, random selection, sampling from the population of interest, such as the general population, and sample size.

The external validity of a programme is greatest when a census of the population is performed. External validity is important in monitoring and surveillance when it is critical that the output from the sample population mirrors the true situation in the study population. For example, estimating disease prevalence would make this important. In the situation of a census, all animals or people in the population of interest are assessed for the presence of disease (Cameron *et al.*, 2003). Populations are groups of interest with a common characteristic. If populations of interest are quite small, this may be manageable, but if they are quite large, such censuses are not feasible. Yet, when the population of interest is too large to census, then representative sampling is needed to estimate the level of disease in that larger population. If the sample is not representative of the population of interest, then the results of the surveillance within that sample will be also invalid. The most reliable method to ensure a representative sample is to use random sampling techniques (Cameron *et al.*, 2003).

Random sampling is sampling which ensures that the probability of being selected is the same for every entity in the study population (Cameron *et al.*, 2003). Random sampling often results in representative samples and allows for statistical evaluation of the results as well (Dohoo *et al.*, 2003a). This is also known as probability sampling. Non-random sampling or non-probability sampling, such as purposive or convenience sampling, can also be used for specific events, but not for measures of disease occurrence. Targeted samples are a type of purposive sampling – sampling of populations at high risk. Convenience sampling can save resources and time. Yet neither of these methods yields representative samples of the greater population (Cameron *et al.*, 2003).

Sample populations are more advantageous than censuses for multiple reasons. They are smaller, less costly and more manageable, while being representative of the study population. The results from an appropriately selected sample population can then be extrapolated to the study population without conducting surveillance on the entire population (Cameron *et al.*, 2003; Dohoo *et al.*, 2003a). For measures of disease such as prevalence, a census gives you a specific and reliable measure of disease, but a sample gives you a value and its variability. The variability is the measure of uncertainty that comes with using a sample of the population to estimate the level of disease rather than using a census to assess the entire population. This variability can be reduced as the sample size increases (Dohoo *et al.*, 2003a). If one is using the surveillance system to assess risk factors, a certain sample size is needed to assure that there is truly no association when the results find no association; this is the concept of the power of a study and is also related to sample size. Sample size is determined through calculations that consider the allowable error, the confidence of the estimate, the variability of the measure of interest in the population and the power wanted in the study (Cameron *et al.*, 2003; Dohoo *et al.*, 2003a). Other factors such as the study population size and the accuracy of tests must also be considered.

However, there are also other considerations when evaluating disease presence. Diseases often cluster based on political, social and environmental factors. This clustering needs to be considered not only in relation to the sample size determination, but also the sampling method. If it is not considered, there are three possibilities. First, the surveillance may accurately estimate the level of disease by chance. Secondly, the sampling method oversamples from clusters without disease, thus underestimating disease. Thirdly, the sampling method may oversample from clusters with disease, and thus overestimate disease (Buehler *et al.*, 2003). In terms of sample size, the greater the similarity of the animals within the clusters in terms of the measure of interest, the greater the sample size (Dohoo *et al.*, 2003a). Clearly, clustering has many impacts on a surveillance system that must be considered.

The results of a surveillance system are used for multiple purposes. The collection of the data should be fitted to the desired analysis instead of fitting the analysis to the data collections. This requires that the purpose of the data is predetermined. It also allows the data to appropriately feed into pertinent statistical analysis, spatio-temporal analysis or simulation models that are often used for risk assessments. The impact of a surveillance system depends upon the usefulness of its output. If the output does not measure variables of interest, the surveillance system is limited in its overall utility.

All systems should be continuously evaluated and improved to meet current needs and threats. Assessment and validation of a system must consider the needs of the stakeholders, the ability to detect disease and the usefulness of the data to the stakeholders. In order to adequately assess a system, the objectives and aims of the system, the disease under surveillance, the legal basis of the programme, the various components of the programme and the responsible organizations need to be defined. By using these definitions and the needs of the stakeholders, even as they change, the baseline purpose of the surveillance assessment can be defined. A needs assessment of the stakeholders can be used to identify areas of weakness in the surveillance system or areas that need to be addressed that currently may not be in the system (Martell-Moran *et al.*, 2011).

There are numerous methods for conducting assessments. These include questionnaires of stakeholders, qualitative methods such as graphical methods, descriptive assessments or classification analysis and quantitative methods such as scoring systems. Graphical methods include fault tree analysis and scenario analysis. Classification analysis is a descriptive assessment of the surveillance system through its component parts. If standardized, classification analysis can compare various surveillance systems, such as those in different regions of countries, or a surveillance system over time. The next step in a classification analysis is to assign a scoring system to the effectiveness of each component, thereby making it a quantitative analysis (Stark, 2003). Lastly, another potential quantitative assessment is via benchmarks or indicators. These can supply quality assurance to the system too by setting benchmarks for the number of tests run or disease reports generated in a period of time (Stark, 2003). Of course, benchmarks are meant to be a tool, and the ability to reach them must adhere to sufficient quality standards as well.

Finally, the results of the analysis from a surveillance system must be communicated to all stakeholders; otherwise the entire programme is for naught. An important component of dissemination of the results is in identifying all of the critical stakeholders in federal, state and local governments, industry, private citizens where appropriate (pet owners, domestic flock owners), veterinarians, physicians, laboratories and the general public (Martell-Moran *et al.*, 2011). Furthermore, the form in which the data are distributed to the various stakeholders may need to differ based on how such stakeholders will use the information. The general public may need information on how to protect themselves from disease; producers may need information on how to enhance their biosecurity in the face of an epidemic; and veterinary public health personnel may need information on the rate of spread and scope of the disease to implement appropriate disease control strategies on a regional or national level.

As different populations with different backgrounds will be using the surveillance results, there need to be various reports that present different types and levels of information: public information requires the statement of easily understandable facts about the disease situation; for those familiar with surveillance systems and epidemiology, reports using tabular summaries that require some level of interpretation can be used; and for stakeholders and policy makers, interpretative reports are needed that include a description of the programme and the results (Wineland and Dargatz, 2003). The latter reports are useful to policy makers and industry key players who have a larger stake in the results of the surveillance system, but do not have the appropriate training to interpret the data alone.

There are numerous components and considerations that go into a surveillance system and its methods. Without consideration of all these parts, from development to assessment and dissemination of the data, the surveillance system will fail to reach its objectives or purpose. A comprehensive surveillance system should be reasonably sensitive and specific for the disease or diseases of interest, have an adequate positive predictive value – especially in the face of endemic diseases, be timely and useful to the stakeholders for developing policy, and be able to dictate when interventions for disease control should be implemented in biosecurity and bio-containment programmes. In order to fully advise stakeholders the programme must be timely, acceptable to those who provide the data and those who use the data, accurate in its depictions of the disease situation/s, and even flexible in its ability to adapt as situations change (Akhtar and White, 2003). There is no right answer or recipe for a surveillance system, but these guidelines can assist in the development of an effective programme.

7.1.3 Types of data collection methods and surveillance techniques

Passive and active data collection systems are both useful, but their usefulness depends upon the purpose of the surveillance programme. Passive surveillance is the collection of laboratory results, tick specimens, slaughterhouse information or reporting by clinicians (veterinarians and physicians) through health departments or similar organizations. The data are second-hand and generated for purposes of diagnosis and not for the sole reason of surveillance. Making critical diseases notifiable can assist in ensuring that they are better reported and recognized by clinicians and veterinarians. Passive data collection is useful for recognizing trends and emerging diseases, but it should not be relied upon as a rapid early warning system. Active collection requires data collection from the apparently normal population for the sole purpose of surveillance and monitoring (Akhtar and White, 2003; Salman, 2003; Thrusfield, 2007a).

In active surveillance, the sampling frame can be derived from different types of populations. These are the general population, sentinel populations and targeted populations. General populations are best used when the disease is very common and can easily be found with a reasonable sample size that is economically feasible within manpower and laboratory constraints. General population surveillance makes use of random sampling and systematic sampling to assess the level of disease in a population (Thrusfield, 2007a; OIE, 2010). General population-based surveillance programmes allow for the calculation of incidence and prevalence if the sampling is representative. This is based on the concept that a representative sample of the population at risk is followed over a time period and all new cases in the representative sample are identified. The results can, therefore, be extrapolated to the general population because there is external validity. The more common a disease is in a population the smaller the sample size that is needed. Sample sizes increase as the disease becomes rare (Salman, 2003). Thus for rare diseases alternative sampling populations are used. For very rare diseases that are being eradicated, a targeted sample population will most likely need to be used. The targeted population should have a higher prevalence of disease than the general population (Cameron *et al.*, 2003). This reduces the sample size and hence the burden on resources needed to successfully eradicate disease.

Surveillance programmes can also use sentinel populations. Sentinel populations are useful in programmes of disease exclusion or prevention. Sentinel populations are populations that are monitored for the presence of a seasonal or foreign animal disease. These populations should have a higher rate of disease than the general population (McCluskey, 2003). If the sentinel population is sufficiently sensitive to disease, it can act as an early warning system to the presence of disease (Rabinowitz *et al.*, 2006). Further, sentinel populations can be used as a public health tool, acting as an early warning system when the sentinel population incubation period is shorter than the human incubation period for the disease. When sentinel populations indicate the presence of a zoonotic disease, this finding can be used by the public health professionals in the area as a way to alert the human population to the disease threat. West Nile fever and eastern equine encephalitis in horses act as an indicator for the human population in this way.

Participatory epidemiology is a specialty that uses local knowledge to understand the local disease situation. Often, when surveillance is done using participatory epidemiology, syndromic surveillance is utilized. The reason for such an approach is that in under-resourced countries there is often poor laboratory support, poor access to remote regions and a lack of appropriate levels of manpower (Thrusfield, 2007a). Syndromic surveillance requires a detailed clinical case definition. The clinical case definition is the only method of capturing all potential suspect cases. Ideally, it is tailored to a specific disease. The purpose of syndromic surveillance is to capture all of these potential cases so that laboratory confirmation (in a field or diagnostic laboratory) can be carried out. Such participatory disease surveillance programmes were used in the Rinderpest eradication campaign, which determined the absence of disease in remote locations (Jost *et al.*, 2007).

7.2 Control Options for Hard Ticks

Tick control has stimulated intense interest for more than a century. There is a variety of strategies for the control of ticks and TBDs (Sonenshine *et al.*, 2006; Willadsen, 2006). Tick prophylaxis should cover the entire period during which ticks are active. Treatment with synthetic chemicals known as acaricides is still the most widely used means to control or prevent hard tick attacks, and has the aim of preventing pathogen transmission (Polar *et al.*, 2005). Control of ticks with acaricides can either be directed against the ticks on the host or against the free-living stages of those ticks in the environment.

7.2.1 Chemical control of ticks on animals

Acaricides include several groups of pesticides: organophosphates (e.g. coumaphos, diazinon), carbamates (e.g. propoxur), pyrethroids (e.g. permethrin, deltamethrin, flumethrin), formamidines (e.g. amitraz), two classes (avermectins and milbemycins) of macrocyclic lactones (e.g. ivermectin, doramectin, moxidectin, eprinomectin), phenylpyrazoles (e.g. fipronil) and natural acaricides such as botanical materials (e.g. extracts of the neem tree containing azadirachtin).

Several methods are used for the application of acaricides to animals, such as dips, sprays using manual or motorized high-pressure sprayers, dusts, pour-ons, spot-ons, the intraruminal bolus and injections. Dipping vats have been used extensively for tick control on livestock including cattle and sheep. When these are used for several animals over extended periods of time, dipping vats need to be maintained with the proper solution levels and percentage of active ingredient. Small numbers of livestock can be treated with hand-held sprayers but manual spraying depends on the skill of the person applying it. Retreatment may be needed several times during a season when sprays and dips are used. Pour-on formulations of acaricides contain high-quality oil that spreads through the greasy hair coat of livestock, and these products can also be used with applicators to treat wild ungulates in game reserves (Latif and Walker, 2004). Pour-on formulations are relatively expensive, but there is little wastage of acaricide and they may be cheaper in the long term. The other topical application methods are often employed owing to their advantage in limiting the amount of acaricide used; the cattle pass through a heavy low pressure spray and become soaked with acaricidal fluid, but body parts such as the ears and groin may not be effectively treated by this method (Latif and Walker, 2004). To achieve long-lasting efficacy, acaricides can be incorporated into plastic or other suitable matrices that provide a slow release of the toxicant over a period of weeks or months. Systemic acaricides offer another means of providing long-lasting and effective tick control. These preparations can be divided into injectable, oral and topically applied products, all of which are delivered to the tick during its feeding activity on the skin. Each application method has its advantages and disadvantages.

Oil, creams or alcohol should not be used for killing the ticks. The application of acaricides is advisable because not all of the ticks, especially the larval and nymphal stages and unengorged adults, may be detected on the animal. When treating dogs and cats against ticks, only approved veterinary medicinal ectoparasiticides should be used. There are several different formulations of acaricides for pets, and both dogs and cats can be sprayed with approved products. Topical spot-on treatments containing fipronil or selamectin can be used on both dogs and cats; however, permethrin can be used only on dogs. Insecticide-impregnated pet collars are also available for dogs and cats, although those containing amitraz cannot be used on cats. These collars offer at least partial protection but may not provide total protection against ticks (Elfassy *et al.*, 2001; Spencer *et al.*, 2003). Dogs and cats travelling to regions with ticks and endemic TBDs should also receive a regular application of acaricidal products, particularly if those TBDs are not endemic in the home country. To advise pet owners and improve owner compliance, the duration of efficacy for an individual product should be established from the relevant product data sheet so that the owners are given the correct retreatment intervals. It is advisable that animals are checked regularly, in particular towards the end of the protection period, to ensure that any visible ticks are removed and that early repeat treatment is considered if appropriate. It should also be remembered that the duration of efficacy of an acaricide may differ between tick species, again highlighting the importance of visual checking to verify that the treatment remains effective (ESCCAP, 2009). If permanent infestations with *Rhipicephalus* (*Rh.*) *sanguineus* occur within a kennel or household environment, regular acaricidal treatment of pet animals coupled with environmental treatment using a compound from a different chemical group can be used. There are special formulations of acaricides for use in kennel or household environments. The active compounds mostly belong to the same chemical classes as the acaricides used for pets.

For wildlife, an important concept called host-targeted tick control has been introduced in recent years as a way to destroy ticks without harming their hosts. The hosts can be attracted (using feed, nesting materials or other attractants) to devices that apply acaricides to them. Several such devices have been developed to attract deer or small mammals to 'feeding stations' where they are treated. Examples include bait boxes, permethrin-treated cotton balls, and so-called four-poster devices (Stafford and Kitron, 2002). One of the most promising of these devices is the 'four-poster' device. It includes a central bin, containing bait to attract deer, with a bait dispenser and applicator station on either side. As the deer inserts its head to feed on the bait, it contaminates its fur with acaricide from the applicators. Four-poster devices have the advantage of providing well-targeted applications using far lower amounts of pesticide than are used in broadcast applications (Pound *et al.*, 2000). The effectiveness of any tick-control approach tends to depend on the ecological conditions at the application site, but these methods can be important tools in integrated pest management programmes, especially when combined with other management techniques appropriate for the local conditions of tick distribution and transmission dynamics. Another novel host-targeted method is the use of rodent-targeted bait boxes containing fipronil. This has proven effective in eliminating immature ticks on mice, thereby reducing the subsequent population of questing nymphs and adults, as well as reducing the proportion of these ticks infected with the Lyme disease spirochaete *Borrelia burgdorferi* (Dolan *et al.*, 2004).

The chemical control of ticks on livestock poses several problems. Besides leaving residues in meat and milk, the use of many acaricides (e.g. organophosphates, carbamates) is associated with risks of side effects or poisoning resulting from overdoses on specific species or from breed sensitivity (de Oliveira Monteiroa *et al.*, 2010). Environmental contamination and effects on non-target animals have also been well documented for organochlorines, organophosphates, carbamates and pyrethroids (Kunz and Kemp, 1994). The development of acaricide resistance by some tick species is an increasing and continuing concern in tropical and subtropical areas, where these ticks have been found to be resistant to chlorinated hydrocarbons, organophosphorus insecticides, pyrethroids and formamidines (Kunz and Kemp, 1994). Acaricide resistance in tick species occurring in Europe has not yet been reported, but to avoid the onset of this problem there are certain rules that should be followed:

- When acaricides are used they should be stored as recommended and used as soon as possible so that they are of full strength as specified by the manufacturer.
- All ticks on treated animals should be destroyed or eliminated.
- The acaricide type recommended by veterinary authorities should be used until official advice to change to another type is given.
- Novel acaricides should only be used when the older types of acaricides have become ineffective (Latif and Walker, 2004).

For the above-mentioned reasons, major alternatives to conventional acaricide treatments have been developed in recent years which can be used in integrated management programmes for hard ticks in order to minimize the environmental impacts and acaricide resistance, and to reduce production costs (Samish *et al.*, 2000). The promising alternative methods described in Sections 7.2.2–7.2.8 are being investigated.

7.2.2 Pheromone or hormone-assisted control

Pheromone-assisted tick control is one of the novel strategies being used for the control of hard ticks. Research with tick pheromones suggests that combinations of pheromones and acaricides can be significantly more effective in controlling ticks than the acaricide alone because ticks are unlikely to develop resistance to their own pheromones. A pheromone–acaricide combination applied to a single spot on cattle can be effective in killing ticks. Another promising device is the 'tick decoy', in which the sex pheromone 2,6-dichlorophenol and an acaricide are impregnated into plastic beads on the surface of which 'mounting' sex pheromone is smeared. Male ticks are attracted to these decoys on the animal's hair coat and killed. This approach also disrupts mating activity so that any surviving females cannot lay viable eggs (Norval *et al.*, 1996). Another novel technology for killing *Ixodes scapularis* ticks in their natural habitats was developed by incorporating the components of the tick arrestment pheromone (guanine, xanthine and haematin) along with permethrin into an oily matrix for dispersal on vegetation. These paste-like droplets attracted and killed ticks before they could infest humans or animals (Sonenshine *et al.*, 2003, 2006; Sonenshine, 2008). Hormones and insect growth regulators (IGRs) such as methoprene have also been used to disrupt tick development in laboratory experiments. Analogues or mimics of ecdysteroids and juvenile hormone are effective in killing ticks by delaying their development, disrupting oviposition or killing the larvae when they hatch from eggs deposited by treated females. However, these compounds do not appear to be uniformly effective against all types of ticks.

7.2.3 Biological control

Biological control of ticks seems an appropriate alternative that may reduce the frequency of use of chemical acaricides and the need for treatment of TBDs. Biological control agents are in principle highly desirable, but their narrow host specificity, often relatively low efficacy, costs of manufacture, certain application problems and sometimes low stability present serious challenges. Numerous pathogens attack ticks, including bacteria, fungi and nematodes (Samish *et al.*, 2004). Several papers have reported the testing of entomopathogens for the control of ticks in laboratories, and the use of these novel biocontrol techniques on animals (Alonso-Díaz *et al.*, 2007) or vegetation (Kaaya, 2000). Among the potential biocontrol agents, entomopathogenic fungi, nematodes and parasitic wasps are the most promising candidates (Samish and Rehacek, 1999).

Entomopathogenic fungi have been studied as control agents of ticks mainly in laboratory assays (Kaaya, 2000; Samish *et al.*, 2001, 2004; Polar *et al.*, 2005). These fungi invade their host by direct penetration of the cuticle. After germination of the attached spore, appressoria are formed which then secrete histolytic enzymes and produce hyphae. After successful penetration by the hyphae, yeast-like blastospores are formed and propagate. After the death of the host, the fungus grows out of the cadaver and sporulates (Kleespies, 1993). The comparative ease with which the spores of these fungi can be produced and artificially disseminated makes them promising potential agents for the control of ticks (Norval and Horak, 2004). It has been reported that temperature and host secretions (e.g. sweat) may affect the virulence of entomopathogenic fungi on animals treated with biopesticides to control ticks (Polar *et al.*, 2005). Alternative targeted methods of applying fungal pathogens to the environment for the control of ticks are needed. Studies of the effects of entomopathogenic fungi on ticks under field conditions are scarce (Benjamin *et al.*, 2002; Alonso-Díaz *et al.*, 2007), but at present one of the best candidates for tick biocontrol is *Metarhizium anisopliae* (Samish *et al.*, 2001). Preliminary field trials have had modest results. The use of an oil-based carrier enhanced tick mortality when compared with a water-based solution (Kaaya and Hassan, 2000). The pathogens that affect ticks typically also affect other arthropods (Brownbridge and Glare, 2007), so effects on non-target arthropods must be considered in strategies for the application of biocontrol materials. There are opportunities to use auto-dissemination devices to deliver pathogens to ticks (Maniania *et al.*, 2007). Such devices use visual cues, pheromones and kairomones to attract host pests to a pathogen source (Vega *et al.*, 2000).

A promising alternative to the use of entomopathogenic fungi is biological control through the use of entomopathogenic nematodes (EPNs) (Samish *et al.*, 2001). The third-stage infective larvae of EPNs of the families Heterorhabditidae and Steinernematidae live freely in the soil. They enter their hosts through natural body openings such as the genital pore and release symbiotic bacteria carried in their intestines (Kocan *et al.*, 1998). These bacteria proliferate in the haemolymph and produce toxins and other metabolites. The tick then dies from septicaemia and the bacteria produce suitable conditions for the nematodes to complete their life cycle (Ehlers, 2001).

Arthropods can also be used for biological control of ticks: chalcid wasps of the genus *Ixodiphagus* are obligatory parasitoids of ixodid ticks and most species will oviposit and develop only in the nymphal stage of the tick. Several wasp larvae can successfully develop in a single engorged nymph, which is killed during this process (Mwangi *et al.*, 1997; Hu *et al.*, 1998; Norval and Horak, 2004).

Domestic chickens are opportunistic predators of ticks and can be used for tick control in rural areas. In particular, the indigenous breeds of Galliformes, if allowed to scavenge among cattle, can consume considerable numbers of ticks (Latif and Walker, 2004).

7.2.4 Genetic resistance

In general, tick tolerance or tick resistance exhibited by certain livestock breeds might vary with the species of infesting tick; heterospecific resistance appears to be low or even absent among different genera of ticks, while a certain degree of cross resistance is expressed to tick species belonging to the same genus (de Castro and Newson, 1993). Genetic resistance has been described in West African N'Dama cattle. This breed has a higher degree of natural resistance to ticks with a long hypostome, such as some *Amblyomma* and *Hyalomma* species, than to tick genera with a short hypostome. In *Bos indicus* breeds, the evidence for a genetic resistance trait to multi-host ticks, such as *Rh. appendiculatus*, is not as strong as for the one-host ticks, i.e. *Rh. microplus* (Mattioli *et al.*, 2000).

7.2.5 Vaccine

In Australia, a commercial recombinant antigen vaccine has been developed for the control of the cattle tick *Rh. (Boophilus) microplus* based on a so-called concealed antigen (Bm86) in cells of the tick gut. A similar recombinant vaccine has been developed in Cuba. Recent reports suggest that the recombinant Bm86 can reduce tick fecundity by as much as 90% (Willadsen, 2006, 2008). Although it is possible that antigen-resistant strains of cattle ticks may appear, large-scale vaccination of cattle herds with these recombinant vaccines offers a promising alternative or supplement to acaricides. Although the vaccine uses an antigen from *Rh. (Bo.) microplus*, it is even more efficacious against *Rh. (Bo.) annulatus* than against the homologous species (Fragoso *et al.*, 1998). Such effects appear not to correlate with the degree of sequence conservation of the antigen across tick species (Willadsen, 2006).

Because vaccines are expensive and involve considerable risk, a high level of efficacy is required to offset these negative aspects of their use. Research on other antigens and other tick species is in progress. Many more potential antigens have been proposed than have been tested. Tick antigen targets studied to date are from a restricted range of functional classes. They include structural proteins, particularly from salivary glands, hydrolytic enzymes and their inhibitors, especially those involved in haemostatic processes, and a range of membrane-associated proteins of unknown function (Willadsen, 2006, 2008). Of special interest is the development of novel combinations using RNA interference (RNAi) to silence subolesin and the tick-protective antigen Rs86 (similar to Bm86) against *Rh. sanguineus*; the synergistic effect of silencing both genes causes a much greater reduction of tick feeding and oviposition than targeting either one alone (de la Fuente and Kocan, 2006). Other promising vaccines target tick-cement protein, disrupting the attachment success as well as causing midgut injury and affecting the tick's ability to transmit pathogens (Labuda *et al.*, 2006); affect the ability to disrupt the male engorgement factor; or apply a combination of anti-tick and anti-pathogen vaccines (Sonenshine *et al.*, 2006, Willadsen, 2008).

7.2.6 Personal protection

Preventive measures are the most effective means of protection from ticks. Ticks can be avoided by avoiding exposure to fields, forests and other tick-infested habitats, especially in areas of known disease foci (Ginsberg and Stafford, 2005). Maintaining a short-clipped lawn in gardens and parks, and establishing barriers to prevent access to woods can minimize human exposure to ticks in these environments. The use of clearly defined paths can help people avoid contact with tick-infested vegetation. People in tick-prone areas should wear boots, socks, long trousers and light-coloured clothing. Trousers should be tucked into the boots, socks drawn over trousers, and the socks taped to form a tight seal. A recent study showed that wearing protective clothing was 40% effective in preventing Lyme disease (Vásquez *et al.*, 2008). Clothing should also be treated with an effective repellent or acaricide such as permethrin before entering tick-infested habitats. Exposed skin also should be treated with repellents or acaricides suitable for use on humans. The most widely used personal protectant is the repellent DEET (diethyl toluamide or N,N-diethyl-3-methyl benzamide), which is available as a lotion or a spray. Repellents should be applied repeatedly as needed to maintain maximum protection, but because of absorption or abrasion the interval should be longer than a few hours. Each person should conduct self-examinations for ticks during and after exposure to tick-infested areas. Early removal of attached ticks is important in minimizing the risk of contracting TBD. There is a large variety of purpose-designed tick removal tools available, and these may be used for the removal of ticks attached to the skin. Oil, creams or alcohol should not be used for killing the ticks.

7.2.7 Habitat modification

Tick control primarily involves the treatment of animals. However, habitat modifications of the infested land can aid in reducing tick abundance. Ticks have species-specific habitat requirements that are often associated with habitats of hosts and the need to avoid desiccation. Therefore, habitats can be manipulated to make them unsuitable for ticks or to minimize encounters between ticks and animals and people (Stafford, 2007). Several acaricides, including organophosphorus and synthetic pyrethroid preparations can be applied directly to vegetation in gardens, parks and other areas that are suitable sites for ticks. Using acaricides off the host is of limited value because ticks commonly occur in microhabitats covered by vegetation, leaf litter and other natural materials, or in the nests, burrows and other cavities used by their hosts. For these reasons, the ticks often do not come into direct contact with acaricides. Therefore, the chemicals must reach the ticks when they are seeking hosts. However, public opposition to the treatment of natural habitats with pesticides has made it unpopular to use this form of tick control. In the European Union (EU) it is prohibited to use any acaricides in the environment.

Selective grazing and pasture rotations often reduce cattle exposure to tick populations. By keeping animals out of certain pastures (e.g. by the removal of deer by hunting and deer-exclusion fences), the number of fed ticks that would serve to build up a population is reduced (Daniels *et al.*, 1993). Although lowering deer populations by hunting can also lower tick numbers, this approach is not generally practical, because deer populations must be reduced to extremely low levels to have a reliable effect on the transmission of tick-borne pathogens (Ginsberg and Stafford, 2005). An extreme case of pasture management is zero grazing by dairy cattle, but there is a risk of the unexpected reintroduction of ticks on cut fodder, on wild mammals or birds, or on newly introduced animals. This practice of zero (or non-continuous) grazing has often been referred to as pasture 'spelling' (Ginsberg and Stafford, 2005). Mowing and burning vegetation in natural areas also lowers tick numbers temporarily, but ticks reinfest treated areas as the vegetation grows back (Wilson, 1986).

For minimizing human and pet exposure to ticks in outdoor areas, efforts should first be made to keep overgrown and heavy vegetation cleared and cut in potentially tick-infested areas. A study on Lyme disease New York state showed that lawns that were cut short and were open to the sun had minimal numbers of ticks, while tick densities increased incrementally in gardens, wood edges and forests (Maupin *et al.*, 1991). Indoor tick control, as well as tick control in and around kennels, with cleaning or changing of pet bedding to prevent or remove ticks, is directed primarily at the brown dog tick, *Rh. sanguineus*.

7.2.8 Integrated tick management

There is no single, ideal solution to the control of ticks, but integrated control scenarios involving increased scientific and practical complexity can be developed and recommended. The integrated approach is probably the most effective way to control ticks (Jongejan and Uilenberg, 1994; Willadsen, 2006). Integrated pest management (IPM) is an approach to the management of arthropod pests that fosters the integration of various pest control methods so as to minimize reliance on individual environmentally damaging approaches and to provide sustained management of pest populations. Given the many tick control techniques currently available and the numerous novel techniques being developed, it is important to refine the theory and practice of integrating these methods so they can be efficiently and effectively applied to prevent human and animal disease.

In conclusion, controlling hard ticks can be an effective way to break the cycle of the tick-borne pathogens and subsequently to control the spread of disease. There are numerous conventional and relatively new approaches to control hard ticks and their spread, and several factors should be considered in determining the most effective options. These factors can be summarized as: the tick species involved, the animal species involved, the environmental and ecological conditions, animal management and the feasibility of the measures. Integration of these factors is required in order to most efficiently control ticks and the potential pathogens that they carry.

7.3 Control Possibilities for Soft Ticks

7.3.1 Argasidae

The control of argasid ticks is a difficult task because they possess a range of biological features (e.g. morphological, physiological and ethological) that make them difficult to eradicate. These soft ticks can form long-term aggregations deep inside the fissures, cracks and crevices of buildings where it is difficult to spray with chemicals to destroy them. Thus, efforts to eradicate these ticks from old buildings can be extremely costly and are generally unsuccessful.

In the case of *Argas reflexus* (the European pigeon tick), a low level of metabolism that allows the resting tick to close its spiracles for several hours, or perhaps even for days, might help this particular species to withstand the application of toxic gases (Dautel *et al.*, 1994). Even when using long-term residual pesticides, *A. reflexus* is difficult to exterminate owing to its long life cycle. A successful control measure of this tick species usually involves three stages (Dautel and Kahl, 1999): (i) the complete uncovering of all potential tick hiding places (in some instances, plaster must be removed from the walls); (ii) the application of a pyrethrum spray to expel ticks; followed by (iii) the use of a propoxur-based contact pesticide. A second control measure is often indicated 3–4 months later because of the resistance of the egg stage to control. Indeed, repeated treatments by a pest control professional are often necessary (Herrmann, 2000).

Several pesticides have been reported to be effective against argasid ticks, including permethrin (Dusbabek *et al.*, 1997), ivermectin (Montasser *et al.*, 2005), and *Bacillus thuringiensis* (Hassanain *et al.*, 1997). It was reported that entomopathogenic fungi have also good potential for the control of *A. persicus* (Sewify and Habib, 2001). Repellents (e.g. permethrin and DEET) have been shown to be effective against *A. persicus*, the fowl or poultry tick (Kumar *et al.*, 1992; Haag-Wackernagel, 2005). No vaccine against argasid ticks exists yet. The various alternatives need further investigation because no effective method for the long-term control of these ticks is known. At present, the only practical measure is to avoid housing domestic fowls in old, infested, buildings and to prevent access to this type of building by free-ranging herds. In order to control argasid ticks in poultry houses, all cracks and crevices – which provide shelter to these ticks – must be eliminated.

7.3.2 *Ornithodoros*

Eradication of *O. erraticus* from old pig sties is invariably unsuccessful. This is because of the tick's long life and long survival without feeding, the existence of accidental hosts other than pigs, and the possibility of penetrating into the cracks and surfaces of the buildings where the ticks are not accessible to acaricides or to the use of a fire torch on these surfaces or on the ground. These factors have been the major reason for abandoning *O. erraticus*-infested pig farms after an outbreak of African swine fever (ASF), and they have led to avoiding the use of this type of building to shelter pigs on the Iberian peninsula (Boinas, 1994). The use of endectocides in pigs can reduce the level of infestation in the premises, but does not prevent infection by the ASF virus in the pigs if they are bitten by an infective tick.

Practical measures include not housing pigs in infested buildings. When such buildings are located near a pig herd, the premises can be isolated with fences to prevent access by the pigs, or the buildings can even be destroyed and new premises built in another location (Arias and Sánchez-Vizcaíno, 2002). No effective vaccine against the ticks exists but there are promising studies on salivary glands extracts and 'concealed' gut antigen extracts (Astigarraga *et al.*, 1995, 1997; Manzano-Román *et al.*, 2006, 2007).

References

Akhtar, S. and White, F. (2003) Animal disease surveillance: prospects for development in Pakistan. *Revue Scientifique et Technique de l'Office International des Epizooties* 22, 977–987.

Alonso-Díaz, M.A., García, L., Galindo-Velasco, E., Lezama-Gutierrez, C.A., Angel-Sahagún, C.A., Rodríguez-Vivas, R.I. and Fragoso-Sánchez, H. (2007) Evaluation of *Metarhizium anisopliae* (Hyphomycetes) for control of *Boophilus microplus* (Acari: Ixodidae) on naturally infested cattle in the Mexican tropics. *Veterinary Parasitology* 147, 336–340.

Arias, M. and Sánchez-Vizcaíno, J.M. (2002) African swine fever eradication: the Spanish model, In: Morilla, A., Yoon, K.J. and Zimmerman, J.J. (eds) *Trends in Emerging Viral Infections of Swine*. Iowa State Press, Ames, Iowa, pp. 133–139.

Astigarraga, A., Oleaga-Pérez, A., Pérez-Sánchez, R. and Encinas-Grandes, A.A. (1995) A study of the vaccinal value of various extracts of concealed antigens and salivary gland extracts against *Ornithodoros erraticus* and *O. moubata*. *Veterinary Parasitology* 60, 133–147.

Astigarraga, A., Oleaga-Pérez, A., Pérez-Sánchez, R., Baranda, J.A., Encinas-Grandes, A. (1997) Host immune response evasion strategies in *Ornithodoros erraticus* and *O. moubata* and their relationship to the development of an antiargasid vaccine. *Parasite Immunology* 19, 401–410.

Benjamin, M.A., Zhioua, E. and Ostfeld, R.S. (2002) Laboratory and field evaluation of the entomopathogenic fungus *Metarhizium anisopliae* (Deuteromycetes) for controlling questing adult *Ixodes scapularis* (Acari: Ixodidae). *Journal of Medical Entomology* 39, 723–728.

Boinas, F.S. (1994) The role of *Ornithodoros erraticus* in the epidemiology of African swine fever in Portugal. PhD thesis, University of Reading, Reading, UK.

Brownbridge, M. and Glare, T. (2007) Impact of entomopathogenic fungus on soil-dwelling invertebrates. In: Ekesi, S. and Maniania, N.K. (eds) *Use of Entomopathogenic Fungi in Biological Pest Management*. Research Signpost, Kerala, India, pp. 295–392.

Buehler, J., Berkelman, R., Hartley, D. and Peters, C. (2003) Syndromic surveillance and bioterrorism-related epidemics. *Emerging Infectious Diseases* 9, 197–204.

Cameron, A., Gardner, I., Doherr, M.G. and Wagner, B. (2003) Sampling considerations in surveys and monitoring and surveillance systems. In: Salman, M.D. (ed.) *Animal Disease Surveillance and Survey Systems: Methods and Applications*. Iowa State Press, Ames, Iowa, pp. 47–66.

Christensen, J. (2003) Application of surveillance and monitoring systems in disease control programs. In: Salman, M.D. (ed.) *Animal Disease Surveillance and Survey Systems: Methods and Applications*. Iowa State Press, Ames, Iowa, pp. 15–34.

Daniels, T.J., Fish, D. and Schwartz, I. (1993) Reduced abundance of *Ixodes scapularis* (Acari: Ixodidae) and Lyme disease risk by deer exclusion. *Journal of Medical Entomology* 30, 1043–1049.

Dautel, H. and Kahl, O. (1999) Ticks (Acari: Ixodoidea) and their medical importance in the urban environment. In: Robinson, W.H., Rettich, F. and Rambo, G.W. (eds) *Proceedings of the 3rd International Conference on Urban Pests, Prague, Czech Republic, 19–22 July 1999*, pp. 73–82.

Dautel, H., Kahl, O., Scheurer, S. and Knülle, W. (1994) Seasonal activities of the pigeon tick, *Argas reflexus* (Acari, Argasidae) in Berlin, Germany. *Folia Parasitologia* 41, 155–160.

de Castro, J.J. and Newson, R.M. (1993) Host resistance in cattle tick control. *Parasitology Today* 9, 13–17.

de la Fuente, J. and Kocan, K.M. (2006) Strategies for development of vaccines for control of ixodid tick species. *Parasite Immunology* 28, 275–283.

de Oliveira Monteiroa, C.M., Furlong, J., de Azevedo Prata, M.C., Soares, A.E., de Paula Batista, E.S. and Dolinskic, C. (2010) Evaluation of the action of *Heterorhabditis bacteriophora* (Rhabditida: Heterorhabditidae) isolate HP88 on the biology of engorged females of *Rhipicephalus* (*Boophilus*) *microplus* (Acari: Ixodidae). *Veterinary Parasitology* 170, 355–358.

Dennis, S.B., Buchanan, R.L. and Miller, A.J. (2002) Microbial risk assessment: achievements and future challenges. *Food Safety Magazine* 7(6), 14–17.

Dohoo, I., Martin, W. and Stryhn, H. (2003a) Sampling. In: Dohoo, I., Martin, W. and Stryhn, H. (eds) *Veterinary Epidemiologic Research* [1st edn]. Atlantic Veterinary College (AVC), University of Prince Edward Island, Charlottetown, Prince Edward Island, Canada, pp. 27–47.

Dohoo, I., Wayne, M. and Henrik, S. (2003b) Screening and Diagnostic Tests. In: Dohoo, I., Martin, W. and Stryhn, H. (eds) *Veterinary Epidemiologic Research* [1st edn]. Atlantic Veterinary College (AVC), University of Prince Edward Island, Charlottetown, Prince Edward Island, Canada, pp. 101–102.

Dohoo, I., Martin, W. and Stryhn, H. (2003c) Validity in observational studies. In: Dohoo, I., Martin, W. and Stryhn, H. (eds) *Veterinary Epidemiologic Research* [1st edn]. Atlantic Veterinary College (AVC), University of Prince Edward Island, Charlottetown, Prince Edward Island, Canada, pp. 207–234.

Dolan, M.C., Maupin, G.O., Schneider, B.S., Denatale, C., Hamon, N., Cole, C., Zeidner, N.S. and Stafford, K.C. III (2004) Control of immature *Ixodes scapularis* (Acari: Ixodidae) on rodent reservoirs of *Borrelia burgdorferi* in a residential community of southeastern Connecticut. *Journal of Medical Entomology* 41, 1043–1054.

Dusbabek, F., Rupes, V., Sbreveimek, P. and Zahradnickova, H. (1997) Enhancement of permethrin efficacy in acaricide-attractant mixtures for control of the fowl tick *Argas persicus* (Acari: Argasidae). *Experimental and Applied Acarology* 21, 293–305.

Ehlers, R.U. (2001) Mass production of entomopathogenic for plant protection. *Applied Microbiology and Biotechnology* 56, 623–633.

Elfassy, O.J., Goodman, F.W., Levy, S.A. and Carter, L.L. (2001) Efficacy of an amitraz-impregnated collar in preventing transmission of *Borrelia burgdorferi* by adult *Ixodes scapularis* to dogs. *Journal of American Veterinary Medical Association* 219, 185–189.

ESCCAP (2009) *Ectoparasites Part 1: Control of Parasitic Insects and Ticks in Dogs and Cats*. ESCCAP Guideline No. 3, European Scientific Counsel Companion Animal Parasites, Malvern, UK.

Fragoso, H., Rad P.H., Ortiz, M., Rodriguez, M., Redondo, M., Herrera, L. and de la Fuente, J. (1998) Protection against *Boophilus annulatus* infestations in cattle vaccinated with the *Boophilus microplus* Bm86-containing vaccine GAVAC. *Vaccine* 16, 1990–1992.

Ginsberg, H.S. and Stafford, K.C. (2005) Forum: management of ticks and tick-borne diseases. In. Goodman, J.L., Dennis, D.T. and Sonenshine, D.E. (eds) *Tick-Borne Diseases of Humans*. ASM Press, Washington, D.C., pp. 65–86.

Haag-Wackernagel, D. (2005) Parasites from feral pigeons as a health hazard for humans. *Annals of Applied Biology* 147, 203–210.

Hassanain, M.A., el-Garhy, M.F., Abdel-Ghaffar, F.A., el-Sharaby, A. and Abdel Megeed, K.N. (1997) Biological control studies of soft and hard ticks in Egypt. I. The effect of *Bacillus thuringiensis* varieties on soft and hard ticks (Ixodidae). *Parasitology Research* 83, 209–213.

Herrmann, J. (2000) Taubenzeckenbekämpfung. *Umweltmedizinischer Informationsdienst* 3, 21–24. Available at: http://www.umweltbundesamt.de/umid/archiv/umid0300.pdf (accessed 4 July 2012).

Hu, R., Hyland, K.E. and Oliver, J.H. (1998) A review on the use of *Ixodiphagus* wasps (Hymenoptera: Encyrtidae) as natural enemies for the control of ticks (Acari: Ixodidae). *Systematic and Applied Acarology* 3, 19–28.

Jebara, K.B. (2004) Surveillance, detection and response: managing emerging diseases at national and international levels. *Revue Scientifique et Technique de l'Office International des Epizooties* 23, 709–715.

Johnson, J.L., Ginsberg, H.S., Zhioua, E., Whitworth, U.G., Markowski, D., Hyland, K.E. and Hu, R. (2004) Passive tick surveillance, dog seropositivity, and incidence of human Lyme disease. *Vector Borne and Zoonotic Diseases* 4, 137–142.

Jongejan, F. and Uilenberg, G. (1994) Ticks and control methods. *Revue Scientifique et Technique de l'Office International des Epizooties* 13, 1201–1226.

Jost, C.C., Mariner, J.C., Roeder, P.L., Sawitri, E. and Macgregor-Skinner, G.J. (2007) Participatory epidemiology in disease surveillance and research. *Revue Scientifique et Technique de l'Office International des Epizooties* 26, 537–549.

Kaaya, G.P. (2000) Laboratory and field evaluation of entomogenous fungi for tick control. *Annals of the New York Academy of Sciences* 916, 559–564.

Kaaya, G.P. and Hassan, S. (2000) Entomogenous fungi as promising biopesticides for tick control. *Experimental and Applied Acarology* 24, 913–926.

Kleespies, R. (1993) Untersuchungen zur Biologie und Anwendung des entomopathogenen Pilzes *Metarhizium anisopliae* (Metsch.) Sorokin zur biologischen Bekämpfung afrikanischer Wanderheuschrecken. PhD Thesis, TH Darmstadt, Germany.

Kocan, K.M., Pidherney, M.S., Blouin, E.F., Claypool, P.L., Samish, M. and Glazer, I. (1998) Interaction of entomopathogenic nematodes (Steinernematidae) with selected species of ixodid ticks (Acari: Ixodidae). *Journal of Medical Entomology* 35, 514–520.

Kumar, S., Prakash, S., Kaushik, M.P. and Rao, K.M. (1992) Comparative activity of three repellents against the ticks *Rhipicephalus sanguineus* and *Argas persicus*. *Medical and Veterinary Entomology* 6, 47–50.

Kunz, S.E. and Kemp, D.H. (1994) Insecticides and acaricides: resistance and environmental impact. *Revue Scientifique et Technique de l'Office International des Epizooties* 13, 1249–1286.

Labuda, M., Trimnell, A.R., Ličková, M., Kazimírová, M., Davies, G.M., Lissina, O., Hails, R.S.and Nuttall, P.A. (2006) An antivector vaccine protects against a lethal vector-borne pathogen. *PLoS Pathogens* 2(4), e27, doi:10.1371/journal.ppat.0020027.

Latif, A.A. and Walker, A.R. (2004) *An Introduction to the Biology and Control of Ticks in Africa*. Produced for the European Union Project International Consortium on Ticks and Tick-borne Diseases (ICTTD-2). Available at: http://www.alanrwalker.com/index/cms-filesystem-action/tick-biology-africa.pdf (accessed 4 July 2012).

Lindenmayer, J.M., Marshall, D. and Onderdonk, A.B. (1991) Dogs as sentinels for Lyme disease in Massachusetts. *American Journal of Public Health* 81, 1448–1455.

Maniania, N.K., Nchu, F. and Ekesi, S. (2007) Fungal pathogen for biocontrol of ticks. In: Ekesi, S. and Maniania, N.K. (eds) *Use of Entomopathogenic Fungi in Biological Pest Management*. Research Signpost, Kerala, India, pp. 295–392.

Manzano-Román, R., Encinas-Grandes, A. and Pérez-Sánchez, R. (2006) Antigens from the midgut membranes of *Ornithodoros erraticus* induce lethal anti-tick immune responses in pigs and mice. *Veterinary Parasitology* 135, 65–79.

Manzano-Román, R., García-Varas, S., Encinas-Grandes, A. and Pérez-Sánchez, R. (2007) Purification and characterization of a 45-kDa concealed antigen from the midgut membranes of *Ornithodoros erraticus* that induces lethal anti-tick immune responses in pigs. *Veterinary Parasitology* 145, 314–325.

Martell-Moran, N.K., Mauer, W.A. and Kaneene, J.B. (2011) Assessment of avian influenza surveillance and reporting needs of stakeholders in Michigan, 2007. *Journal of the American Veterinary Medical Association* 238, 1570–1583.

Mattioli, R.C., Pandey, V.S., Murray, M. and Fitzpatrick, J.L. (2000) Immunogenetic influences on tick resistance in African cattle with particular reference to trypanotolerant N'Dama (*Bos taurus*) and trypanosusceptible Gobra zebu (*Bos indicus*) cattle. *Acta Tropica* 75, 263–277.

Maupin, G.O., Fish, D., Zultowsky, J., Campos, E.G. and Piesman, J. (1991) Landscape ecology of Lyme disease in a residential area of Westchester County, New York. *American Journal of Epidemiology* 133, 1105–1113.

McCluskey, B.J. (2003) Use of sentinel herds in monitoring and surveillance systems. In: Salman, M.D. (ed.) *Animal Disease Surveillance and Survey Systems: Methods and Applications*. Iowa State Press, Ames, Iowa, pp. 119–134.

Montasser, A.A., Gadelhak, G.G. and Tariq, S. (2005) Impact of ivermectin on the ultrastructure of the testis of *Argas (Persicargas) persicus* (Ixodoidea: Argasidae). *Experimental and Applied Acarology* 36, 119–129.

Mwangi, E.N., Hassan, S.M., Kaaya, G.P. and Essijman, S. (1997) The impact of *Ixodiphagus hookeri*, a tick parasitoid, on *Amblyomma variegatum* (Acari: Ixodidae) in a field trial in Kenya. *Experimental and Applied Acarology* 21, 117–126.

Nicholson, M.C. and Mather, T.N. (1996) Methods for evaluating Lyme disease risks using geographic information systems and geospatial analysis. *Journal of Medical Entomology* 33, 711–720.

Norval, R.A.I. and Horak, I.G. (2004) Vectors: ticks. In: Coetzer, J.A.W. and Tustin, R.C. (eds) *Infectious Diseases of Livestock, Volume 1*, 2nd edn. Oxford University Press Southern Africa, Cape Town, South Africa, pp. 3–42.

Norval, R.A.I., Sonenshine, D.E., Allan, S.A. and Burridge, M.J. (1996) Efficacy of pheromone–acaricide impregnated tail-tag decoys for controlling the bont tick, *Amblyomma hebraeum* (Acari: Ixodidae), on cattle in Zimbabwe. *Experimental and Applied Acarology* 20, 31–46.

OIE (2010) *Animal Health Code*, 19th edn. World Organisation for Animal Health, Paris.

Polar, P., Kairo, M.T.K., Peterkin, D., Moore, D., Pegram, R. and John, S.A. (2005) Assessment of fungal isolates for development of a myco-acaricide for cattle tick control. *Vector-Borne Zoonotic Diseases* 5, 276–284.

Pound, J.M., Miller, J.A., George, J.E. and Lemeilleur, C.A. (2000) The "4-poster" passive topical treatment device to apply acaricide for controlling ticks (Acari: Ixodidae) feeding on white-tailed deer. *Journal of Medical Entomology* 37, 588–594.

Rabinowitz, P., Gordon, Z., Chudnov, D., Wilcox, M., Odofin, L., Liu, A. and Dein, J. (2006) Animals as sentinels of bioterrorism agents. *Emerging Infectious Diseases* 12, 647–652.

Salman, M.D. (2003) Surveillance and monitoring systems for animal health programs and disease surveys. In: Salman, M.D. (ed.) *Animal Disease Surveillance and Survey Systems: Methods and Applications*. Iowa State Press, Ames, Iowa, pp. 1–14.

Samish, M. and Rehacek, J. (1999) Pathogens and predators of ticks and their potential in biological control. *Annual Review of Entomology* 44, 159–182.

Samish, M., Alekseev, E. and Glazer, I. (2000) Biocontrol of ticks by entomopathogenic nematodes: research update. *Annals of the New York Academy of Sciences* 916, 589–594.

Samish, M., Alekseev, E. and Glazer, I. (2001) Entomopathogenic nematodes for the biocontrol of ticks. *Trends in Parasitology* 17, 368–371.

Samish, M., Ginsberg, H. and Glazer, I. (2004) Biological control of ticks. *Parasitology* 129, 389–403.

Sewify, H. and Habib, S.M. (2001) Biological control of the tick fowl *Argas persicargas persicus* by the entomopathogenic fungi *Beauveria bassiana* and *Metarhizium anisopliae*. *Anzeiger für Schädlingskunde* 74, 121–123.

Sonenshine, D.E. (2008) Pheromones and other semiochemicals of ticks and their use in tick control. In: Bowman, A.S. and Nuttall, P.A. (eds) *Ticks. Biology, Disease and Control*. Cambridge University Press, Cambridge, UK, pp. 470–491.

Sonenshine, D.E., Adams, T., Allan, S.A., McLaughlin, J. and Webster, F.X. (2003) Chemical composition of some components of the arrestment pheromone of the black-legged tick, *Ixodes scapularis* (Acari: Ixodidae) and their use in tick control. *Journal of Medical Entomology* 40, 849–859.

Sonenshine, D.E., Kocan, K. and de la Fuente, J. (2006) Tick control: further thoughts on a research agenda. *Trends in Parasitology* 22, 550–551.

Spencer, J.A., Butler, J.M., Stafford, K.C., Pough, M.B., Levy, S.A., Bledsoe, D.L. and Blagburn, B.L. (2003) Evaluation of permethrin and imidacloprid for prevention of *Borrelia burgdorferi* transmission from blacklegged ticks (*Ixodes scapularis*) to *Borrelia burgdorferi*-free dogs. *Parasitology Research* 90, 106–107.

Stafford, K.C. III (2007) *Tick Management Handbook: An Integrated Guide for Homeowners, Pest Control Operators, and Public Health Officials for the Prevention of Tick-associated Disease*, revised edn. Bulletin No. 1010, The Connecticut Agricultural Experiment Station, New Haven, Connecticut. Available at: http://www.ct.gov/caes/lib/caes/documents/publications/bulletins/b1010.pdf (accessed 4 July 2012).

Stafford, K.C. and Kitron, U. (2002) Environmental management for Lyme borreliosis control. In: Gray, J.S., Kahl, O., Lane, R.S. and Stanek, G. (eds) *Lyme borreliosis: Biology, Epidemiology and Control*. CAB International, Wallingford, UK, pp. 301–334.

Stark, K.D.C. (2003) Quality assessment of animal disease surveillance and survey systems. In: Salman, M.D. (ed.) *Animal Disease Surveillance and Survey Systems: Methods and Applications*. Iowa State Press, Ames, Iowa, pp. 169–176.

Thrusfield, M. (2007a) Surveillance. In: Thrusfield, M. *Veterinary Epidemiology*, 3rd edn. Blackwell Science, Oxford, UK, pp. 168–186.

Thrusfield, M. (2007b) Data collection and management. In: Thrusfield, M. *Veterinary Epidemiology*, 3rd edn. Blackwell Science, Oxford, UK, pp. 187–213.

Vázquez, M., Muehlenbein, C., Cartter, M., Hayes, E.B., Ertel, S. and Shapiro, E.D. (2008) Effectiveness of personal protective measures to prevent Lyme disease. *Emerging Infectious Diseases* 14, 210–216.

Vega, F.E., Dowd, P.F., Lacey, L.A., Pell, J.K., Jackson, D.M. and Klein, M.G. (2000) Dissemination of beneficial microbial agents by insects. In: Lacey, L.A. and Kaya, H.K. (eds) *Field Manual of Techniques in Invertebrate Pathology*. Kluwer Academic, Dordrecht, The Netherlands, pp. 153–177.

White, D.J. (1993) Lyme disease surveillance and personal protection against ticks. In: Ginsberg, H.S. (ed.) *Ecology and Environmental Management of Lyme Disease*. Rutgers University Press, New Brunswick, New Jersey, pp. 99–125.

Willadsen, P. (2006) Tick control: thoughts on a research agenda. *Veterinary Parasitology* 138, 161–168.

Willadsen, P. (2008) Anti-tick vaccines. In: Bowman A.S. and Nuttall P.A. (eds) *Ticks: Biology, Disease and Control*. Cambridge University Press, Cambridge, UK, pp. 424–446.

Wilson, M.L. (1986) Reduced abundance of adult *Ixodes dammini* (Acari: Ixodidae) following destruction of vegetation. *Journal of Economic Entomology* 79, 693–696.

Wineland, N.E. and Dargatz, D.A. (2003) Dissemination of surveillance findings. In: Salman, M.D. (ed.) *Animal Disease Surveillance and Survey Systems: Methods and Applications*. Iowa State Press, Ames, Iowa, pp. 177–184.

Zepeda, C. and Salman, M.D. (2003) Planning survey, surveillance, and monitoring systems in disease control programs. In: Salman, M.D. (ed.) *Animal Disease Surveillance and Survey Systems: Methods and Applications*. Iowa State Press, Ames, Iowa, pp. 35–46.

Glossary

Argasid ticks: soft ticks.

Co-feeding: a phenomenon in which ticks become infected with a pathogen during feeding adjacent to infective ticks on the same vertebrate host, even when the vertebrate host has not developed a systemic infection.

Competence: the ability of a vector to transmit a pathogen to a susceptible host in such a way that the host becomes infected.

Conscutum: the hard, sclerotized protective shield (plate) which covers most of the dorsal surface of Ixodidae males.

Diapause: a neurohormonally mediated dynamic state of low metabolic activity. Associated with this is reduced morphogenesis, increased resistance to environmental extremes, and altered or reduced behavioural activity. Diapause occurs during a genetically determined stage(s) of metamorphosis, and its full expression develops in a species-specific manner, usually in response to a number of environmental stimuli that precede unfavourable conditions. Once diapause has begun, metabolic activity is suppressed even if conditions favourable for development prevail.

Enamelling: is often called ornamentation. It is most conspicuous on the conscutum of males. The colour is mainly pink, orange or red; the enamel looks like paint on the surface of the integument.

Endophilic and Exophilic: when not feeding, endophilic (= nidiculous) ticks live in the nest, burrow or den of the host. Exophilic ticks live in the open environment away from the host's nest or burrow.

Ixodid ticks: hard ticks.

Kairomone: a compound (chemical messenger) emanating from a potential host, e.g. carbon dioxide in ox breath that may induce appetitive behaviour in blood-feeding arthropods (ticks, mosquitoes, etc.). Kairomones enable an interspecific system of chemical communication.

Monotropic: ticks which as immature stages or adults feed on the same type of host.

Nidiculous: endophilic and nidiculous are synonymous.

Questing: many exophilic ticks cling to plant stems or similar substrates and await passing animals. The tick's front legs, which have chemosensillae, are held out. Such questing ticks may be collected by flagging or dragging a cloth to which they try to attach.

Reservoir: an animate or inanimate object on or in which an infectious agent usually lives, and which therefore is often a source of such an infectious agent.

Scutum: the hard plate on the anterior dorsal surface of the larva, nymph and adult female ixodid tick. Soft ticks (Argasidae) do not have a scutum.

Telotropic: when the immature stages of a tick are able to feed on both different types of hosts and the same types of host as the adult ticks. For instance, both rodents and ruminants can support the feeding of immature stages of a certain telotropic tick.

Transovarial transmission: the transmission of microorganisms, including viruses, from mother to offspring via the ovaries.

Trans-stadial transmission: the passage of microorganisms and viruses in arthropods from one stage (stadium or instar) to the next.

Appendix 1 Tables of Geographic Distribution Data for Ticks and Tick-borne Diseases

Table A1.1. Geographic distribution data for African swine fever (ASFV). See Appendix 2 (hosted on http://www.cabi.org/vetmedresource/) for complete reference indicated by the Reference ID.

Country	Administrative region 1	Administrative region 2	Reference ID
Armenia	Lori		4000
Armenia	Tavush		4000
Georgia	Ajaria	Batumi	4000
Georgia	Guria	Chokhatauri	4000
Georgia	Imereti	Bagdati	4000
Georgia	Kakheti	Akhmeta	4000
Georgia	Kvemo Kartli	Bolnisi	4000
Georgia	Mtskheta-Mtianeti	Akhalgori	4000
Georgia	Racha-Lechkhumi-Kvemo Svaneti	Ambrolauri	4000
Georgia	Samegrelo-Zemo Svaneti	Abasha	4000
Georgia	Samtskhe-Javakheti	Adigeni	4000
Georgia	Shida Kartli	Gori	4000
Georgia	Tbilisi	Tbilisi	4000
Italy	Sardegna	Cagliari	4000
Portugal	Beja	Aljustrel	122
Portugal	Beja		122
Portugal	Évora	Alandroal	122
Portugal	Évora		122
Portugal	Faro	Albufeira	122
Portugal	Portalegre	Alter do Chão	122
Portugal	Portalegre		122
Russia	Astrakhan	Akhtubinsk	4000
Russia	Kalmyk	Chernozemel'skiy rayon	4000
Russia	Krasnodar	Abinskiy rayon	4000
Russia	Rostov	Aksayskiy rayon	4000
Russia	Stavropol	Aleksandrovskiy rayon	4000
Russia	Volgograd	Alekseevskiy rayon	4000
Spain	Andalucía		0
Spain	Extremadura		0

Table A1.2. Geographic distribution data for Crimean-Congo haemorrhagic fever virus (CCHFV). See Appendix 2 (hosted on http://www.cabi.org/vetmedresource/) for complete reference indicated by the Reference ID.

Country	Administrative region 1	Administrative region 2	Reference ID	No. entries[a]
Albania	Kukës		723	
Bulgaria	Blagoevgrad	Gotse Delchev	1109	
Bulgaria	Blagoevgrad	Gotse Delchev	1412	
Greece	Anatoliki Makedonia kai Thraki		1413	
Greece	Anatoliki Makedonia kai Thraki	Rodopi	721	
Kosovo	Pristina		1402	
Russia	Astrakhan		1068	
Russia	Astrakhan		1404	
Russia	Astrakhan		635	
Russia	Dagestan		1403	
Russia	Dagestan		1411	
Russia	Ingush		1411	
Russia	Kalmyk		1401	
Russia	Kalmyk		1404	
Russia	Kalmyk		1408	
Russia	Rostov		1068	
Russia	Rostov		634	
Russia	Rostov	Dubovskiy rayon	1407	
Russia	Rostov	Martynovskiy rayon	1414	
Russia	Rostov	Orlovskiy rayon	1405	
Russia	Rostov	Orlovskiy rayon	1414	
Russia	Rostov	Peschanokopskiy rayon	1407	
Russia	Rostov	Proletarskiy rayon	1407	
Russia	Rostov	Remontnenskiy rayon	1407	
Russia	Rostov	Sal'skSal'skiy rayon	1405	
Russia	Rostov	Tselinnyy rayon	1406	
Russia	Rostov	Tsimlyanskiy rayon	1406	
Russia	Rostov	Zimovnikovskiy rayon	1406	
Russia	Rostov	Zimovnikovskiy rayon	1414	
Russia	Stavropol		1404	
Russia	Stavropol		1404	
Russia	Stavropol		635	
Russia	Stavropol	Apanasenkovskiy rayon	1410	
Russia	Stavropol	Apanasenkovskiy rayon	1414	
Russia	Stavropol	Arzgirskiy rayon	1414	
Russia	Stavropol	Blagodarnenskiy rayon	1410	
Russia	Stavropol	Budennovsk	1408	
Russia	Stavropol	Budennovskiy rayon	1414	
Russia	Stavropol	Ipatovskiy rayon	1414	
Russia	Stavropol	Izobilnenskiy rayon	1410	
Russia	Stavropol	Krasnogorodskiy rayon	1409	
Russia	Stavropol	Krasnogorodskiy rayon	1410	
Russia	Stavropol	Novoaleksandrovskiy rayon	1414	
Russia	Stavropol	Novoselitskiy rayon	1410	
Russia	Stavropol	Petropavlovskiy rayon	1414	
Russia	Stavropol	Rostov	1400	
Russia	Stavropol	Stepnovskiy rayon	1414	
Russia	Volgograd		1068	
Russia	Volgograd		1404	

Continued

Table A1.2. Continued.

Country	Administrative region 1	Administrative region 2	Reference ID	No. entries[a]
Russia	Volgograd	Oktyabr'skiy	1408	
Turkey	Adana		1399	3
Turkey	Adiyaman		1399	
Turkey	Afyon		1399	4
Turkey	Aksaray		1399	
Turkey	Amasya		1342	44
Turkey	Ankara		1399	20
Turkey	Artvin		1399	31
Turkey	Artvin		708	3
Turkey	Aydın		1399	14
Turkey	Balıkesir		1399	2
Turkey	Bayburt		1399	3
Turkey	Bilecik		1399	3
Turkey	Bingöl		1399	
Turkey	Bolu		1399	14
Turkey	Bursa		1399	
Turkey	Çanakkale		1399	
Turkey	Çankırı		1399	38
Turkey	Çorum		1399	117
Turkey	Erzincan		1399	21
Turkey	Erzurum		1399	73
Turkey	Erzurum		708	2
Turkey	Eskişehir		1399	6
Turkey	Giresun		1102	8
Turkey	Giresun		1399	25
Turkey	Gümüshane		1399	70
Turkey	Gümüshane		991	
Turkey	Hatay		1399	
Turkey	İçel		1399	2
Turkey	Isparta		1399	2
Turkey	Istanbul		1399	
Turkey	Istanbul		649	4
Turkey	İzmir		1399	4
Turkey	Karabuk		1399	18
Turkey	Kars		1399	
Turkey	Kastamonu		1399	42
Turkey	Kayseri		1399	5
Turkey	Kırıkkale		1399	4
Turkey	Kırklareli		1399	2
Turkey	Kırşehir		1399	3
Turkey	Konya		1399	4
Turkey	Kütahya		1399	8
Turkey	Malatya		1399	4
Turkey	Manisa		1399	2
Turkey	Mugla		1399	3
Turkey	Mus		1399	2
Turkey	Nevşehir		1399	2
Turkey	Ordu		1102	15
Turkey	Osmaniye		1399	
Turkey	Samsun		1399	11
Turkey	Şanlıurfa		1399	3
Turkey	Sinop		1399	3

Continued

Table A1.2. Continued.

Country	Administrative region 1	Administrative region 2	Reference ID	No. entries[a]
Turkey	Sivas		1399	114
Turkey	Tokat		1102	219
Turkey	Tokat		708	2
Turkey	Trabzon		708	2
Turkey	Tunceli		1399	2
Turkey	Van		1399	
Turkey	Yozgat		1399	95
Turkey	Zinguldak		1399	

[a]If more than one.

Table A1.3. Geographic distribution data for the tick-borne encephalitis (TBE) group. See Appendix 2 (hosted on http://www.cabi.org/vetmedresource/) for complete reference indicated by the Reference ID.

Country	Administrative region 1	Administrative region 2	Reference ID	No. entries[a]
Åland Islands			1329	
Czech Republic	Jihočeský		1045	
Czech Republic	Jihočeský		226	8
Czech Republic	Jihočeský		525	
Czech Republic	Jihomoravský		1338	
Czech Republic	Jihomoravský		227	
Czech Republic	Jihomoravský		525	
Czech Republic	Kraj Vysočina		227	4
Czech Republic	Královéhradecký		225	
Czech Republic	Královéhradecký		230	
Czech Republic	Plzeňský		230	
Denmark	Århus		888	
Denmark	Bornholm		1290	
Denmark	Bornholm		345	2
Denmark	Bornholm	Hasle	1325	
Denmark	Fyn	Otterup	582	
Denmark	Ringkøbing	Ringkøbing	582	
Denmark	South Jutland		888	
Denmark	Storstrøm		888	
Denmark	Storstrøm	Fakse	582	
Denmark	Vejle		888	
Denmark	Vejle	Vejle	582	
Estonia	Harju		1317	
Estonia	Harju		377	2
Estonia	Hiiu		1317	
Estonia	Ida-Viru		1317	
Estonia	Ida-Viru		376	
Estonia	Ida-Viru		377	
Estonia	Järva		1317	
Estonia	Jõgeva		1317	
Estonia	Lääne		1317	
Estonia	Lääne		377	2
Estonia	Lääne-Viru		1317	
Estonia	Pärnu		1317	
Estonia	Põlva		1317	
Estonia	Rapla		1317	

Continued

Table A1.3. Continued.

Country	Administrative region 1	Administrative region 2	Reference ID	No. entries[a]
Estonia	Saare		1317	
Estonia	Tartu		1317	
Estonia	Tartu		1348	2
Estonia	Tartu		376	
Estonia	Valga		1317	
Estonia	Valga		376	
Estonia	Valga		377	
Estonia	Viljandi		1317	
Estonia	Võru		1317	
Finland			409	
Finland	Lapland	Lapland	1341	2
Finland	Southern Finland		409	
Finland	Southern Finland	Uusimaa	1329	
Finland	Western Finland		34	
Finland	Western Finland	Central Ostrobothnia	465	
France	Alsace		1285	
France	Bourgogne		1285	
France	Champagne-Ardenne		1285	
France	Franche-Comté		1285	
France	Lorraine		1285	
Germany	Baden-Württemberg		1058	
Germany	Baden-Württemberg		1327	
Germany	Baden-Württemberg		969	
Germany	Baden-Württemberg	Freiburg	968	
Germany	Bayern		1058	
Germany	Bayern		1327	
Germany	Bayern		969	
Germany	Bayern	Mittelfranken	1287	
Germany	Bayern	Niederbayern	1287	
Germany	Bayern	Niederbayern	968	
Germany	Bayern	Oberbayern	1287	
Germany	Bayern	Oberpfalz	1287	
Germany	Bayern	Unterfranken	439	
Germany	Brandenburg		1058	
Germany	Brandenburg	Brandenburg	424	
Germany	Hessen		1058	
Germany	Hessen		1327	
Germany	Hessen		968	
Germany	Hessen		969	
Germany	Mecklenburg-Vorpommern		1121	
Germany	Mecklenburg-Vorpommern		523	
Germany	Nordrhein-Westfalen		1064	
Germany	Rheinland-Pfalz		1058	
Germany	Rheinland-Pfalz		1327	
Germany	Rheinland-Pfalz		968	
Germany	Rheinland-Pfalz		969	
Germany	Rheinland-Pfalz	Rheinhessen-Pfalz	1059	
Germany	Sachsen		969	
Germany	Sachsen-Anhalt		1121	
Germany	Sachsen-Anhalt		523	
Germany	Thüringen		1121	
Germany	Thüringen		969	
Greece	Anatoliki Makedonia kai Thraki		736	

Continued

Table A1.3. Continued.

Country	Administrative region 1	Administrative region 2	Reference ID	No. entries[a]
Greece	Dytiki Makedonia		736	
Greece	Kentriki Makedonia		736	
Hungary	Borsod-Abaúj-Zemplén		1286	
Hungary	Fejér		784	
Hungary	Gyor-Moson-Sopron		784	
Hungary	Heves		1286	7
Hungary	Heves		784	
Hungary	Komárom-Esztergom		784	
Hungary	Nógrád		1286	8
Hungary	Nógrád		784	
Hungary	Somogy		784	
Hungary	Tolna		784	
Hungary	Vas		784	
Hungary	Veszprém		784	
Hungary	Zala		784	
Italy	Friuli-Venezia Giulia		336	
Italy	Piemonte	Torino	780	
Italy	Trentino-Alto Adige	Trento	1304	
Italy	Trentino-Alto Adige	Trento	160	
Italy	Trentino-Alto Adige	Trento	804	3
Italy	Trentino-Alto Adige	Trento	806	3
Italy	Veneto	Belluno	1304	
Italy	Veneto	Belluno	201	4
Latvia	Kurzeme	Dobele	968	
Latvia	Kurzeme	Kuldiga	1317	
Latvia	Kurzeme	Kuldiga	968	
Latvia	Kurzeme	Liepaja	1301	
Latvia	Kurzeme	Liepaja	968	
Latvia	Kurzeme	Saldus	968	
Latvia	Kurzeme	Tukums	968	
Latvia	Kurzeme	Ventspils	968	
Latvia	Latgale	Balvi	1317	
Latvia	Latgale	Balvi	968	
Latvia	Latgale	Daugavpils	968	
Latvia	Latgale	Jekabpils	968	
Latvia	Latgale	Kraslava	968	
Latvia	Latgale	Ludza	968	
Latvia	Latgale	Madona	968	
Latvia	Latgale	Preili	968	
Latvia	Latgale	Rezekne	968	
Latvia	Riga	Limbaži	1317	
Latvia	Riga	Riga	126	
Latvia	Riga	Riga	1301	
Latvia	Riga	Riga	968	2
Latvia	Vidzeme		1301	
Latvia	Vidzeme	Aizkraukle	968	
Latvia	Vidzeme	Aluksne	1317	
Latvia	Vidzeme	Aluksne	968	
Latvia	Vidzeme	Gulbene	968	
Latvia	Vidzeme	Limbaži	968	
Latvia	Vidzeme	Madona	968	
Latvia	Vidzeme	Valka	968	
Latvia	Vidzeme	Valmiera	968	

Continued

Table A1.3. Continued.

Country	Administrative region 1	Administrative region 2	Reference ID	No. entries[a]
Latvia	Zemgale	Aizkraukle	1317	
Latvia	Zemgale	Bauska	968	
Latvia	Zemgale	Jelgava	968	
Latvia	Zemgale	Ogre	968	
Liechtenstein			541	
Lithuania			489	
Lithuania	Alytaus		489	
Lithuania	Alytaus	Alytaus	1317	
Lithuania	Kauno		489	4
Lithuania	Kauno	Birštono	1317	
Lithuania	Klaipėdos	Klaipėdos	1317	
Lithuania	Marijampoles	Marijampoles	1317	
Lithuania	Panevėžio		413	
Lithuania	Panevėžio	Biržu	1317	
Lithuania	Šiauliai	Akmenes	1317	
Lithuania	Šiauliai		413	2
Lithuania	Taurages		489	2
Lithuania	Taurages	Jurbarko	1317	
Lithuania	Telšiai	Mažeikiu	1317	
Lithuania	Telšiai		489	
Lithuania	Utenos	Anykščiu	1317	
Lithuania	Vilniaus		489	2
Lithuania	Vilniaus	Šalčininku	1317	
Norway	Aust-Agder		887	
Norway	Aust-Agder	Arendal	209	
Norway	Aust-Agder	Arendal	210	
Norway	Sør-Trøndelag		1290	
Poland	Greater Poland		942	
Poland	Kuyavian-Pomeranian		942	
Poland	Lesser Poland		942	
Poland	Lódz		942	
Poland	Lower Silesian		942	
Poland	Lublin		193	
Poland	Lublin		942	
Poland	Lublin	Lublin	403	
Poland	Lublin	Radzyn	192	
Poland	Lubusz		942	
Poland	Masovian		942	
Poland	Opole		942	
Poland	Podlachian		532	
Poland	Podlachian		942	
Poland	Podlachian	Białystok	1350	2
Poland	Pomeranian		942	
Poland	Silesian		942	
Poland	Subcarpathian		942	
Poland	Świętokrzyskie		942	
Poland	Warmian-Masurian		942	
Poland	West Pomeranian		942	
Russia			1293	
Sweden	Blekinge		1054	
Sweden	Blekinge		1289	
Sweden	Blekinge	Karlskrona	1289	
Sweden	Blekinge	Olofström	1289	

Continued

Table A1.3. Continued.

Country	Administrative region 1	Administrative region 2	Reference ID	No. entries[a]
Sweden	Blekinge	Ronneby	1289	
Sweden	Skåne		483	
Sweden	Stockholm	Söderhamn	636	
Sweden	Stockholm	Söderhamn	636	
Sweden	Stockholm	Upplands-Bro	636	
Switzerland	Aargau		1328	
Switzerland	Bern		1328	
Switzerland	Bern		165	
Switzerland	Neuchâtel		1328	
Switzerland	Sant Gallen		1328	
Switzerland	Schaffhausen		1328	
Switzerland	Thurgau		1328	
Switzerland	Thurgau		541	
Switzerland	Thurgau		855	
Switzerland	Zürich		556	
Turkey	Tokat		308	
Turkey	Yozgat		1351	

[a]If more than one.

Table A1.4. Geographic distribution data for *Anaplasma* spp. and *Ehrlichia* spp. now included in *Anaplasma* spp. See Appendix 2 (hosted on http://www.cabi.org/vetmedresource/) for complete reference indicated by the Reference ID.

Species	Country	Administrative region 1	Administrative region 2	Reference ID	No. entries[a]
Anaplasma bovis	Italy	Basilicata	Potenza	156	
Anaplasma centrale	Egypt	Al Wadi al Jadid		1343	2
Anaplasma centrale	Italy	Basilicata	Matera	156	
Anaplasma centrale	Italy	Basilicata	Potenza	156	
Anaplasma centrale	Italy	Calabria	Crotone	156	
Anaplasma centrale	Italy	Calabria	Vibo Valentia	156	
Anaplasma centrale	Italy	Sicily	Palermo	366	2
Anaplasma marginale	Hungary	Borsod-Abaúj-Zemplén		443	2
Anaplasma marginale	Italy	Abruzzo	Bari	156	
Anaplasma marginale	Italy	Apulia		207	
Anaplasma marginale	Italy	Apulia	Foggia	156	
Anaplasma marginale	Italy	Basilicata		207	
Anaplasma marginale	Italy	Basilicata		996	
Anaplasma marginale	Italy	Basilicata	Matera	156	
Anaplasma marginale	Italy	Basilicata	Potenza	156	
Anaplasma marginale	Italy	Calabria		996	
Anaplasma marginale	Italy	Calabria	Crotone	156	
Anaplasma marginale	Italy	Calabria	Vibo Valentia	156	
Anaplasma marginale	Italy	Campania		207	
Anaplasma marginale	Italy	Campania		996	
Anaplasma marginale	Italy	Lazio		996	
Anaplasma marginale	Italy	Lombardia		996	
Anaplasma marginale	Italy	Marche		996	
Anaplasma marginale	Italy	Sicily		994	
Anaplasma marginale	Italy	Sicily		996	
Anaplasma marginale	Italy	Sicily	Agrigento	992	
Anaplasma marginale	Italy	Sicily	Caltanissetta	992	
Anaplasma marginale	Italy	Sicily	Catania	992	
Anaplasma marginale	Italy	Sicily	Enna	992	
Anaplasma marginale	Italy	Sicily	Messina	992	
Anaplasma marginale	Italy	Sicily	Palermo	242	
Anaplasma marginale	Italy	Sicily	Palermo	244	

Continued

Table A1.4. Continued.

Species	Country	Administrative region 1	Administrative region 2	Reference ID	No. entries[a]
Anaplasma marginale	Italy	Sicily	Palermo	366	2
Anaplasma marginale	Italy	Sicily	Palermo	992	
Anaplasma marginale	Italy	Sicily	Trapani	992	
Anaplasma marginale	Italy	Toscana		996	
Anaplasma marginale	Italy	Umbria		996	
Anaplasma marginale	Spain	Castilla-La Mancha	Ciudad Real	246	
Anaplasma marginale	Switzerland	Graubünden		1211	
Anaplasma marginale	Switzerland	Graubünden		270	
Anaplasma marginale	Switzerland	Graubünden		438	
Anaplasma ovis	Cyprus	Famagusta		775	2
Anaplasma ovis	Cyprus	Larnaca		775	2
Anaplasma ovis	Cyprus	Limassol		775	2
Anaplasma ovis	Cyprus	Nicosia		775	2
Anaplasma ovis	Cyprus	Paphos		775	2
Anaplasma ovis	Hungary	Borsod-Abaúj-Zemplén		443	2
Anaplasma ovis	Italy	Sicily		994	2
Anaplasma ovis	Italy	Sicily	Agrigento	992	2
Anaplasma ovis	Italy	Sicily	Caltanissetta	992	2
Anaplasma ovis	Italy	Sicily	Catania	992	2
Anaplasma ovis	Italy	Sicily	Enna	992	2
Anaplasma ovis	Italy	Sicily	Messina	992	2
Anaplasma ovis	Italy	Sicily	Palermo	242	
Anaplasma ovis	Italy	Sicily	Palermo	992	2
Anaplasma ovis	Italy	Sicily	Trapani	992	2
Anaplasma ovis	Romania	Tulcea		59	
Anaplasma ovis	Spain	Andalucía	Cádiz	241	
Anaplasma ovis	Turkey	Elâziğ		24	
Anaplasma phagocytophilum	Albania	Tiranë		409	
Anaplasma phagocytophilum	Belarus	Minsk		1336	
Anaplasma phagocytophilum	Bulgaria	Sofia		448	
Anaplasma phagocytophilum	Bulgaria	Sofia	Stolichna	185	
Anaplasma phagocytophilum	Croatia	Koprivnicko-Krizevacka		656	
Anaplasma phagocytophilum	Cyprus	Famagusta		1139	
Anaplasma phagocytophilum	Czech Republic	Jihomoravský		1338	3

Anaplasma phagocytophilum	Czech Republic	Moravskoslezský		1338	
Anaplasma phagocytophilum	Czech Republic	Středočský		1091	
Anaplasma phagocytophilum	Czech Republic	Středočský		1092	3
Anaplasma phagocytophilum	Denmark	Copenhagen		888	
Anaplasma phagocytophilum	Denmark	Frederiksborg		888	
Anaplasma phagocytophilum	Denmark	Fyn		888	
Anaplasma phagocytophilum	Denmark	North Jutland		888	
Anaplasma phagocytophilum	Denmark	South Jutland		1120	
Anaplasma phagocytophilum	Denmark	South Jutland		888	
Anaplasma phagocytophilum	Egypt	Al Jizah		1343	
Anaplasma phagocytophilum	Estonia	Pärnu		1348	3
Anaplasma phagocytophilum	Estonia	Saare		1336	
Anaplasma phagocytophilum	France	Alsace	Bas-Rhin	484	
Anaplasma phagocytophilum	France	Alsace	Bas-Rhin	719	
Anaplasma phagocytophilum	France	Aquitaine	Dordogne	484	
Anaplasma phagocytophilum	France	Aquitaine	Gironde	484	
Anaplasma phagocytophilum	France	Aquitaine	Landes	484	
Anaplasma phagocytophilum	France	Aquitaine	Lot-et-Garonne	484	
Anaplasma phagocytophilum	France	Aquitaine	Pyrénées-Atlantiques	484	2
Anaplasma phagocytophilum	France	Auvergne	Allier	484	
Anaplasma phagocytophilum	France	Auvergne	Cantal	484	
Anaplasma phagocytophilum	France	Auvergne	Haute-Loire	484	
Anaplasma phagocytophilum	France	Auvergne	Puy-de-Dôme	484	
Anaplasma phagocytophilum	France	Basse-Normandie	Calvados	484	
Anaplasma phagocytophilum	France	Basse-Normandie	Manche	484	
Anaplasma phagocytophilum	France	Basse-Normandie	Orne	484	
Anaplasma phagocytophilum	France	Bourgogne	Côte-d'Or	484	
Anaplasma phagocytophilum	France	Bourgogne	Nièvre	484	
Anaplasma phagocytophilum	France	Bourgogne	Saône-et-Loire	484	
Anaplasma phagocytophilum	France	Bourgogne	Saône-et-Loire	719	
Anaplasma phagocytophilum	France	Bourgogne	Yonne	484	
Anaplasma phagocytophilum	France	Bretagne	Côtes-d'Armor	484	
Anaplasma phagocytophilum	France	Bretagne	Finistère	484	
Anaplasma phagocytophilum	France	Bretagne	Ille-et-Vilaine	484	
Anaplasma phagocytophilum	France	Bretagne	Morbihan	484	
Anaplasma phagocytophilum	France	Centre	Cher	484	

Continued

Table A1.4. Continued.

Species	Country	Administrative region 1	Administrative region 2	Reference ID	No. entries[a]
Anaplasma phagocytophilum	France	Centre	Eure-et-Loir	484	
Anaplasma phagocytophilum	France	Centre	Eure-et-Loir	719	
Anaplasma phagocytophilum	France	Centre	Indre	484	
Anaplasma phagocytophilum	France	Centre	Indre-et-Loire	484	
Anaplasma phagocytophilum	France	Centre	Indre-et-Loire	719	
Anaplasma phagocytophilum	France	Centre	Loir-et-Cher	484	
Anaplasma phagocytophilum	France	Centre	Loiret	484	
Anaplasma phagocytophilum	France	Champagne-Ardenne	Ardennes	484	
Anaplasma phagocytophilum	France	Champagne-Ardenne	Haute-Marne	484	
Anaplasma phagocytophilum	France	Champagne-Ardenne	Marne	484	
Anaplasma phagocytophilum	France	Corse	Corse-du-Sud	484	
Anaplasma phagocytophilum	France	Franche-Comté	Doubs	484	
Anaplasma phagocytophilum	France	Franche-Comté	Haute-Saône	484	
Anaplasma phagocytophilum	France	Franche-Comté	Haute-Saône	719	
Anaplasma phagocytophilum	France	Franche-Comté	Jura	484	
Anaplasma phagocytophilum	France	Franche-Comté	Territoire de Belfort	484	
Anaplasma phagocytophilum	France	Haute-Normandie	Eure	484	
Anaplasma phagocytophilum	France	Haute-Normandie	Seine-Maritime	484	
Anaplasma phagocytophilum	France	Île-de-France	Essonne	719	
Anaplasma phagocytophilum	France	Île-de-France	Seine-et-Marne	484	
Anaplasma phagocytophilum	France	Île-de-France	Seine-et-Marne	719	
Anaplasma phagocytophilum	France	Île-de-France	Yvelines	484	
Anaplasma phagocytophilum	France	Languedoc-Roussillon	Bouches-du-Rhône	561	
Anaplasma phagocytophilum	France	Languedoc-Roussillon	Lozère	484	
Anaplasma phagocytophilum	France	Limousin	Corrèze	484	
Anaplasma phagocytophilum	France	Limousin	Corrèze	719	
Anaplasma phagocytophilum	France	Limousin	Creuse	484	
Anaplasma phagocytophilum	France	Limousin	Haute-Vienne	484	
Anaplasma phagocytophilum	France	Limousin	Haute-Vienne	719	
Anaplasma phagocytophilum	France	Lorraine	Meurthe-et-Moselle	484	
Anaplasma phagocytophilum	France	Lorraine	Meurthe-et-Moselle	719	
Anaplasma phagocytophilum	France	Lorraine	Meuse	484	
Anaplasma phagocytophilum	France	Lorraine	Meuse	484	
Anaplasma phagocytophilum	France	Lorraine	Moselle	484	

Anaplasma phagocytophilum	France	Lorraine	Moselle	719	
Anaplasma phagocytophilum	France	Lorraine	Vod	484	
Anaplasma phagocytophilum	France	Lorraine	Vosges	719	
Anaplasma phagocytophilum	France	Midi-Pyrénées	Aveyron	484	
Anaplasma phagocytophilum	France	Midi-Pyrénées	Haute-Garonne	484	
Anaplasma phagocytophilum	France	Midi-Pyrénées	Haute-Garonne	719	
Anaplasma phagocytophilum	France	Midi-Pyrénées	Hautes-Pyrénées	484	2
Anaplasma phagocytophilum	France	Midi-Pyrénées	Lot	484	
Anaplasma phagocytophilum	France	Midi-Pyrénées	Tarn	484	
Anaplasma phagocytophilum	France	Midi-Pyrénées	Tarn	719	
Anaplasma phagocytophilum	France	Midi-Pyrénées	Tarn-et-Garonne	484	
Anaplasma phagocytophilum	France	Nord-Pas-de-Calais	Nord	484	
Anaplasma phagocytophilum	France	Nord-Pas-de-Calais	Pas-De-Calais	484	
Anaplasma phagocytophilum	France	Nord-Pas-de-Calais	Pas-De-Calais	719	
Anaplasma phagocytophilum	France	Pays de la Loire	Loire-Atlantique	484	
Anaplasma phagocytophilum	France	Pays de la Loire	Maine-et-Loire	484	
Anaplasma phagocytophilum	France	Pays de la Loire	Maine-et-Loire	719	
Anaplasma phagocytophilum	France	Pays de la Loire	Mayenne	484	
Anaplasma phagocytophilum	France	Pays de la Loire	Sarthe	484	
Anaplasma phagocytophilum	France	Pays de la Loire	Vendée	484	
Anaplasma phagocytophilum	France	Picardie	Aisne	719	
Anaplasma phagocytophilum	France	Picardie	Oise	484	
Anaplasma phagocytophilum	France	Picardie	Oise	719	
Anaplasma phagocytophilum	France	Picardie	Somme	484	
Anaplasma phagocytophilum	France	Poitou-Charentes	Charente	484	
Anaplasma phagocytophilum	France	Poitou-Charentes	Deux-Sèvres	484	
Anaplasma phagocytophilum	France	Poitou-Charentes	Vienne	484	
Anaplasma phagocytophilum	France	Provence-Alpes-Côte-d'Azur	Alpes-Maritimes	484	
Anaplasma phagocytophilum	France	Provence-Alpes-Côte-d'Azur	Bouches Du-Rhône	484	
Anaplasma phagocytophilum	France	Provence-Alpes-Côte-d'Azur	Bouches-Du-Rhône	719	
Anaplasma phagocytophilum	France	Provence-Alpes-Côte-d'Azur	Var	484	
Anaplasma phagocytophilum	France	Provence-Alpes-Côte-d'Azur	Var	719	
Anaplasma phagocytophilum	France	Provence-Alpes-Côte-d'Azur	Vaucluse	484	
Anaplasma phagocytophilum	France	Rhône-Alpes	Ain	484	
Anaplasma phagocytophilum	France	Rhône-Alpes	Haute-Savoie	484	
Anaplasma phagocytophilum	France	Rhône-Alpes	Isère	484	

Continued

Table A1.4. Continued.

Species	Country	Administrative region 1	Administrative region 2	Reference ID	No. entries[a]
Anaplasma phagocytophilum	France	Rhône-Alpes	Loire	484	
Anaplasma phagocytophilum	France	Rhône-Alpes	Rhône	1032	
Anaplasma phagocytophilum	France	Rhône-Alpes	Rhône	484	
Anaplasma phagocytophilum	France	Rhône-Alpes	Savoie	484	
Anaplasma phagocytophilum	Germany	Baden-Württemberg		718	
Anaplasma phagocytophilum	Germany	Bayern		1352	
Anaplasma phagocytophilum	Germany	Berlin		540	
Anaplasma phagocytophilum	Germany	Berlin	Berlin	1294	
Anaplasma phagocytophilum	Germany	Brandenburg		540	
Anaplasma phagocytophilum	Germany	Thüringen		1352	
Anaplasma phagocytophilum	Greece	Kriti		1108	
Anaplasma phagocytophilum	Italy	Basilicata	Potenza	156	
Anaplasma phagocytophilum	Italy	Friuli-Venezia Giulia		191	8
Anaplasma phagocytophilum	Italy	Friuli-Venezia Giulia		639	
Anaplasma phagocytophilum	Italy	Friuli-Venezia Giulia		93	2
Anaplasma phagocytophilum	Italy	Lazio		832	
Anaplasma phagocytophilum	Italy	Lazio		833	
Anaplasma phagocytophilum	Italy	Sardegna		28	2
Anaplasma phagocytophilum	Italy	Sardegna		619	
Anaplasma phagocytophilum	Italy	Sicily		999	
Anaplasma phagocytophilum	Italy	Sicily	Agrigento	992	4
Anaplasma phagocytophilum	Italy	Sicily	Caltanissetta	992	4
Anaplasma phagocytophilum	Italy	Sicily	Catania	992	
Anaplasma phagocytophilum	Italy	Sicily	Enna	992	
Anaplasma phagocytophilum	Italy	Sicily	Messina	992	
Anaplasma phagocytophilum	Italy	Sicily	Palermo	242	4
Anaplasma phagocytophilum	Italy	Sicily	Palermo	243	
Anaplasma phagocytophilum	Italy	Sicily	Palermo	992	4
Anaplasma phagocytophilum	Italy	Sicily	Ragusa	992	
Anaplasma phagocytophilum	Italy	Sicily	Syracuse	992	
Anaplasma phagocytophilum	Italy	Sicily	Trapani	992	4
Anaplasma phagocytophilum	Italy	Trentino-Alto Adige	Trento	158	
Anaplasma phagocytophilum	Italy	Trentino-Alto Adige	Trento	806	4
Anaplasma phagocytophilum	Lithuania	Klaipedos		1336	

Anaplasma phagocytophilum	Moldova			1134	
Anaplasma phagocytophilum	Moldova	Chisinau		528	
Anaplasma phagocytophilum	Moldova	Chisinau		666	
Anaplasma phagocytophilum	Morocco	Taza-Al Hoceima-Taounate		643	
Anaplasma phagocytophilum	Norway	Aust-Agder		1323	2
Anaplasma phagocytophilum	Norway	Sør-Trøndelag	Hitra	1323	
Anaplasma phagocytophilum	Norway	Telemark	Kragerø	1323	
Anaplasma phagocytophilum	Poland	Greater Poland		1131	4
Anaplasma phagocytophilum	Poland	Lublin		1062	
Anaplasma phagocytophilum	Poland	Lublin		11	
Anaplasma phagocytophilum	Poland	Lublin		179	6
Anaplasma phagocytophilum	Poland	Lublin		990	
Anaplasma phagocytophilum	Poland	Lublin	Lublin	194	2
Anaplasma phagocytophilum	Poland	Masovian		1119	
Anaplasma phagocytophilum	Poland	Masovian	Warsaw	1098	
Anaplasma phagocytophilum	Poland	Podlachian		1129	
Anaplasma phagocytophilum	Poland	Podlachian	Augustów	0	2
Anaplasma phagocytophilum	Poland	Podlachian	Hajnówka	385	
Anaplasma phagocytophilum	Poland	Podlachian	Hajnówka	386	
Anaplasma phagocytophilum	Poland	Pomeranian		931	
Anaplasma phagocytophilum	Poland	Warmian-Masurian	Białystok	0	
Anaplasma phagocytophilum	Poland	West Pomeranian		822	
Anaplasma phagocytophilum	Portugal	Faro		836	
Anaplasma phagocytophilum	Portugal	Madeira		1143	
Anaplasma phagocytophilum	Portugal	Madeira		238	
Anaplasma phagocytophilum	Portugal	Madeira		838	
Anaplasma phagocytophilum	Russia	Vologda		35	
Anaplasma phagocytophilum	Serbia	Branicevski		652	
Anaplasma phagocytophilum	Serbia	Grad Beograd		652	
Anaplasma phagocytophilum	Serbia	Macvanski		652	
Anaplasma phagocytophilum	Serbia	Ni'avski		652	
Anaplasma phagocytophilum	Serbia	Pomoravski		652	
Anaplasma phagocytophilum	Serbia	Ra'ki		652	
Anaplasma phagocytophilum	Serbia	Sremski		652	
Anaplasma phagocytophilum	Serbia	Zajecarski		652	
Anaplasma phagocytophilum	Slovakia	Banskobystrický		529	6

Continued

Table A1.4. Continued.

Species	Country	Administrative region 1	Administrative region 2	Reference ID	No. entries[a]
Anaplasma phagocytophilum	Slovakia	Banskobystrický		941	12
Anaplasma phagocytophilum	Slovakia	Bratislavský		1349	
Anaplasma phagocytophilum	Slovakia	Košický		256	
Anaplasma phagocytophilum	Slovakia	Košický		499	
Anaplasma phagocytophilum	Slovakia	Košický		936	2
Anaplasma phagocytophilum	Slovakia	Nitrianský		1349	
Anaplasma phagocytophilum	Slovakia	Presov		936	
Anaplasma phagocytophilum	Slovakia	Zilinsky		1347	
Anaplasma phagocytophilum	Slovenia			191	
Anaplasma phagocytophilum	Spain	Andalucía	Cádiz	241	
Anaplasma phagocytophilum	Spain	Castilla y León		327	2
Anaplasma phagocytophilum	Spain	Comunidad de Madrid	Madrid	18	2
Anaplasma phagocytophilum	Spain	Galicia		360	
Anaplasma phagocytophilum	Spain	Galicia	Ourense	52	3
Anaplasma phagocytophilum	Spain	Galicia	Ourense	53	
Anaplasma phagocytophilum	Spain	Galicia	Pontevedra	52	4
Anaplasma phagocytophilum	Spain	Galicia	Pontevedra	53	
Anaplasma phagocytophilum	Spain	La Rioja	La Rioja	701	2
Anaplasma phagocytophilum	Spain	La Rioja	La Rioja	773	
Anaplasma phagocytophilum	Spain	País Vasco		78	2
Anaplasma phagocytophilum	Spain	País Vasco		79	4
Anaplasma phagocytophilum	Switzerland	Bern		586	3
Anaplasma phagocytophilum	Switzerland	Graubünden		1211	
Anaplasma phagocytophilum	Switzerland	Graubünden		438	
Anaplasma phagocytophilum	Switzerland	Neuchâtel		586	3
Anaplasma phagocytophilum	Tunisia	Jendouba		643	
Anaplasma phagocytophilum	Turkey	Giresun		26	
Anaplasma phagocytophilum	Turkey	Rize		26	
Anaplasma phagocytophilum	Turkey	Trabzon		26	
Anaplasma phagocytophilum	Ukraine	Kharkiv		1134	
Anaplasma phagocytophilum	UK	England	Cheshire	134	
Anaplasma phagocytophilum	UK	England	Hampshire	808	
Anaplasma phagocytophilum	UK	England	Northumberland	131	
Anaplasma platys	Italy	Abruzzo		912	

Anaplasma platys	Italy	Sicily		994	
Anaplasma platys	Italy	Sicily		999	
Anaplasma platys	Italy	Sicily	Palermo	245	
Anaplasma platys	Tunisia	Jendouba		643	
Anaplasma platys	Turkey	Muğla		1012	
Anaplasma spp.	Albania	Durrës	Durrësit	185	
Anaplasma spp.	Albania	Kukës	Kukësit	185	2
Anaplasma spp.	Cyprus			1118	
Anaplasma spp.	Germany	Berlin	Berlin	1330	4
Anaplasma spp.	Slovakia	Bratislavsky		124	
Anaplasma spp.	Slovakia	Nitriansky		124	
Anaplasma spp.	Turkey	Antalya		185	2
Anaplasma spp.	Turkey	Kayseri		185	3
Anaplasma spp.	Turkey	Malatya		185	

[a]If more than one.

Table A1.5. Geographic distribution data for *Rickettsia* spp. See Appendix 2 (hosted on http://www.cabi.org/vetmedresource/) for complete reference indicated by the Reference ID.

Species	Country	Administrative region 1	Administrative region 2	Reference ID	No. entries[a]
Rickettsia aeschlimanni	Algeria			110	4
Rickettsia aeschlimanni	Croatia	Splitsko-Dalmatinska		781	
Rickettsia aeschlimanni	Egypt	Shamal Sina'		1343	
Rickettsia aeschlimanni	France	Corse	Corse-du-Sud	624	
Rickettsia aeschlimanni	Russia	Stavropol		1144	
Rickettsia aeschlimanni	Spain	Castilla y León		327	
Rickettsia aeschlimanni	Spain	Castilla y León		327	
Rickettsia aeschlimanni	Spain	Castilla y León		328	6
Rickettsia aeschlimanni	Spain	La Rioja	La Rioja	700	
Rickettsia aeschlimanni	Spain	La Rioja	La Rioja	772	
Rickettsia conorii	Albania	Durrës	Durrësit	185	2
Rickettsia conorii	Albania	Kukës	Kukësit	185	3
Rickettsia conorii	Algeria			110	
Rickettsia conorii	Algeria	Oran		664	
Rickettsia conorii	France	Languedoc-Roussillon	Hérault	1292	
Rickettsia conorii	France	Provence-Alpes-Côte-d'Azur	Var	1299	
Rickettsia conorii	Greece	Anatoliki Makedonia kai Thraki		720	
Rickettsia conorii	Greece	Anatoliki Makedonia kai Thraki	Drama	222	
Rickettsia conorii	Greece	Anatoliki Makedonia kai Thraki	Kavala	222	
Rickettsia conorii	Greece	Dytiki Makedonia	Florina	222	
Rickettsia conorii	Greece	Dytiki Makedonia	Grevena	222	
Rickettsia conorii	Greece	Dytiki Makedonia	Kastoria	222	
Rickettsia conorii	Greece	Dytiki Makedonia	Kozani	222	
Rickettsia conorii	Greece	Kentriki Makedonia	Imathia	222	
Rickettsia conorii	Greece	Kentriki Makedonia	Khalkidiki	222	
Rickettsia conorii	Greece	Kentriki Makedonia	Kilkis	222	
Rickettsia conorii	Greece	Kentriki Makedonia	Pella	222	
Rickettsia conorii	Greece	Kentriki Makedonia	Pieria	222	
Rickettsia conorii	Greece	Kentriki Makedonia	Serrai	222	
Rickettsia conorii	Greece	Kentriki Makedonia	Thessaloniki	222	
Rickettsia conorii	Greece	Kriti		367	
Rickettsia conorii	Greece	Stereá Elláda	Fokis	778	
Rickettsia conorii	Greece	Thessalia	Trikala	222	

Rickettsia conorii	Israel			416	4
Rickettsia conorii	Italy	Piemonte	Cuneo	607	4
Rickettsia conorii	Italy	Piemonte	Cuneo	608	4
Rickettsia conorii	Italy	Sicily		999	2
Rickettsia conorii	Kosovo	Kosovska Mitrovica	Kosovska Mitrovica	346	2
Rickettsia conorii	Kosovo	Kosovska Mitrovica	Kosovska Mitrovica	827	
Rickettsia conorii	Kosovo	Pecki	Pec	827	
Rickettsia conorii	Morocco	Grand Casablanca		1104	
Rickettsia conorii	Serbia	Grad Beograd	Stari Grad	827	
Rickettsia conorii	Serbia	Pcinjski	Vranje	827	
Rickettsia conorii	Serbia	Raski	Tutin	827	
Rickettsia conorii	Serbia	Srednje-Banatski	Zrenjanin	827	
Rickettsia conorii	Serbia	Zlatiborski	Uzice	827	
Rickettsia conorii	Spain	Andalucía	Sevilla	95	
Rickettsia conorii	Spain	Castilla y León	Soria	681	
Rickettsia conorii	Spain	Cataluña	Barcelona	907	
Rickettsia conorii	Spain	Cataluña	Tarragona	907	
Rickettsia conorii	Spain	Galicia	Ourense	53	
Rickettsia conorii	Spain	Galicia	Pontevedra	53	
Rickettsia conorii	Spain	Islas Baleares		907	
Rickettsia conorii	Spain	La Rioja	La Rioja	1295	
Rickettsia conorii	Turkey	Antalya		185	
Rickettsia conorii	Turkey	Edirne		1122	
Rickettsia conorii	Turkey	Kayseri		185	3
Rickettsia conorii	Turkey	Kırklareli		1122	
Rickettsia conorii	Turkey	Malatya		185	
Rickettsia conorii	Turkey	Tekirdağ		1122	
Rickettsia conorii israelensis (*Rickettsia conorii* Israeli spotted fever strain)	Italy	Sicily		369	
Rickettsia helvetica	Albania	Durrës	Durrësit	185	2
Rickettsia helvetica	Albania	Kukës	Kukësit	185	3
Rickettsia helvetica	Austria	Burgenland		265	
Rickettsia helvetica	Bulgaria	Sofia	Stolichna	185	
Rickettsia helvetica	Croatia	Medimurska		264	
Rickettsia helvetica	Denmark	North Jutland		0	

Continued

Table A1.5. Continued.

Species	Country	Administrative region 1	Administrative region 2	Reference ID	No. entries[a]
Rickettsia helvetica	Denmark	Vestsjælland		1146	
Rickettsia helvetica	Denmark	Vestsjælland	Korsør	973	
Rickettsia helvetica	France	Auvergne	Puy-de-Dôme	727	
Rickettsia helvetica	Germany	Bayern		1063	
Rickettsia helvetica	Germany	Berlin	Berlin	1294	
Rickettsia helvetica	Italy	Toscana		1281	
Rickettsia helvetica	Italy	Trentino-Alto Adige	Trento	1281	
Rickettsia helvetica	Italy	Veneto		1281	
Rickettsia helvetica	Moldova			1134	
Rickettsia helvetica	Netherlands	Flevoland		917	
Rickettsia helvetica	Netherlands	Friesland		917	
Rickettsia helvetica	Netherlands	Noord-Holland		917	
Rickettsia helvetica	Netherlands	Overijssel		917	
Rickettsia helvetica	Poland	Greater Poland		1145	3
Rickettsia helvetica	Poland	Greater Poland		932	7
Rickettsia helvetica	Poland	Lesser Poland		932	3
Rickettsia helvetica	Poland	Pomeranian		932	2
Rickettsia helvetica	Slovakia	Banskobystrický		941	12
Rickettsia helvetica	Spain	Andalucía	Granada	1127	
Rickettsia helvetica	Spain	Andalucía	Huelva	1127	
Rickettsia helvetica	Spain	Andalucía	Jaén	611	
Rickettsia helvetica	Switzerland	Zürich		125	2
Rickettsia helvetica	Tunisia	Jendouba		0	
Rickettsia helvetica	Turkey	Antalya		185	2
Rickettsia helvetica	Turkey	Kayseri		185	2
Rickettsia helvetica	Turkey	Malatya		185	
Rickettsia helvetica	Ukraine	Kharkiv		1134	
Rickettsia massiliae	Algeria			110	3
Rickettsia massiliae	Greece	Stereá Elláda	Fokis	778	
Rickettsia massiliae	Morocco	Gharb-Chrarda-Béni Hssen		1104	
Rickettsia massiliae	Spain	Andalucía	Granada	1127	
Rickettsia massiliae	Spain	Andalucía	Jaén	611	3
Rickettsia massiliae	Spain	Andalucía	Sevilla	614	
Rickettsia massiliae	Spain	La Rioja	La Rioja	700	

Rickettsia massiliae	Switzerland	Ticino		96	
Rickettsia monacensis	Algeria	El Tarf		1111	
Rickettsia monacensis	Germany	Bayern		1112	
Rickettsia monacensis	Germany	Bayern		878	
Rickettsia monacensis	Moldova			1134	
Rickettsia monacensis	Morocco	Taza-Al Hoceima-Taounate		1104	
Rickettsia monacensis	Portugal	Madeira		238	
Rickettsia monacensis	Spain	Andalucía	Granada	1127	2
Rickettsia monacensis	Spain	Andalucía	Huelva	1127	2
Rickettsia monacensis	Spain	Andalucía	Jaén	611	2
Rickettsia monacensis	Spain	La Rioja		468	
Rickettsia monacensis	Spain	País Vasco		468	
Rickettsia monacensis	Tunisia	Jendouba		0	
Rickettsia monacensis	Ukraine	Kharkiv		1134	
Rickettsia mongolotimonae[b]	France	Provence-Alpes-Côte-d'Azur	Bouches-du-Rhône	348	
Rickettsia mongolotimonae	Greece	Kriti		776	2
Rickettsia raoultii	Germany	Baden-Württemberg	Freiburg	1137	
Rickettsia raoultii	Germany	Berlin	Berlin	1330	
Rickettsia raoultii	Italy	Toscana	Lucca	1332	
Rickettsia raoultii	Italy	Toscana	Lucca	862	
Rickettsia raoultii	Poland	Podlachian		1129	
Rickettsia raoultii	Spain	Andalucía	Granada	1127	
Rickettsia raoultii	Spain	Andalucía	Huelva	1127	2
Rickettsia raoultii	Spain	Andalucía	Jaén	611	2
Rickettsia rhipicephali	Greece	Stereá Elláda	Fokis	778	
Rickettsia slovaca	Croatia	Medimurska		264	
Rickettsia slovaca	Croatia	Splitsko-Dalmatinska		781	
Rickettsia slovaca	France	Bretagne		1130	
Rickettsia slovaca	Germany	Bayern		1137	
Rickettsia slovaca	Greece	Kentriki Makedonia	Khalkidiki	1303	
Rickettsia slovaca	Italy	Toscana	Lucca	1332	3
Rickettsia slovaca	Italy	Toscana	Lucca	862	
Rickettsia slovaca	Spain	Andalucía	Granada	1127	
Rickettsia slovaca	Spain	Andalucía	Jaén	611	
Rickettsia slovaca	Spain	Castilla y León		330	2
Rickettsia slovaca	Spain	Castilla y León	Burgos	587	

Continued

Table A1.5. Continued.

Species	Country	Administrative region 1	Administrative region 2	Reference ID	No. entries[a]
Rickettsia slovaca	Spain	Castilla y León	Palencia	587	
Rickettsia slovaca	Spain	La Rioja	La Rioja	1295	
Rickettsia slovaca	Spain	La Rioja	La Rioja	700	
Rickettsia slovaca	Switzerland	Ticino		96	
Rickettsia spp.	Albania	Tiranë		409	
Rickettsia spp.	Cyprus			1118	
Rickettsia spp.	Germany	Bayern		236	
Rickettsia spp.	Germany	Bayern		236	
Rickettsia spp.	Germany	Berlin	Berlin	1330	5
Rickettsia spp.	Germany	Brandenburg		236	2
Rickettsia spp.	Germany	Sachsen		236	2
Rickettsia spp.	Germany	Sachsen-Anhalt		236	
Rickettsia spp.	Italy	Friuli-Venezia Giulia		337	
Rickettsia spp.	Italy	Veneto	Belluno	752	
Rickettsia spp.	Poland	Podlachian		928	4
Rickettsia spp.	Poland	Warmian-Masurian		928	
Rickettsia spp.	Portugal	Madeira		238	
Rickettsia spp.	Russia	Vologda		35	
Rickettsia spp.	Slovakia	Banskobystrický		124	
Rickettsia spp.	Slovakia	Banskobystrický		529	15
Rickettsia spp.	Slovakia	Bratislavský		124	
Rickettsia spp.	Slovakia	Nitrianský		124	3
Rickettsia spp.	Spain	Andalucía	Granada	1127	2
Rickettsia spp.	Spain	País Vasco		79	

[a]If more than one.
[b]*Rickettsia mongolotimonae* is the name first proposed for this species. It was later identified as a member of the *R. sibirica* species complex, and is now called both *R. sibirica mongolotimonae* and *R. sibirica mongolitimonae*.

Table A1.6. Geographic distribution data for *Borrelia* spp. See Appendix 2 (hosted on http://www.cabi.org/vetmedresource/) for complete reference indicated by the Reference ID.

Species	Country	Administrative region 1	Administrative region 2	Reference ID	No. entries[a]
Borrelia afzelii	Austria	Kärnten		114	
Borrelia afzelii	Austria	Niederösterreich		511	8
Borrelia afzelii	Austria	Oberösterreich		114	2
Borrelia afzelii	Austria	Steiermark		114	
Borrelia afzelii	Austria	Steiermark		963	
Borrelia afzelii	Austria	Tirol		114	2
Borrelia afzelii	Austria	Vorarlberg		114	2
Borrelia afzelii	Austria	Wien		511	4
Borrelia afzelii	Bulgaria	Sofia	Stolichna	185	
Borrelia afzelii	Croatia	Primorsko-Goranska		1011	
Borrelia afzelii	Czech Republic	Jihočeský		224	2
Borrelia afzelii	Czech Republic	Jihočeský		255	
Borrelia afzelii	Czech Republic	Jihomoravský		550	6
Borrelia afzelii	Czech Republic	Jihomoravský		742	
Borrelia afzelii	Czech Republic	Královéhradecký		225	
Borrelia afzelii	Czech Republic	Královéhradecký		230	
Borrelia afzelii	Czech Republic	Liberecký		271	8
Borrelia afzelii	Czech Republic	Moravskoslezský		220	
Borrelia afzelii	Czech Republic	Plzeňský		230	
Borrelia afzelii	Czech Republic	Ústecký		550	5
Borrelia afzelii	Czech Republic	Zlínský		550	3
Borrelia afzelii	Denmark	North Jutland		1026	
Borrelia afzelii	Finland	Western Finland		34	
Borrelia afzelii	France	Alsace	Haut-Rhin	331	
Borrelia afzelii	France	Île-de-France	Yvelines	753	
Borrelia afzelii	Germany	Baden-Württemberg	Tübingen	794	
Borrelia afzelii	Germany	Bayern		332	2
Borrelia afzelii	Germany	Bayern		333	5
Borrelia afzelii	Germany	Berlin	Berlin	1294	
Borrelia afzelii	Germany	Niedersachsen	Hannover	431	
Borrelia afzelii	Germany	Nordrhein-Westfalen		600	

Continued

Table A1.6. Continued.

Species	Country	Administrative region 1	Administrative region 2	Reference ID	No. entries[a]
Borrelia afzelii	Hungary	Pest		344	3
Borrelia afzelii	Ireland	Galway		380	2
Borrelia afzelii	Ireland	Kerry		754	
Borrelia afzelii	Italy	Friuli-Venezia Giulia		639	
Borrelia afzelii	Italy	Lazio		831	
Borrelia afzelii	Italy	Lazio		833	
Borrelia afzelii	Italy	Toscana		1331	
Borrelia afzelii	Italy	Toscana	Pisa	98	
Borrelia afzelii	Italy	Trentino-Alto Adige	Trento	609	3
Borrelia afzelii	Italy	Trentino-Alto Adige	Trento	741	
Borrelia afzelii	Latvia	Riga		319	3
Borrelia afzelii	Moldova	Chisinau		528	
Borrelia afzelii	Moldova	Chisinau		666	
Borrelia afzelii	Netherlands	Flevoland		1291	
Borrelia afzelii	Netherlands	Gelderland	Arnhem	365	
Borrelia afzelii	Netherlands	Noord-Holland		1050	
Borrelia afzelii	Netherlands	Overijssel		1050	
Borrelia afzelii	Norway	Aust-Agder		1323	2
Borrelia afzelii	Norway	Sør-Trøndelag	Hitra	1323	
Borrelia afzelii	Norway	Telemark		475	2
Borrelia afzelii	Norway	Telemark	Kragerø	1323	
Borrelia afzelii	Poland	Greater Poland		603	
Borrelia afzelii	Poland	Lublin	Lublin City	194	
Borrelia afzelii	Poland	Lublin	Lublin City	1337	
Borrelia afzelii	Poland	Silesian	Tarnowskie	1288	
Borrelia afzelii	Poland	Warmian-Masurian		384	
Borrelia afzelii	Poland	Warmian-Masurian		738	2
Borrelia afzelii	Poland	Warmian-Masurian		739	2
Borrelia afzelii	Russia	Vologda		35	
Borrelia afzelii	Serbia	Branicevski		652	
Borrelia afzelii	Serbia	Grad Beograd		652	
Borrelia afzelii	Serbia	Macvanski		652	
Borrelia afzelii	Serbia	Ni'avski		652	

Borrelia afzelii	Serbia	Pomoravski		652	
Borrelia afzelii	Serbia	Ra'ki		652	
Borrelia afzelii	Serbia	Sremski		652	
Borrelia afzelii	Serbia	Zajecarski		652	
Borrelia afzelii	Slovakia			1320	
Borrelia afzelii	Slovakia	Bratislavský		1349	
Borrelia afzelii	Slovakia	Košický		602	2
Borrelia afzelii	Slovakia	Nitrianský		1349	
Borrelia afzelii	Slovakia	Trencianský		1349	
Borrelia afzelii	Slovakia	Trencianský		603	
Borrelia afzelii	Slovakia	Zilinský		1347	
Borrelia afzelii	Spain	La Rioja	La Rioja	315	
Borrelia afzelii	Spain	País Vasco		78	
Borrelia afzelii	Sweden	Gävleborg		350	
Borrelia afzelii	Sweden	Kalmar		350	
Borrelia afzelii	Sweden	Skåne		350	
Borrelia afzelii	Sweden	Skåne		696	
Borrelia afzelii	Sweden	Stockholm	Nynäshamn	470	
Borrelia afzelii	Sweden	Västerbotten	Umeå	470	
Borrelia afzelii	Switzerland	Graubünden		163	
Borrelia afzelii	Switzerland	Neuchâtel		143	
Borrelia afzelii	Switzerland	Neuchâtel		163	
Borrelia afzelii	Switzerland	Neuchâtel		368	
Borrelia afzelii	Switzerland	Ticino		1284	
Borrelia afzelii	Switzerland	Ticino		163	16
Borrelia afzelii	Switzerland	Valais		1284	
Borrelia afzelii	Switzerland	Valais		163	2
Borrelia afzelii	Turkey	Çankiri		396	
Borrelia afzclii	Turkey	Istanbul		395	
Borrelia afzelii	Turkey	Ordu		396	
Borrelia burgdorferi s.l.	Austria	Burgenland		114	2
Borrelia burgdorferi s.l.	Austria	Burgenland		258	
Borrelia burgdorferi s.l.	Austria	Burgenland	Eisenstadt	172	
Borrelia burgdorferi s.l.	Austria	Burgenland	Eisenstadt Umgebung	172	
Borrelia burgdorferi s.l.	Austria	Burgenland	Güssing	172	

Continued

Table A1.6. Continued.

Species	Country	Administrative region 1	Administrative region 2	Reference ID	No. entries[a]
Borrelia burgdorferi s.l.	Austria	Burgenland	Jennersdorf	172	
Borrelia burgdorferi s.l.	Austria	Burgenland	Mattersburg	172	
Borrelia burgdorferi s.l.	Austria	Burgenland	Neusiedl am See	172	
Borrelia burgdorferi s.l.	Austria	Burgenland	Oberpullendorf	172	
Borrelia burgdorferi s.l.	Austria	Burgenland	Oberwart	172	
Borrelia burgdorferi s.l.	Austria	Kärnten		114	2
Borrelia burgdorferi s.l.	Austria	Niederösterreich		114	2
Borrelia burgdorferi s.l.	Austria	Oberösterreich		114	2
Borrelia burgdorferi s.l.	Austria	Salzburg		114	2
Borrelia burgdorferi s.l.	Austria	Steiermark		258	
Borrelia burgdorferi s.l.	Austria	Tirol		114	2
Borrelia burgdorferi s.l.	Austria	Vorarlberg		114	2
Borrelia burgdorferi s.l.	Austria	Wien		114	2
Borrelia burgdorferi s.l.	Belgium	Antwerp		580	
Borrelia burgdorferi s.l.	Belgium	Brussels		580	
Borrelia burgdorferi s.l.	Belgium	East Flanders		580	
Borrelia burgdorferi s.l.	Belgium	East Flanders		668	
Borrelia burgdorferi s.l.	Belgium	Flemish Brabant		580	
Borrelia burgdorferi s.l.	Belgium	Flemish Brabant		615	
Borrelia burgdorferi s.l.	Belgium	Hainaut		580	
Borrelia burgdorferi s.l.	Belgium	Liege		580	
Borrelia burgdorferi s.l.	Belgium	Limburg		580	
Borrelia burgdorferi s.l.	Belgium	Luxembourg		580	
Borrelia burgdorferi s.l.	Belgium	Namur		580	
Borrelia burgdorferi s.l.	Belgium	Namur	Namur	1298	4
Borrelia burgdorferi s.l.	Belgium	Walloon Brabant		1333	
Borrelia burgdorferi s.l.	Belgium	Walloon Brabant		580	
Borrelia burgdorferi s.l.	Belgium	West Flanders		580	
Borrelia burgdorferi s.l.	Belgium	West Flanders		668	
Borrelia burgdorferi s.l.	Bulgaria			184	
Borrelia burgdorferi s.l.	Bulgaria	Sofia		448	
Borrelia burgdorferi s.l.	Bulgaria	Sofia	Stolichna	1319	2
Borrelia burgdorferi s.l.	Bulgaria	Stara Zagora		1007	2

Borrelia burgdorferi s.l.	Croatia	Koprivnicko-Krizevacka		656	
Borrelia burgdorferi s.l.	Czech Republic	Jihočeský		491	
Borrelia burgdorferi s.l.	Czech Republic	Jihomoravský		1077	
Borrelia burgdorferi s.l.	Czech Republic	Jihomoravský		1338	4
Borrelia burgdorferi s.l.	Czech Republic	Jihomoravský		490	2
Borrelia burgdorferi s.l.	Czech Republic	Jihomoravský		491	
Borrelia burgdorferi s.l.	Czech Republic	Jihomoravský		669	
Borrelia burgdorferi s.l.	Czech Republic	Jihomoravský	Brno	472	
Borrelia burgdorferi s.l.	Czech Republic	Kraj Vysočina		491	
Borrelia burgdorferi s.l.	Czech Republic	Královéhradeck		491	
Borrelia burgdorferi s.l.	Czech Republic	Libereck		491	
Borrelia burgdorferi s.l.	Czech Republic	Libereck	Liberec	472	
Borrelia burgdorferi s.l.	Czech Republic	Moravskoslezský		1035	3
Borrelia burgdorferi s.l.	Czech Republic	Moravskoslezský		491	
Borrelia burgdorferi s.l.	Czech Republic	Moravskoslezský		1338	
Borrelia burgdorferi s.l.	Czech Republic	Olomouck		491	
Borrelia burgdorferi s.l.	Czech Republic	Pardubický		491	
Borrelia burgdorferi s.l.	Czech Republic	Pardubický	Ústí nad Orlicí	472	
Borrelia burgdorferi s.l.	Czech Republic	Plzeňský		491	
Borrelia burgdorferi s.l.	Czech Republic	Prague		669	
Borrelia burgdorferi s.l.	Czech Republic	Ústecký		550	4
Borrelia burgdorferi s.l.	Czech Republic	Zlínsk		491	
Borrelia burgdorferi s.l.	Czech Republic	Zlínsk		550	4
Borrelia burgdorferi s.l.	Denmark	Copenhagen		888	
Borrelia burgdorferi s.l.	Denmark	Frederiksborg		888	
Borrelia burgdorferi s.l.	Denmark	Fyn		888	
Borrelia burgdorferi s.l.	Denmark	North Jutland		888	
Borrelia burgdorferi s.l.	Denmark	North Jutland		946	
Borrelia burgdorferi s.l.	Denmark	South Jutland		888	
Borrelia burgdorferi s.l.	Finland	Western Finland		34	2
Borrelia burgdorferi s.l.	France	Alsace		1285	
Borrelia burgdorferi s.l.	France	Alsace		572	
Borrelia burgdorferi s.l.	France	Alsace		851	
Borrelia burgdorferi s.l.	France	Alsace	Bas-Rhin	719	
Borrelia burgdorferi s.l.	France	Aquitaine		572	
Borrelia burgdorferi s.l.	France	Basse-Normandie		572	

Continued

Table A1.6. Continued.

Species	Country	Administrative region 1	Administrative region 2	Reference ID	No. entries[a]
Borrelia burgdorferi s.l.	France	Bourgogne		1285	
Borrelia burgdorferi s.l.	France	Bourgogne		572	
Borrelia burgdorferi s.l.	France	Bretagne		572	
Borrelia burgdorferi s.l.	France	Bretagne	Morbihan	626	
Borrelia burgdorferi s.l.	France	Centre		572	
Borrelia burgdorferi s.l.	France	Centre	Eure-et-Loir	719	
Borrelia burgdorferi s.l.	France	Centre	Indre-et-Loire	626	
Borrelia burgdorferi s.l.	France	Centre	Indre-et-Loire	719	
Borrelia burgdorferi s.l.	France	Centre	Loiret	626	
Borrelia burgdorferi s.l.	France	Champagne-Ardenne		1285	
Borrelia burgdorferi s.l.	France	Champagne-Ardenne		572	
Borrelia burgdorferi s.l.	France	Champagne-Ardenne	Ardennes	626	
Borrelia burgdorferi s.l.	France	Franche-Comté		1285	
Borrelia burgdorferi s.l.	France	Franche-Comté		572	
Borrelia burgdorferi s.l.	France	Haute-Normandie		572	
Borrelia burgdorferi s.l.	France	Haute-Normandie	Eure	719	
Borrelia burgdorferi s.l.	France	Île-de-France		262	
Borrelia burgdorferi s.l.	France	Île-de-France		572	
Borrelia burgdorferi s.l.	France	Île-de-France	Seine-et-Marne	626	
Borrelia burgdorferi s.l.	France	Limousin		572	
Borrelia burgdorferi s.l.	France	Limousin		851	
Borrelia burgdorferi s.l.	France	Lorraine		1285	
Borrelia burgdorferi s.l.	France	Lorraine		572	
Borrelia burgdorferi s.l.	France	Lorraine	Meurthe-et-Moselle	719	
Borrelia burgdorferi s.l.	France	Lorraine	Meuse	626	
Borrelia burgdorferi s.l.	France	Lorraine	Moselle	626	3
Borrelia burgdorferi s.l.	France	Midi-Pyrénées		572	
Borrelia burgdorferi s.l.	France	Nord-Pas-de-Calais	Nord	408	
Borrelia burgdorferi s.l.	France	Pays de la Loire		572	
Borrelia burgdorferi s.l.	France	Pays de la Loire	Maine-et-Loire	626	
Borrelia burgdorferi s.l.	France	Picardie		572	
Borrelia burgdorferi s.l.	France	Provence-Alpes-Côte-d'Azur	Alpes-Maritimes	719	

Borrelia burgdorferi s.l.	France	Provence-Alpes-Côte-d'Azur	Bouches-du-Rhône	626	2
Borrelia burgdorferi s.l.	France	Rhône-Alpes		572	
Borrelia burgdorferi s.l.	France	Rhône-Alpes		851	
Borrelia burgdorferi s.l.	France	Rhône-Alpes	Ain	782	
Borrelia burgdorferi s.l.	France	Rhône-Alpes	Haute-Savoie	719	
Borrelia burgdorferi s.l.	France	Rhône-Alpes	Loir-et-Cher	782	
Borrelia burgdorferi s.l.	France	Rhône-Alpes	Rhône	782	
Borrelia burgdorferi s.l.	Germany	Baden-Württemberg	Freiburg	691	
Borrelia burgdorferi s.l.	Germany	Baden-Württemberg	Stuttgart	691	
Borrelia burgdorferi s.l.	Germany	Bayern		564	
Borrelia burgdorferi s.l.	Germany	Bayern	Unterfranken	439	
Borrelia burgdorferi s.l.	Germany	Berlin		540	
Borrelia burgdorferi s.l.	Germany	Berlin		633	
Borrelia burgdorferi s.l.	Germany	Berlin	Berlin	1330	5
Borrelia burgdorferi s.l.	Germany	Brandenburg		540	
Borrelia burgdorferi s.l.	Germany	Brandenburg		633	
Borrelia burgdorferi s.l.	Germany	Hessen		930	
Borrelia burgdorferi s.l.	Germany	Hessen	Darmstadt	496	
Borrelia burgdorferi s.l.	Germany	Mecklenburg-Vorpommern		633	
Borrelia burgdorferi s.l.	Germany	Sachsen-Anhalt		633	
Borrelia burgdorferi s.l.	Germany	Thüringen		268	
Borrelia burgdorferi s.l.	Germany	Thüringen		633	
Borrelia burgdorferi s.l.	Greece	Attiki	Attica	476	
Borrelia burgdorferi s.l.	Italy	Calabria		381	
Borrelia burgdorferi s.l.	Italy	Friuli-Venezia Giulia		191	10
Borrelia burgdorferi s.l.	Italy	Friuli-Venezia Giulia		49	
Borrelia burgdorferi s.l.	Italy	Friuli-Venezia Giulia		93	3
Borrelia burgdorferi s.l.	Italy	Lazio		832	
Borrelia burgdorferi s.l.	Italy	Marche		325	
Borrelia burgdorferi s.l.	Italy	Molise		325	
Borrelia burgdorferi s.l.	Italy	Toscana	Arezzo	1283	
Borrelia burgdorferi s.l.	Italy	Toscana	Florence	1283	
Borrelia burgdorferi s.l.	Italy	Toscana	Siena	1283	
Borrelia burgdorferi s.l.	Italy	Trentino-Alto Adige	Trento	806	4
Borrelia burgdorferi s.l.	Italy	Veneto	Belluno	201	3

Continued

Table A1.6. Continued.

Species	Country	Administrative region 1	Administrative region 2	Reference ID	No. entries[a]
Borrelia burgdorferi s.l.	Italy	Veneto	Belluno	752	
Borrelia burgdorferi s.l.	Lithuania		Birzu	1100	
Borrelia burgdorferi s.l.	Lithuania		Joniskio	1100	
Borrelia burgdorferi s.l.	Lithuania		Kupiskio	1100	
Borrelia burgdorferi s.l.	Lithuania		Pakruojo	1100	
Borrelia burgdorferi s.l.	Lithuania		Pasvalio	1100	
Borrelia burgdorferi s.l.	Lithuania		Rokiskio	1100	
Borrelia burgdorferi s.l.	Lithuania	Alytaus		1009	
Borrelia burgdorferi s.l.	Lithuania	Alytaus	Varenos	732	
Borrelia burgdorferi s.l.	Lithuania	Kauno		1009	2
Borrelia burgdorferi s.l.	Lithuania	Klaipėdos		1009	3
Borrelia burgdorferi s.l.	Lithuania	Klaipėdos	Šilutes	732	
Borrelia burgdorferi s.l.	Lithuania	Klaipėdos	Klaipėdos	1100	
Borrelia burgdorferi s.l.	Lithuania	Klaipėdos	Klaipėdos	732	
Borrelia burgdorferi s.l.	Lithuania	Klaipėdos	Neringos	732	
Borrelia burgdorferi s.l.	Lithuania	Klaipėdos	Skuodo	1100	
Borrelia burgdorferi s.l.	Lithuania	Marijampoles		1009	
Borrelia burgdorferi s.l.	Lithuania	Marijampoles	Marijampoles	732	
Borrelia burgdorferi s.l.	Lithuania	Panevezio		1009	
Borrelia burgdorferi s.l.	Lithuania	Panevezio	Paneveûio	732	
Borrelia burgdorferi s.l.	Lithuania	Panevezio	Rokiökio	732	
Borrelia burgdorferi s.l.	Lithuania	Šiauliai	Šiauliu	732	
Borrelia burgdorferi s.l.	Lithuania	Šiauliai	Joniökio	732	
Borrelia burgdorferi s.l.	Lithuania	Šiauliu		1009	4
Borrelia burgdorferi s.l.	Lithuania	Šiauliu		1009	
Borrelia burgdorferi s.l.	Lithuania	Šiauliu		1009	
Borrelia burgdorferi s.l.	Lithuania	Utenos		1009	2
Borrelia burgdorferi s.l.	Lithuania	Utenos	Utenos	732	
Borrelia burgdorferi s.l.	Lithuania	Utenos	Utenos	732	
Borrelia burgdorferi s.l.	Lithuania	Vilniaus		1009	2
Borrelia burgdorferi s.l.	Lithuania	Vilniaus	Ukmerges	732	
Borrelia burgdorferi s.l.	Netherlands	Friesland	Ameland	467	
Borrelia burgdorferi s.l.	Netherlands	Noord-Holland		1050	

Borrelia burgdorferi s.l.	Norway	Aust-Agder	Birkenes	732	
Borrelia burgdorferi s.l.	Norway	Hordaland	Kvinnherad	732	
Borrelia burgdorferi s.l.	Norway	Rogaland		707	
Borrelia burgdorferi s.l.	Norway	Telemark	Kragerø	732	
Borrelia burgdorferi s.l.	Norway	Vest-Agder	Kristiansand	732	
Borrelia burgdorferi s.l.	Poland	Greater Poland		645	2
Borrelia burgdorferi s.l.	Poland	Greater Poland	Czarnków-Trzcianka	107	
Borrelia burgdorferi s.l.	Poland	Greater Poland	Jarocin	107	
Borrelia burgdorferi s.l.	Poland	Greater Poland	Krotoszyn	107	
Borrelia burgdorferi s.l.	Poland	Greater Poland	Leszno	107	
Borrelia burgdorferi s.l.	Poland	Greater Poland	Miedzychód	107	
Borrelia burgdorferi s.l.	Poland	Greater Poland	Poznań	107	
Borrelia burgdorferi s.l.	Poland	Greater Poland	Poznań	644	
Borrelia burgdorferi s.l.	Poland	Kuyavian-Pomeranian		1324	
Borrelia burgdorferi s.l.	Poland	Kuyavian-Pomeranian	Naklo	283	
Borrelia burgdorferi s.l.	Poland	Kuyavian-Pomeranian	Sepólno	283	
Borrelia burgdorferi s.l.	Poland	Kuyavian-Pomeranian	Włocławek	284	
Borrelia burgdorferi s.l.	Poland	Lódz		1324	
Borrelia burgdorferi s.l.	Poland	Lódz	Lowicz	284	
Borrelia burgdorferi s.l.	Poland	Lower Silesian		267	
Borrelia burgdorferi s.l.	Poland	Lower Silesian		272	
Borrelia burgdorferi s.l.	Poland	Lower Silesian		513	6
Borrelia burgdorferi s.l.	Poland	Lublin		105	2
Borrelia burgdorferi s.l.	Poland	Lublin		1062	
Borrelia burgdorferi s.l.	Poland	Lublin		178	4
Borrelia burgdorferi s.l.	Poland	Lublin		193	
Borrelia burgdorferi s.l.	Poland	Lublin		925	
Borrelia burgdorferi s.l.	Poland	Lublin		990	
Borrelia burgdorferi s.l.	Poland	Lublin	Lubartów	192	
Borrelia burgdorferi s.l.	Poland	Lublin	Lublin City	192	
Borrelia burgdorferi s.l.	Poland	Lublin	Lublin City	194	
Borrelia burgdorferi s.l.	Poland	Lublin	Lublin City	196	6
Borrelia burgdorferi s.l.	Poland	Lublin	Radzyn	192	
Borrelia burgdorferi s.l.	Poland	Lublin	Włodawa	284	
Borrelia burgdorferi s.l.	Poland	Lubusz		1324	
Borrelia burgdorferi s.l.	Poland	Masovian		1324	

Continued

Table A1.6. Continued.

Species	Country	Administrative region 1	Administrative region 2	Reference ID	No. entries[a]
Borrelia burgdorferi s.l.	Poland	Masovian		531	7
Borrelia burgdorferi s.l.	Poland	Masovian	Kozienice	284	
Borrelia burgdorferi s.l.	Poland	Masovian	Kozienice	285	
Borrelia burgdorferi s.l.	Poland	Masovian	Przasnysz	284	
Borrelia burgdorferi s.l.	Poland	Masovian	Warsaw	1098	
Borrelia burgdorferi s.l.	Poland	Opole		1324	
Borrelia burgdorferi s.l.	Poland	Podlachian		1324	
Borrelia burgdorferi s.l.	Poland	Podlachian		715	
Borrelia burgdorferi s.l.	Poland	Podlachian	Hajnówka	334	
Borrelia burgdorferi s.l.	Poland	Pomeranian		1324	
Borrelia burgdorferi s.l.	Poland	Pomeranian		931	
Borrelia burgdorferi s.l.	Poland	Silesian		956	2
Borrelia burgdorferi s.l.	Poland	Świętokrzyskie		1324	
Borrelia burgdorferi s.l.	Poland	Warmian-Masurian		1324	
Borrelia burgdorferi s.l.	Poland	Warmian-Masurian		738	2
Borrelia burgdorferi s.l.	Poland	Warmian-Masurian		739	2
Borrelia burgdorferi s.l.	Poland	Warmian-Masurian	Lidzbark	283	
Borrelia burgdorferi s.l.	Poland	West Pomeranian		1324	
Borrelia burgdorferi s.l.	Poland	West Pomeranian		687	
Borrelia burgdorferi s.l.	Poland	West Pomeranian	Stargard	1296	
Borrelia burgdorferi s.l.	Poland	West Pomeranian	Szczecin	138	
Borrelia burgdorferi s.l.	Poland	West Pomeranian	Szczecin	891	5
Borrelia burgdorferi s.l.	Portugal	Lisboa	Mafra	76	
Borrelia burgdorferi s.l.	Russia	Leningrad		533	2
Borrelia burgdorferi s.l.	Russia	Moskva		522	
Borrelia burgdorferi s.l.	Russia	Novgorod		533	2
Borrelia burgdorferi s.l.	Russia	Vologda		35	
Borrelia burgdorferi s.l.	Serbia	Grad Beograd		654	3
Borrelia burgdorferi s.l.	Serbia	Grad Beograd		655	
Borrelia burgdorferi s.l.	Serbia	Toplicki		655	
Borrelia burgdorferi s.l.	Serbia	Zapadno-Backi		655	
Borrelia burgdorferi s.l.	Slovakia	Banskobystrický		529	4
Borrelia burgdorferi s.l.	Slovakia	Banskobystrický		87	

Borrelia burgdorferi s.l.	Slovakia	Banskobystrický		939	
Borrelia burgdorferi s.l.	Slovakia	Bratislavský		87	
Borrelia burgdorferi s.l.	Slovakia	Bratislavský		939	
Borrelia burgdorferi s.l.	Slovakia	Košický		256	
Borrelia burgdorferi s.l.	Slovakia	Košický		382	22
Borrelia burgdorferi s.l.	Slovakia	Košický		87	
Borrelia burgdorferi s.l.	Slovakia	Košický		934	18
Borrelia burgdorferi s.l.	Slovakia	Košický		936	3
Borrelia burgdorferi s.l.	Slovakia	Košický		940	4
Borrelia burgdorferi s.l.	Slovakia	Košický		945	
Borrelia burgdorferi s.l.	Slovakia	Nitrianský		87	
Borrelia burgdorferi s.l.	Slovakia	Prešovský		87	
Borrelia burgdorferi s.l.	Slovakia	Prešovský		939	
Borrelia burgdorferi s.l.	Slovakia	Prešovský		936	
Borrelia burgdorferi s.l.	Slovakia	Prešovský		940	6
Borrelia burgdorferi s.l.	Slovakia	Trenčianský		87	
Borrelia burgdorferi s.l.	Slovakia	Trenčianský		939	3
Borrelia burgdorferi s.l.	Slovakia	Trnavský		87	
Borrelia burgdorferi s.l.	Slovakia	Zilinský		1347	
Borrelia burgdorferi s.l.	Slovakia	Zilinský		87	
Borrelia burgdorferi s.l.	Slovenia			191	3
Borrelia burgdorferi s.l.	Spain	Galicia	Ourense	53	
Borrelia burgdorferi s.l.	Spain	Galicia	Pontevedra	53	
Borrelia burgdorferi s.l.	Spain	La Rioja	La Rioja	705	
Borrelia burgdorferi s.l.	Spain	País Vasco		78	
Borrelia burgdorferi s.l.	Spain	País Vasco		79	2
Borrelia burgdorferi s.l.	Sweden	Dalarna		1312	
Borrelia burgdorferi s.l.	Sweden	Gävleborg		1312	
Borrelia burgdorferi s.l.	Sweden	Gotland		1312	
Borrelia burgdorferi s.l.	Sweden	Halland		1312	
Borrelia burgdorferi s.l.	Sweden	Halland	Varberg	470	
Borrelia burgdorferi s.l.	Sweden	Jämtland		1312	
Borrelia burgdorferi s.l.	Sweden	Jönköping		1312	
Borrelia burgdorferi s.l.	Sweden	Kalmar		1312	
Borrelia burgdorferi s.l.	Sweden	Kronoberg		1312	
Borrelia burgdorferi s.l.	Sweden	Orebro		1312	

Continued

Table A1.6. Continued.

Species	Country	Administrative region 1	Administrative region 2	Reference ID	No. entries[a]
Borrelia burgdorferi s.l.	Sweden	Östergötland		1312	
Borrelia burgdorferi s.l.	Sweden	Skåne		1312	
Borrelia burgdorferi s.l.	Sweden	Södermanland		1312	
Borrelia burgdorferi s.l.	Sweden	Stockholm		1312	
Borrelia burgdorferi s.l.	Sweden	Stockholm	Norrtälje	470	
Borrelia burgdorferi s.l.	Sweden	Uppsala		1312	
Borrelia burgdorferi s.l.	Sweden	Uppsala	Uppsala	470	
Borrelia burgdorferi s.l.	Sweden	Värmland		1312	
Borrelia burgdorferi s.l.	Sweden	Västerbotten		1318	
Borrelia burgdorferi s.l.	Sweden	Västerbotten	Umeå	1322	
Borrelia burgdorferi s.l.	Sweden	Västerbotten	Umeå	1322	
Borrelia burgdorferi s.l.	Sweden	Västernorrland		1312	
Borrelia burgdorferi s.l.	Sweden	Västmanland		1312	
Borrelia burgdorferi s.l.	Sweden	Västra Götaland		1312	
Borrelia burgdorferi s.l.	Switzerland	Genève		798	
Borrelia burgdorferi s.l.	Switzerland	Neuchâtel		1326	
Borrelia burgdorferi s.l.	Switzerland	Valais		1321	
Borrelia burgdorferi s.l.	Turkey	Bursa		101	
Borrelia burgdorferi s.l.	Turkey	Düzce		507	
Borrelia burgdorferi s.l.	Turkey	Isparta		252	
Borrelia burgdorferi s.l.	Turkey	Izmir		101	
Borrelia burgdorferi s.s.	Austria	Niederösterreich		114	
Borrelia burgdorferi s.s.	Austria	Niederösterreich		511	7
Borrelia burgdorferi s.s.	Austria	Oberösterreich		114	
Borrelia burgdorferi s.s.	Austria	Salzburg		114	
Borrelia burgdorferi s.s.	Austria	Tirol		114	
Borrelia burgdorferi s.s.	Austria	Wien		511	4
Borrelia burgdorferi s.s.	Bulgaria	Sofia	Stolichna	185	
Borrelia burgdorferi s.s.	Croatia	Primorsko-Goranska		1011	
Borrelia burgdorferi s.s.	Czech Republic	Jihočeský		224	2
Borrelia burgdorferi s.s.	Czech Republic	Jihočeský		255	
Borrelia burgdorferi s.s.	Czech Republic	Jihomoravský		550	3
Borrelia burgdorferi s.s.	Czech Republic	Jihomoravský		742	

Borrelia burgdorferi s.s.	Czech Republic	Královéhradeck		225	
Borrelia burgdorferi s.s.	Czech Republic	Královéhradeck		230	
Borrelia burgdorferi s.s.	Czech Republic	Libereck		271	4
Borrelia burgdorferi s.s.	Czech Republic	Moravskoslezský		220	
Borrelia burgdorferi s.s.	Czech Republic	Plzeňský		230	
Borrelia burgdorferi s.s.	Denmark	North Jutland		1026	
Borrelia burgdorferi s.s.	France	Alsace	Haut-Rhin	331	
Borrelia burgdorferi s.s.	France	Île-de-France	Yvelines	753	
Borrelia burgdorferi s.s.	Germany	Baden-Württemberg	Tübingen	794	
Borrelia burgdorferi s.s.	Germany	Bayern		332	2
Borrelia burgdorferi s.s.	Germany	Bayern		333	4
Borrelia burgdorferi s.s.	Germany	Berlin	Berlin	1294	
Borrelia burgdorferi s.s.	Germany	Niedersachsen	Hannover	431	
Borrelia burgdorferi s.s.	Germany	Nordrhein-Westfalen		600	
Borrelia burgdorferi s.s.	Hungary	Pest		344	3
Borrelia burgdorferi s.s.	Ireland	Kerry		754	
Borrelia burgdorferi s.s.	Italy	Friuli-Venezia Giulia		639	
Borrelia burgdorferi s.s.	Italy	Lazio		831	
Borrelia burgdorferi s.s.	Italy	Toscana		1331	
Borrelia burgdorferi s.s.	Italy	Toscana	Pisa	98	
Borrelia burgdorferi s.s.	Italy	Trentino-Alto Adige	Trento	741	
Borrelia burgdorferi s.s.	Latvia	Riga		319	
Borrelia burgdorferi s.s.	Latvia	Riga		319	
Borrelia burgdorferi s.s.	Moldova	Chisinau		528	
Borrelia burgdorferi s.s.	Moldova	Chisinau		666	
Borrelia burgdorferi s.s.	Morocco	Taza-Al Hoceima-Taounate	Taza	642	
Borrelia burgdorferi s.s.	Norway	Telemark		475	
Borrelia burgdorferi s.s.	Poland	Greater Poland		603	
Borrelia burgdorferi s.s.	Poland	Greater Poland	Poznań	645	
Borrelia burgdorferi s.s.	Poland	Lublin	Lublin City	194	
Borrelia burgdorferi s.s.	Poland	Lublin	Lublin City	1337	
Borrelia burgdorferi s.s.	Poland	Silesian	Tarnowskie	1288	
Borrelia burgdorferi s.s.	Poland	Warmian-Masurian		384	
Borrelia burgdorferi s.s.	Serbia	Branicevski		652	
Borrelia burgdorferi s.s.	Serbia	Grad Beograd		652	
Borrelia burgdorferi s.s.	Serbia	Macvanski		652	

Continued

Table A1.6. Continued.

Species	Country	Administrative region 1	Administrative region 2	Reference ID	No. entries[a]
Borrelia burgdorferi s.s.	Serbia	Ni'avski		652	
Borrelia burgdorferi s.s.	Serbia	Pomoravski		652	
Borrelia burgdorferi s.s.	Serbia	Ra'ki		652	
Borrelia burgdorferi s.s.	Serbia	Sremski		652	
Borrelia burgdorferi s.s.	Serbia	Zajecarski		652	
Borrelia burgdorferi s.s.	Slovakia	Kosicky		602	2
Borrelia burgdorferi s.s.	Spain	País Vasco		80	
Borrelia burgdorferi s.s.	Sweden	Blekinge		350	
Borrelia burgdorferi s.s.	Sweden	Gävleborg		350	
Borrelia burgdorferi s.s.	Sweden	Kalmar		350	
Borrelia burgdorferi s.s.	Sweden	Skåne		350	
Borrelia burgdorferi s.s.	Sweden	Stockholm	Stockholm	350	
Borrelia burgdorferi s.s.	Switzerland	Neuchâtel		143	
Borrelia burgdorferi s.s.	Switzerland	Neuchâtel		163	
Borrelia burgdorferi s.s.	Switzerland	Valais		163	2
Borrelia burgdorferi s.s.	Turkey	Istanbul		395	
Borrelia garinii	Austria	Burgenland		114	2
Borrelia garinii	Austria	Kärnten		114	2
Borrelia garinii	Austria	Niederösterreich		114	
Borrelia garinii	Austria	Niederösterreich		511	6
Borrelia garinii	Austria	Oberösterreich		114	2
Borrelia garinii	Austria	Salzburg		114	2
Borrelia garinii	Austria	Steiermark		963	
Borrelia garinii	Austria	Tirol		114	
Borrelia garinii	Austria	Vorarlberg		114	2
Borrelia garinii	Austria	Wien		114	2
Borrelia garinii	Austria	Wien		511	3
Borrelia garinii	Bulgaria	Sofia	Stolichna	185	
Borrelia garinii	Croatia	Primorsko-Goranska		1011	
Borrelia garinii	Czech Republic	Jihočeský		224	2
Borrelia garinii	Czech Republic	Jihočeský		255	
Borrelia garinii	Czech Republic	Jihomoravský		742	
Borrelia garinii	Czech Republic	Královéhradeck		225	

Borrelia garinii	Czech Republic	Královéhradeck		230	
Borrelia garinii	Czech Republic	Libereck		271	9
Borrelia garinii	Czech Republic	Moravskoslezský		220	
Borrelia garinii	Czech Republic	Ústečký		550	
Borrelia garinii	Denmark	North Jutland		1026	
Borrelia garinii	Finland	Western Finland		34	
Borrelia garinii	France	Alsace	Haut-Rhin	331	
Borrelia garinii	France	Île-de-France	Yvelines	753	
Borrelia garinii	Germany	Baden-Württemberg	Tübingen	794	
Borrelia garinii	Germany	Bayern		332	2
Borrelia garinii	Germany	Bayern		333	5
Borrelia garinii	Germany	Berlin	Berlin	1294	
Borrelia garinii	Germany	Niedersachsen	Hannover	431	
Borrelia garinii	Germany	Nordrhein-Westfalen		600	
Borrelia garinii	Hungary	Baranya		341	
Borrelia garinii	Hungary	Somogy		341	
Borrelia garinii	Hungary	Veszprém		341	
Borrelia garinii	Ireland	Galway		380	
Borrelia garinii	Ireland	Kerry		754	
Borrelia garinii	Italy	Friuli-Venezia Giulia		639	
Borrelia garinii	Italy	Lazio		831	
Borrelia garinii	Italy	Lazio		833	
Borrelia garinii	Italy	Toscana		1331	
Borrelia garinii	Italy	Toscana	Pisa	98	
Borrelia garinii	Italy	Trentino-Alto Adige	Trento	1344	
Borrelia garinii	Italy	Trentino-Alto Adige	Trento	609	
Borrelia garinii	Italy	Trentino-Alto Adige	Trento	741	
Borrelia garinii	Italy	Veneto	Belluno	1344	
Borrelia garinii	Latvia	Riga		319	3
Borrelia garinii	Moldova	Chisinau		528	
Borrelia garinii	Moldova	Chisinau		666	
Borrelia garinii	Morocco	Taza-Al Hoceima-Taounate	Taza	642	
Borrelia garinii	Netherlands	Gelderland	Arnhem	365	
Borrelia garinii	Netherlands	Noord-Holland		1050	
Borrelia garinii	Netherlands	Overijssel		1050	
Borrelia garinii	Norway	Aust-Agder		1323	2

Continued

Table A1.6. Continued.

Species	Country	Administrative region 1	Administrative region 2	Reference ID	No. entries[a]
Borrelia garinii	Norway	Finnmark		558	
Borrelia garinii	Norway	Sør-Trøndelag	Hitra	1323	
Borrelia garinii	Norway	Telemark		475	
Borrelia garinii	Poland	Greater Poland		603	2
Borrelia garinii	Poland	Greater Poland	Poznań	645	
Borrelia garinii	Poland	Lublin	Lublin City	1337	
Borrelia garinii	Poland	Silesian	Tarnowskie	1288	
Borrelia garinii	Poland	Warmian-Masurian		384	
Borrelia garinii	Poland	Warmian-Masurian		738	2
Borrelia garinii	Poland	Warmian-Masurian		739	2
Borrelia garinii	Portugal	Aveiro		597	
Borrelia garinii	Portugal	Azores		597	
Borrelia garinii	Portugal	Beja		597	
Borrelia garinii	Portugal	Braga		597	
Borrelia garinii	Portugal	Castelo Branco		597	
Borrelia garinii	Portugal	Coimbra		597	
Borrelia garinii	Portugal	Évora		597	
Borrelia garinii	Portugal	Faro		597	
Borrelia garinii	Portugal	Guarda		597	
Borrelia garinii	Portugal	Leiria		597	
Borrelia garinii	Portugal	Lisboa		597	
Borrelia garinii	Portugal	Lisboa	Mafra	76	
Borrelia garinii	Portugal	Madeira		597	
Borrelia garinii	Portugal	Portalegre		597	
Borrelia garinii	Portugal	Porto		597	
Borrelia garinii	Portugal	Santarém		597	
Borrelia garinii	Portugal	Setúbal		597	
Borrelia garinii	Portugal	Viana do Castelo		597	
Borrelia garinii	Portugal	Viseu		597	
Borrelia garinii	Russia	Vologda		35	
Borrelia garinii	Serbia	Branicevski		652	
Borrelia garinii	Serbia	Grad Beograd		652	
Borrelia garinii	Serbia	Macvanski		652	

Borrelia garinii	Serbia	Ni'avski		652	
Borrelia garinii	Serbia	Pomoravski		652	
Borrelia garinii	Serbia	Ra'ki		652	
Borrelia garinii	Serbia	Sremski		652	
Borrelia garinii	Serbia	Zajecarski		652	
Borrelia garinii	Slovakia	Bratislavsky		1349	
Borrelia garinii	Slovakia	Kosicky		1349	
Borrelia garinii	Slovakia	Kosicky		602	2
Borrelia garinii	Slovakia	Nitriansky		1349	
Borrelia garinii	Slovakia	Trenciansky		1349	
Borrelia garinii	Slovakia	Zilinsky		1349	
Borrelia garinii	Spain	La Rioja	La Rioja	315	
Borrelia garinii	Spain	País Vasco		80	
Borrelia garinii	Sweden			695	
Borrelia garinii	Sweden	Gävleborg		350	
Borrelia garinii	Sweden	Gotland	Gotland	470	
Borrelia garinii	Sweden	Halland	Kungsbacka	470	
Borrelia garinii	Sweden	Jönköping		426	
Borrelia garinii	Sweden	Kalmar		350	
Borrelia garinii	Sweden	Skåne		350	
Borrelia garinii	Sweden	Skåne		696	
Borrelia garinii	Sweden	Skåne	Båstad	470	
Borrelia garinii	Sweden	Uppsala	Älvkarleby	470	
Borrelia garinii	Sweden	Uppsala	Tierp	470	
Borrelia garinii	Switzerland	Neuchâtel		143	
Borrelia garinii	Switzerland	Neuchâtel		163	
Borrelia garinii	Switzerland	Ticino		1284	2
Borrelia garinii	Switzerland	Ticino		163	
Borrelia garinii	Switzerland	Ticino		485	
Borrelia garinii	Switzerland	Valais		1284	2
Borrelia garinii	Switzerland	Valais		163	
Borrelia garinii	Turkey	Istanbul		395	
Borrelia lusitaniae	Bulgaria	Sofia	Stolichna	185	
Borrelia lusitaniae	Denmark	North Jutland		1026	
Borrelia lusitaniae	Hungary			1320	3
Borrelia lusitaniae	Hungary	Pest		344	3

Continued

Table A1.6. Continued.

Species	Country	Administrative region 1	Administrative region 2	Reference ID	No. entries[a]
Borrelia lusitaniae	Italy	Toscana	Pisa	98	
Borrelia lusitaniae	Italy	Trentino-Alto Adige	Trento	741	
Borrelia lusitaniae	Moldova	Chisinau		528	
Borrelia lusitaniae	Moldova	Chisinau		666	
Borrelia lusitaniae	Morocco	Taza-Al Hoceima-Taounate		425	
Borrelia lusitaniae	Morocco	Taza-Al Hoceima-Taounate	Taza	642	
Borrelia lusitaniae	Poland	Greater Poland		603	2
Borrelia lusitaniae	Poland	Lublin	Lublin City	1337	
Borrelia lusitaniae	Portugal	Lisboa	Mafra	76	
Borrelia lusitaniae	Portugal	Madeira		238	
Borrelia lusitaniae	Portugal	Setúbal	Grândola	76	
Borrelia lusitaniae	Romania			1320	
Borrelia lusitaniae	Romania	Tulcea		603	
Borrelia lusitaniae	Serbia	Branicevski		652	
Borrelia lusitaniae	Serbia	Macvanski		652	
Borrelia lusitaniae	Serbia	Ni'avski		652	
Borrelia lusitaniae	Serbia	Pomoravski		652	
Borrelia lusitaniae	Serbia	Ra'ki		652	
Borrelia lusitaniae	Serbia	Sremski		652	
Borrelia lusitaniae	Serbia	Zajecarski		652	
Borrelia lusitaniae	Serbia	Zajecarski		652	
Borrelia lusitaniae	Slovakia			1320	2
Borrelia lusitaniae	Slovakia	Banskobystricky		602	
Borrelia lusitaniae	Slovakia	Bratislavsky		1349	
Borrelia lusitaniae	Slovakia	Kosicky		1349	
Borrelia lusitaniae	Slovakia	Kosicky		602	2
Borrelia lusitaniae	Slovakia	Nitriansky		1349	
Borrelia lusitaniae	Slovakia	Trenciansky		1349	
Borrelia lusitaniae	Slovakia	Trenciansky		603	
Borrelia lusitaniae	Slovakia	Zilinsky		1347	
Borrelia lusitaniae	Slovakia	Zilinsky		1349	
Borrelia lusitaniae	Spain	País Vasco		80	
Borrelia lusitaniae	Switzerland	Ticino		1284	

Borrelia lusitaniae	Switzerland	Ticino		163	7
Borrelia lusitaniae	Switzerland	Valais		1284	
Borrelia lusitaniae	Tunisia	Jendouba		425	2
Borrelia lusitaniae	Turkey	Istanbul		395	
Borrelia miyamotoi	Sweden	Gävleborg		350	
Borrelia miyamotoi	Sweden	Kalmar		350	
Borrelia miyamotoi	Sweden	Skåne		350	
Borrelia spielmanii	Germany	Bayern		333	2
Borrelia spielmanii	Germany	Bayern		333	
Borrelia spielmanii	Germany	Niedersachsen	Hannover	431	
Borrelia valaisiana	Austria	Burgenland		114	
Borrelia valaisiana	Austria	Niederösterreich		114	2
Borrelia valaisiana	Austria	Oberösterreich		114	2
Borrelia valaisiana	Austria	Salzburg		114	
Borrelia valaisiana	Austria	Vorarlberg		114	2
Borrelia valaisiana	Austria	Wien		114	
Borrelia valaisiana	Bulgaria	Sofia	Stolichna	185	
Borrelia valaisiana	Czech Republic	Jihočeský		255	
Borrelia valaisiana	Czech Republic	Jihomoravský		742	
Borrelia valaisiana	Czech Republic	Královéhradeck		225	
Borrelia valaisiana	Czech Republic	Královéhradeck		230	
Borrelia valaisiana	Czech Republic	Libereck		271	9
Borrelia valaisiana	Czech Republic	Moravskoslezský		220	
Borrelia valaisiana	Czech Republic	Plzeňský		230	
Borrelia valaisiana	France	Alsace	Haut-Rhin	331	
Borrelia valaisiana	Germany	Bayern		333	5
Borrelia valaisiana	Germany	Berlin	Berlin	1294	
Borrelia valaisiana	Germany	Niedersachsen	Hannover	431	
Borrelia valaisiana	Germany	Nordrhein Westfalen		600	
Borrelia valaisiana	Ireland	Galway		380	
Borrelia valaisiana	Ireland	Kerry		754	
Borrelia valaisiana	Italy	Lazio		831	
Borrelia valaisiana	Italy	Lazio		833	
Borrelia valaisiana	Italy	Toscana		1331	
Borrelia valaisiana	Italy	Trentino-Alto Adige	Trento	609	3
Borrelia valaisiana	Italy	Trentino-Alto Adige	Trento	741	
Borrelia valaisiana	Italy	Veneto	Belluno	1344	

Continued

Table A1.6. Continued.

Species	Country	Administrative region 1	Administrative region 2	Reference ID	No. entries[a]
Borrelia valaisiana	Italy	Veneto	Belluno	324	
Borrelia valaisiana	Latvia	Riga		319	3
Borrelia valaisiana	Moldova	Chisinau		528	
Borrelia valaisiana	Moldova	Chisinau		666	
Borrelia valaisiana	Netherlands	Flevoland		1291	
Borrelia valaisiana	Netherlands	Gelderland	Arnhem	365	
Borrelia valaisiana	Netherlands	Noord-Holland		1050	
Borrelia valaisiana	Netherlands	Overijssel		1050	2
Borrelia valaisiana	Poland	Greater Poland		603	2
Borrelia valaisiana	Poland	Lublin	Lublin City	1337	
Borrelia valaisiana	Portugal	Lisboa	Mafra	76	
Borrelia valaisiana	Serbia	Branicevski		652	
Borrelia valaisiana	Serbia	Grad Beograd		652	
Borrelia valaisiana	Serbia	Macvanski		652	
Borrelia valaisiana	Serbia	Ni'avski		652	
Borrelia valaisiana	Serbia	Pomoravski		652	
Borrelia valaisiana	Serbia	Ra'ki		652	
Borrelia valaisiana	Serbia	Sremski		652	
Borrelia valaisiana	Serbia	Zajecarski		652	
Borrelia valaisiana	Slovakia	Kosicky		602	2
Borrelia valaisiana	Slovakia	Zilinsky		1347	
Borrelia valaisiana	Spain	La Rioja	La Rioja	315	
Borrelia valaisiana	Spain	País Vasco		80	
Borrelia valaisiana	Switzerland	Neuchâtel		143	
Borrelia valaisiana	Switzerland	Neuchâtel		163	
Borrelia valaisiana	Switzerland	Neuchâtel		368	
Borrelia valaisiana	Switzerland	Ticino		1284	
Borrelia valaisiana	Switzerland	Ticino		163	
Borrelia valaisiana	Switzerland	Ticino		485	
Borrelia valaisiana	Switzerland	Valais		1284	
Borrelia valaisiana	Switzerland	Valais		163	
Borrelia valaisiana	Turkey	Istanbul		395	

[a]If more than one.

Table A1.7. Geographic distribution data for *Babesia* spp. See Appendix 2 (hosted on http://www.cabi.org/vetmedresource/) for complete reference indicated by the Reference ID.

Species	Country	Administrative region 1	Administrative region 2	Reference ID	No. entries[a]
Babesia bigemina	Egypt	Al Jizah		1101	2
Babesia bigemina	Italy	Apulia		207	
Babesia bigemina	Italy	Basilicata		207	
Babesia bigemina	Italy	Calabria		997	
Babesia bigemina	Italy	Calabria	Crotone	156	
Babesia bigemina	Italy	Campania		207	
Babesia bigemina	Italy	Emilia-Romagna	Bologna	755	
Babesia bigemina	Italy	Lazio		997	
Babesia bigemina	Italy	Lazio	Roma	755	
Babesia bigemina	Italy	Marche		997	
Babesia bigemina	Italy	Sicily		997	
Babesia bigemina	Italy	Sicily	Palermo	366	2
Babesia bigemina	Italy	Sicily	Ragusa	366	
Babesia bigemina	Italy	Umbria		997	
Babesia bigemina	Italy	Umbria	Perugia	755	
Babesia bigemina	Italy	Veneto		997	
Babesia bigemina	Italy	Veneto	Padua	755	
Babesia bigemina	Morocco	Doukkala-Abda		291	
Babesia bigemina	Morocco	Gharb-Chrarda-Béni Hssen		291	
Babesia bigemina	Morocco	Marrakech-Tensift-Al Haouz		291	
Babesia bigemina	Morocco	Tadla-Azilal		291	
Babesia bigemina	Spain	Islas Baleares		41	
Babesia bigemina	Spain	País Vasco		363	2
Babesia bigemina	Switzerland	Graubünden		434	
Babesia bigemina	Turkoy	Antalya		508	
Babesia bigemina	Turkey	Kayseri		457	
Babesia bovis	Egypt	Al Jizah		1101	2
Babesia bovis	Italy	Calabria		997	3
Babesia bovis	Italy	Calabria	Crotone	156	2
Babesia bovis	Italy	Campania		997	3
Babesia bovis	Italy	Emilia-Romagna	Bologna	755	
Babesia bovis	Italy	Lazio		997	3

Continued

Table A1.7. Continued.

Species	Country	Administrative region 1	Administrative region 2	Reference ID	No. entries[a]
Babesia bovis	Italy	Lazio	Roma	755	
Babesia bovis	Italy	Marche		997	3
Babesia bovis	Italy	Sicily		997	3
Babesia bovis	Italy	Sicily	Palermo	366	3
Babesia bovis	Italy	Umbria		997	3
Babesia bovis	Italy	Umbria	Perugia	755	
Babesia bovis	Italy	Veneto		997	3
Babesia bovis	Italy	Veneto	Padua	755	
Babesia bovis	Spain	Islas Baleares		41	
Babesia bovis	Spain	País Vasco		363	
Babesia bovis	Spain	País Vasco		364	
Babesia bovis	Tunisia	Jendouba		641	3
Babesia bovis	Tunisia	Nabeul		641	
Babesia bovis	Tunisia	Nabeul		641	
Babesia bovis	Turkey	Kayseri		457	
Babesia bovis	Turkey	Tekirdag		43	4
Babesia caballi	Hungary	Hajdú-Bihar		441	
Babesia caballi	Italy	Calabria		997	2
Babesia caballi	Italy	Emilia-Romagna		997	2
Babesia caballi	Italy	Emilia-Romagna	Bologna	755	
Babesia caballi	Italy	Lazio		997	2
Babesia caballi	Italy	Lazio	Roma	755	
Babesia caballi	Italy	Marche		997	2
Babesia caballi	Italy	Sardegna		997	2
Babesia caballi	Italy	Sicily		997	2
Babesia caballi	Italy	Toscana		606	
Babesia caballi	Italy	Umbria		997	2
Babesia caballi	Italy	Umbria	Perugia	755	
Babesia caballi	Italy	Veneto	Padua	755	
Babesia caballi	Spain	Galicia		146	
Babesia caballi	Spain	País Vasco		363	3
Babesia caballi	Turkey	Ankara		390	2
Babesia caballi	Turkey	Malatya		25	
Babesia canis	Albania	Tiranë		409	

Babesia canis	Egypt	Asyut		510	
Babesia canis	Germany	Baden-Württemberg		83	
Babesia canis	Germany	Baden-Württemberg	Freiburg	576	
Babesia canis	Germany	Bayern		576	
Babesia canis	Germany	Bayern		83	
Babesia canis	Germany	Berlin		83	
Babesia canis	Germany	Brandenburg		83	
Babesia canis	Germany	Bremen		83	
Babesia canis	Germany	Hamburg		83	
Babesia canis	Germany	Hessen		83	
Babesia canis	Germany	Mecklenburg-Vorpommern		83	
Babesia canis	Germany	Niedersachsen		83	
Babesia canis	Germany	Nordrhein-Westfalen		83	
Babesia canis	Germany	Rheinland-Pfalz		83	
Babesia canis	Germany	Saarland		83	
Babesia canis	Germany	Sachsen		83	
Babesia canis	Germany	Sachsen-Anhalt		83	
Babesia canis	Germany	Schleswig-Holstein		83	
Babesia canis	Germany	Thüringen		83	
Babesia canis	Greece	Attiki	Attica	476	
Babesia canis	Hungary	Hajdú-Bihar		441	
Babesia canis	Italy	Emilia-Romagna		1107	
Babesia canis	Italy	Emilia-Romagna	Bologna	755	
Babesia canis	Italy	Emilia-Romagna	Bologna	755	
Babesia canis	Italy	Friuli-Venezia Giulia		1107	
Babesia canis	Italy	Lazio		1107	
Babesia canis	Italy	Lazio		997	
Babesia canis	Italy	Lazio	Roma	755	3
Babesia canis	Italy	Lombardia		997	
Babesia canis	Italy	Marche		997	
Babesia canis	Italy	Sicily		997	2
Babesia canis	Italy	Sicily		999	2
Babesia canis	Italy	Toscana		1107	
Babesia canis	Italy	Umbria		1107	
Babesia canis	Italy	Umbria		997	
Babesia canis	Italy	Umbria	Perugia	755	3

Continued

Table A1.7. Continued.

Species	Country	Administrative region 1	Administrative region 2	Reference ID	No. entries[a]
Babesia canis	Italy	Veneto		1107	
Babesia canis	Italy	Veneto	Padua	755	
Babesia canis	Poland	Masovian	Warsaw	1098	
Babesia canis	UK	England	Kent	440	
Babesia canis canis	Croatia	Bjelovarska-Bilogorska		88	
Babesia canis canis	Croatia	Grad Zagreb		88	2
Babesia canis canis	Croatia	Medimurska		88	
Babesia canis canis	Croatia	Varaždinska		88	
Babesia canis canis	Hungary	Budapest		340	
Babesia canis canis	Hungary	Gyor-Moson-Sopron		341	
Babesia canis canis	Hungary	Pest		341	
Babesia canis canis	Hungary	Veszprém		341	
Babesia canis canis	Italy	Friuli-Venezia Giulia		1107	
Babesia canis canis	Italy	Lazio		1107	
Babesia canis canis	Italy	Umbria		1107	
Babesia canis canis	Italy	Veneto		1107	
Babesia canis canis	Poland	Pomeranian		190	
Babesia canis canis	Spain	Cataluña	Barcelona	978	
Babesia canis vogeli	Croatia	Bjelovarska-Bilogorska		88	
Babesia canis vogeli	Croatia	Grad Zagreb		88	2
Babesia canis vogeli	Croatia	Medimurska		88	
Babesia canis vogeli	Croatia	Varaždinska		88	
Babesia canis vogeli	Italy	Emilia-Romagna		1107	
Babesia canis vogeli	Spain	Cataluña	Barcelona	978	
Babesia divergens	Belgium	Flemish Brabant		321	2
Babesia divergens	Estonia	Ida-Viru		1335	
Babesia divergens	France	Rhône-Alpes	Rhône	261	
Babesia divergens	Germany	Bayern		564	
Babesia divergens	Hungary	Borsod-Abaúj-Zemplén		442	
Babesia divergens	Hungary	Heves		442	
Babesia divergens	Italy	Sicily	Palermo	366	
Babesia divergens	Italy	Veneto		997	
Babesia divergens	Poland	West Pomeranian	Szczecin	890	
Babesia divergens	Slovakia	Bratislavský		1349	

Babesia divergens	Slovakia	Nitrianský		1349	
Babesia divergens	Spain	País Vasco		364	
Babesia divergens	Switzerland	Graubünden		435	
Babesia divergens	Switzerland	Graubünden		622	
Babesia divergens	Switzerland	Ticino		435	2
Babesia divergens	Turkey	Kayseri		457	
Babesia equi	Italy	Emilia-Romagna	Bologna	755	
Babesia equi	Italy	Lazio	Roma	755	
Babesia equi	Italy	Umbria	Perugia	755	
Babesia equi	Italy	Veneto	Padua	755	
Babesia equi	Turkey	Malatya		25	
Babesia gibsoni	Croatia	Bjelovarska-Bilogorska		88	
Babesia gibsoni	Croatia	Grad Zagreb		88	2
Babesia gibsoni	Croatia	Medimurska		88	
Babesia gibsoni	Croatia	Varaždinska		88	
Babesia gibsoni	Germany	Baden-Württemberg	Tübingen	417	
Babesia gibsoni	Hungary	Budapest		323	
Babesia gibsoni	Spain	Cataluña	Barcelona	978	
Babesia major	Italy	Apulia		157	
Babesia major	Italy	Basilicata		157	
Babesia major	Italy	Calabria		157	
Babesia major	Spain	País Vasco		364	
Babesia major	Turkey	Amasya		46	
Babesia major	Turkey	Giresun		46	
Babesia major	Turkey	Gümüshane		46	
Babesia major	Turkey	Tokat		46	
Babesia major	Turkey	Trabzon		46	
Babesia microti	Czech Republic	Jihomoravský		820	
Babesia microti	Estonia	Harju		1335	
Babesia microti	Estonia	Ida-Viru		1335	
Babesia microti	Estonia	Pärnu		1335	2
Babesia microti	Estonia	Tartu		1335	2
Babesia microti	Estonia	Tartu		1348	2
Babesia microti	Finland	Western Finland		34	
Babesia microti	Italy	Emilia-Romagna		1107	2
Babesia microti	Poland			549	

Continued

Table A1.7. Continued.

Species	Country	Administrative region 1	Administrative region 2	Reference ID	No. entries[a]
Babesia microti	Poland	Lower Silesian	Katowice City	504	
Babesia microti	Poland	Lublin		1062	
Babesia microti	Poland	Lublin	Parczew	1061	
Babesia microti	Poland	Lublin	Puławy	1061	
Babesia microti	Poland	Lublin	Włodawa	1061	
Babesia microti	Poland	Masovian		504	
Babesia microti	Poland	Podlachian		504	
Babesia microti	Poland	Podlachian	Białystok	72	
Babesia microti	Poland	Pomeranian	Sopot	504	
Babesia microti	Poland	Warmian-Masurian		1046	8
Babesia microti	Poland	Warmian-Masurian		737	
Babesia microti	Poland	Warmian-Masurian		879	4
Babesia microti	Poland	West Pomeranian		504	
Babesia microti	Poland	West Pomeranian	Szczecin	890	
Babesia microti	Poland	West Pomeranian	Szczecin	891	5
Babesia microti	Slovakia	Bratislavský		1349	
Babesia microti	Slovakia	Nitrianský		1349	
Babesia microti	Spain	La Rioja	La Rioja	315	
Babesia microti	Switzerland	Graubünden		1305	
Babesia microti	Turkey	Niğde		188	
Babesia microti	UK	England	Northumberland	131	
Babesia motasi	Romania	Tulcea		59	
Babesia motasi	Spain	País Vasco		677	
Babesia ovis	Romania	Tulcea		59	
Babesia ovis	Spain	País Vasco		677	
Babesia ovis	Turkey	Afyon		186	
Babesia ovis	Turkey	Diyarbakır		23	2
Babesia ovis	Turkey	Elâziğ		23	2
Babesia ovis	Turkey	Elâziğ		45	4
Babesia ovis	Turkey	Erzincan		23	2
Babesia ovis	Turkey	Erzurum		23	2
Babesia ovis	Turkey	Iğdır		23	2
Babesia ovis	Turkey	Kayseri		458	2
Babesia ovis	Turkey	Malatya		23	2

Babesia ovis	Turkey	Mardin		23	2
Babesia ovis	Turkey	Mus		23	2
Babesia ovis	Turkey	Şanlıurfa		303	
Babesia spp.	Austria	Burgenland	Oberwart	115	
Babesia spp.	Austria	Kärnten	Wolfsberg	115	
Babesia spp.	Austria	Niederösterreich		115	
Babesia spp.	Austria	Oberösterreich	Voitsberg	115	
Babesia spp.	Austria	Salzburg	Hallein	115	
Babesia spp.	Austria	Salzburg	Sankt Johann im Pongau	115	
Babesia spp.	Austria	Steiermark	Graz Umgebung	115	
Babesia spp.	Austria	Steiermark	Mürzzuschlag	115	
Babesia spp.	Austria	Tirol	Imst	115	
Babesia spp.	Austria	Vorarlberg	Bludenz	115	
Babesia spp.	Austria	Wien	Wien	115	
Babesia spp.	Estonia	Harju		1335	
Babesia spp.	Estonia	Ida-Viru		1335	
Babesia spp.	Estonia	Pärnu		1335	2
Babesia spp.	Estonia	Tartu		1335	2
Babesia spp.	France	Nord-Pas-de-Calais	Nord	408	
Babesia spp.	Greece	Ipeiros	Ioannina	985	2
Babesia spp.	Greece	Thessalia	Trikala	985	2
Babesia spp.	Italy	Calabria		997	
Babesia spp.	Italy	Veneto	Belluno	752	
Babesia spp.	Netherlands	Gelderland	Arnhem	447	
Babesia spp.	Netherlands	Zuid-Holland		447	
Babesia spp.	Switzerland	Graubünden		1211	
Babesia spp.	Switzerland	Graubünden		438	
Babesia spp.	Switzerland	Neuchâtel		166	
Babesia spp.	Switzerland	Ticino		166	
Babesia spp.	Switzerland	Valais		166	
Babesia spp.	Switzerland	Zürich		166	
Babesia spp.	Turkey	Diyarbakır		1066	
Babesia vogeli	Turkey	Istanbul		394	

[a]If more than one.

Table A1.8. Geographic distribution data for *Theileria* spp. See Appendix 2 (hosted on http://www.cabi.org/vetmedresource/) for complete reference indicated by the Reference ID.

Species	Country	Administrative region 1	Administrative region 2	Reference ID	No. entries[a]
Theileria annae	Spain	País Vasco		363	
Theileria annulata	Egypt	Ad Daqahliyah		295	
Theileria annulata	Egypt	Ad Daqahliyah		301	
Theileria annulata	Egypt	Al Gharbiyah		295	
Theileria annulata	Italy	Sicily	Palermo	366	4
Theileria annulata	Italy	Sicily	Ragusa	366	
Theileria annulata	Morocco	Doukkala-Abda		701	
Theileria annulata	Spain	Islas Baleares		40	
Theileria annulata	Spain	Islas Baleares		41	
Theileria annulata	Spain	País Vasco		363	
Theileria annulata	Spain	País Vasco		364	
Theileria annulata	Tunisia	Béja		641	
Theileria annulata	Tunisia	Bizerte		641	
Theileria annulata	Tunisia	Bizerte		641	
Theileria annulata	Tunisia	Jendouba		641	3
Theileria annulata	Tunisia	Nabeul		641	2
Theileria annulata	Tunisia	Zaghouan		641	2
Theileria annulata	Turkey	Adıyaman		282	3
Theileria annulata	Turkey	Amasya		46	
Theileria annulata	Turkey	Ankara		844	
Theileria annulata	Turkey	Ankara		845	3
Theileria annulata	Turkey	Antalya		508	
Theileria annulata	Turkey	Bingöl		22	
Theileria annulata	Turkey	Bingöl		282	3
Theileria annulata	Turkey	Elâziğ		22	
Theileria annulata	Turkey	Elâziğ		282	3
Theileria annulata	Turkey	Erzincan		282	3
Theileria annulata	Turkey	Erzurum		22	
Theileria annulata	Turkey	Erzurum		282	3
Theileria annulata	Turkey	Giresun		46	
Theileria annulata	Turkey	Gümüshane		46	
Theileria annulata	Turkey	Kars		282	
Theileria annulata	Turkey	Kayseri		456	
Theileria annulata	Turkey	Malatya		282	3
Theileria annulata	Turkey	Mus		22	
Theileria annulata	Turkey	Mus		282	3
Theileria annulata	Turkey	Şanlıurfa		282	3
Theileria annulata	Turkey	Tekirdağ		43	4
Theileria annulata	Turkey	Tokat		46	
Theileria annulata	Turkey	Trabzon		46	
Theileria annulata	Turkey	Van		282	3
Theileria buffeli	Italy	Apulia	Foggia	156	
Theileria buffeli	Italy	Basilicata	Matera	156	
Theileria buffeli	Italy	Basilicata	Potenza	156	
Theileria buffeli	Italy	Sicily	Palermo	366	4
Theileria buffeli	Spain	País Vasco		363	
Theileria buffeli	Spain	País Vasco		364	
Theileria buffeli	Tunisia	Bizerte		641	2
Theileria buffeli	Tunisia	Nabeul		641	2
Theileria buffeli	Tunisia	Zaghouan		641	2
Theileria buffeli	Turkey	Ankara		254	

Continued

Table A1.8. Continued.

Species	Country	Administrative region 1	Administrative region 2	Reference ID	No. entries[a]
Theileria buffeli	Turkey	Bingöl		22	
Theileria buffeli	Turkey	Elâziğ		22	
Theileria buffeli	Turkey	Erzurum		22	
Theileria buffeli	Turkey	Kayseri		456	
Theileria buffeli	Turkey	Mus		22	
Theileria caballi	Croatia	Bjelovarska-Bilogorska		88	
Theileria caballi	Croatia	Grad Zagreb		88	2
Theileria caballi	Croatia	Medimurska		88	
Theileria caballi	Croatia	Varaždinska		88	
Theileria equi	Croatia	Bjelovarska-Bilogorska		88	
Theileria equi	Croatia	Grad Zagreb		88	2
Theileria equi	Croatia	Medimurska		88	
Theileria equi	Croatia	Varaždinska		88	
Theileria equi	Italy	Calabria		997	2
Theileria equi	Italy	Lazio		997	2
Theileria equi	Italy	Marche		997	2
Theileria equi	Italy	Sardegna		997	2
Theileria equi	Italy	Sicily		997	2
Theileria equi	Italy	Toscana		606	
Theileria equi	Italy	Umbria		997	2
Theileria equi	Italy	Veneto		997	2
Theileria equi	Morocco	Fès-Boulemane	Fès	800	
Theileria equi	Morocco	Grand Casablanca	Rabat	800	
Theileria equi	Morocco	Marrakech-Tensift-Al Haouz	Marrakech	800	
Theileria equi	Morocco	Taza-Al Hoceima-Taounate	Taza	800	
Theileria equi	Spain	Galicia		146	
Theileria equi	Spain	País Vasco		363	2
Theileria equi	Turkey	Ankara		390	2
Theileria orientalis	Turkey	Bingöl		22	
Theileria orientalis	Turkey	Elâziğ		22	
Theileria orientalis	Turkey	Erzurum		22	
Theileria orientalis	Turkey	Mus		22	
Theileria ovis	Croatia	Splitsko-Dalmatinska		277	
Theileria ovis	Spain	País Vasco		363	
Theileria ovis	Spain	País Vasco		677	
Theileria ovis	Turkey	Aksaray		1353	4
Theileria ovis	Turkey	Bingöl		1353	
Theileria ovis	Turkey	Çankırı		1353	
Theileria ovis	Turkey	Çankırı		1353	
Theileria ovis	Turkey	Diyarbakır		44	2
Theileria ovis	Turkey	Elâziğ		1353	
Theileria ovis	Turkey	Elâziğ		21	6
Theileria ovis	Turkey	Erzincan		44	2
Theileria ovis	Turkey	Erzurum		44	2
Theileria ovis	Turkey	Kars		44	2
Theileria ovis	Turkey	Malatya		44	2
Theileria ovis	Turkey	Mardin		44	2
Theileria ovis	Turkey	Mersin		1353	2
Theileria ovis	Turkey	Mus		44	2
Theileria ovis	Turkey	Van		1353	
Theileria sergenti	Turkey	Bingöl		22	
Theileria sergenti	Turkey	Elâziğ		22	

Continued

Table A1.8. Continued.

Species	Country	Administrative region 1	Administrative region 2	Reference ID	No. entries[a]
Theileria sergenti	Turkey	Erzurum		22	
Theileria sergenti	Turkey	Mus		22	
Theileria spp.	Switzerland	Graubünden		1211	
Theileria spp.	Switzerland	Graubünden		438	
Theileria spp.	Turkey	Amasya		46	
Theileria spp.	Turkey	Giresun		46	
Theileria spp.	Turkey	Gümüshane		46	
Theileria spp.	Turkey	Tokat		46	
Theileria spp.	Turkey	Trabzon		46	

[a]If more than one.

Table A1.9. Geographic distribution data for equine piroplasmoses. See Appendix 2 (hosted on http://www.cabi.org/vetmedresource/) for complete reference indicated by the Reference ID.

Species	Country	Administrative region 1	Administrative region 2	Reference ID	No. entries[a]
Theileria caballi	Croatia	Bjelovarska-Bilogorska		88	
Theileria equi	Croatia	Bjelovarska-Bilogorska		88	
Theileria caballi	Croatia	Grad Zagreb		88	2
Theileria equi	Croatia	Grad Zagreb		88	2
Theileria caballi	Croatia	Medimurska		88	
Theileria equi	Croatia	Medimurska		88	
Theileria caballi	Croatia	Varazdinska		88	
Theileria equi	Croatia	Varazdinska		88	
Ehrlichia equi	Denmark	North Jutland		946	
Babesia caballi	Hungary	Hajdú-Bihar		441	
Babesia caballi	Italy	Calabria		997	2
Theileria equi	Italy	Calabria		997	2
Babesia caballi	Italy	Emilia-Romagna		997	2
Babesia caballi	Italy	Emilia-Romagna	Bologna	755	
Babesia equi	Italy	Emilia-Romagna	Bologna	755	
Babesia caballi	Italy	Lazio		997	2
Theileria equi	Italy	Lazio		997	2
Babesia caballi	Italy	Lazio	Roma	755	
Babesia equi	Italy	Lazio	Roma	755	
Babesia caballi	Italy	Marche		997	2
Theileria equi	Italy	Marche		997	2
Babesia caballi	Italy	Sardegna		997	2
Theileria equi	Italy	Sardegna		997	2
Babesia caballi	Italy	Sicily		997	2
Theileria equi	Italy	Sicily		997	2
Babesia caballi	Italy	Toscana		606	2
Theileria equi	Italy	Toscana		606	
Babesia caballi	Italy	Umbria		997	2
Theileria equi	Italy	Umbria		997	2
Babesia caballi	Italy	Umbria	Perugia	755	
Babesia equi	Italy	Umbria	Perugia	755	
Theileria equi	Italy	Veneto		997	2
Babesia caballi	Italy	Veneto	Padua	755	

Continued

Table A1.9. Continued.

Species	Country	Administrative region 1	Administrative region 2	Reference ID	No. entries[a]
Babesia equi	Italy	Veneto	Padua	755	
Theileria equi	Morocco	Fès-Boulemane	Fès	800	
Theileria equi	Morocco	Grand Casablanca	Rabat	800	
Theileria equi	Morocco	Marrakech-Tensift-Al Haouz	Marrakech	800	
Theileria equi	Morocco	Taza-Al Hoceima-Taounate	Taza	800	
Babesia caballi	Spain	Galicia		146	
Theileria equi	Spain	Galicia		146	
Babesia caballi	Spain	País Vasco		363	3
Theileria equi	Spain	País Vasco		363	2
Babesia caballi	Turkey	Ankara		390	2
Theileria equi	Turkey	Ankara		390	2
Babesia caballi	Turkey	Malatya		25	
Babesia equi	Turkey	Malatya		25	

[a]If more than one.

Table A1.10. Geographic distribution data for *Bartonella* spp. See Appendix 2 (hosted on http://www.cabi.org/vetmedresource/) for complete reference indicated by the Reference ID.

Species	Country	Administrative region 1	Administrative region 2	Reference ID	No. entries[a]
Bartonella henselae	Croatia	Zadarska		1194	
Bartonella henselae	Denmark	Frederiksborg	Helsingør	304	
Bartonella henselae	Denmark	Fyn	Søndersø	304	
Bartonella henselae	France	Île-de-France		1179	
Bartonella henselae	France	Lorraine	Meurthe-et-Moselle	1179	
Bartonella henselae	France	Provence-Alpes-Côte-d'Azur	Bouches-du-Rhône	1179	
Bartonella henselae	France	Rhône-Alpes	Rhône	1179	
Bartonella henselae	Greece	Kriti		1246	
Bartonella henselae	Italy	Emilia-Romagna		1182	2
Bartonella henselae	Italy	Emilia-Romagna	Modena	1181	
Bartonella henselae	Italy	Friuli-Venezia Giulia		830	
Bartonella henselae	Italy	Lombardia		1182	2
Bartonella henselae	Italy	Sardegna		1182	
Bartonella henselae	Italy	Toscana		1182	
Bartonella henselae	Italy	Toscana		1195	
Bartonella henselae	Italy	Toscana		1237	4
Bartonella henselae	Norway	Sør-Trøndelag		1161	
Bartonella henselae	Poland			1177	
Bartonella henselae	Poland	Lódz		1138	
Bartonella henselae	Poland	Lublin		1138	
Bartonella henselae	Poland	Masovian		1138	
Bartonella henselae	Poland	Masovian	Warsaw	763	2
Bartonella henselae	Spain	Andalucía		1128	4
Bartonella henselae	Spain	Cataluña	Barcelona	1261	
Bartonella henselae	Turkey	Ankara		1175	

Continued

Table A1.10. Continued.

Species	Country	Administrative region 1	Administrative region 2	Reference ID	No. entries[a]
Bartonella quintana	France	Île-de-France		1206	
Bartonella quintana	France	Provence-Alpes-Côte-d'Azur	Bouches-du-Rhône	1265	
Bartonella quintana	Italy	Toscana		1195	
Bartonella quintana	Poland			1177	
Bartonella quintana	Poland	Masovian		1138	
Bartonella spp.	Albania	Tiranë		409	
Bartonella spp.	Algeria	Alger		1167	4
Bartonella spp.	Algeria	Biskra		1167	4
Bartonella spp.	Algeria	Mascara		1167	4
Bartonella spp.	Algeria	Tiaret		1167	4
Bartonella spp.	Czech Republic	Jihomoravský		427	4
Bartonella spp.	Czech Republic	Prague		427	2
Bartonella spp.	France	Nord-Pas-de-Calais	Nord	408	
Bartonella spp.	Greece	Thessalia	Thessaloniki	1191	
Bartonella spp.	Italy	Apulia	Taranto	1191	
Bartonella spp.	Italy	Basilicata		1191	
Bartonella spp.	Spain	Andalucía		1128	2
Bartonella vinsonii	Denmark	Fyn	Søndersø	304	2

[a]If more than one.

Table A1.11. Geographic distribution data for *Francisella tularensis*. See Appendix 2 (hosted on http://www.cabi.org/vetmedresource/) for complete reference indicated by the Reference ID.

Country	Administrative region 1	Administrative region 2	Reference ID	No. entries[a]
Austria	Burgenland		259	
Austria	Niederösterreich		259	
Austria	Niederösterreich	Gänserndorf	1053	
Austria	Niederösterreich	Gänserndorf	401	5
Austria	Niederösterreich	Korneuburg	401	
Austria	Niederösterreich	Mistelbach	1053	
Austria	Steiermark		259	
Croatia	Koprivnicko-Krizevacka		656	
Czech Republic	Jihomoravský		1053	
Czech Republic	Jihomoravský		171	
France	Pays de la Loire	Vendée	882	
Germany	Hessen		509	2
Germany	Niedersachsen		509	7
Slovakia	Banskobystrický		401	2
Slovakia	Trnavský		401	4
Spain	Castilla y León		39	3
Spain	Castilla y León	Soria	31	
Spain	Castilla y León	Valladolid	617	
Spain	Castilla y León	Valladolid	91	
Sweden	Dalarna		300	
Sweden	Gävleborg		300	

Continued

Table A1.11. Continued.

Country	Administrative region 1	Administrative region 2	Reference ID	No. entries[a]
Sweden	Jämtland		300	
Sweden	Västernorrland		300	
Turkey	Amasya		560	
Turkey	Ankara		19	
Turkey	Antalya		19	
Turkey	Bursa		19	
Turkey	Düzce		710	
Turkey	Edirne		19	
Turkey	Edirne		250	
Turkey	Kars		19	
Turkey	Kars		825	
Turkey	Kastamonu		19	
Turkey	Kırklareli		19	
Turkey	Kırklareli		250	
Turkey	Kocaeli		19	
Turkey	Samsun		19	
Turkey	Tekirdağ		19	
Turkey	Tekirdağ		250	
Turkey	Van		19	
Turkey	Zinguldak		19	

[a]If more than one.

Table A1.12. Geographic distribution data for *Coxiella burnetii*. See Appendix 2 (hosted on http://www.cabi.org/vetmedresource/) for complete reference indicated by the Reference ID.

Country	Administrative region 1	Administrative region 2	Reference ID	No. entries[a]
Albania	Berat		1174	2
Albania	Dibër	Dibrës	1174	3
Albania	Dibër	Matit	1174	
Albania	Elbasan		1174	3
Albania	Elbasan	Gramshit	1174	3
Albania	Elbasan	Librazhdit	1174	2
Albania	Fier		1174	
Albania	Fier	Lushnjës	1174	2
Albania	Gjirokastër		1174	
Albania	Gjirokastër	Përmetit	1174	3
Albania	Korçë	Kolonjës	1174	
Albania	Lezhë		1174	2
Albania	Tiranë	Kavajës	1174	
Albania	Tiranë	Tiranës	1174	3
Algeria	Sétif		1221	
Belgium	Antwerp		1198	
Belgium	East Flanders		1198	
Belgium	Flemish Brabant		1198	
Belgium	Hainaut		1198	
Belgium	West Flanders		1198	
Bosnia and Herzegovina	Federacija Bosna i Hercegovina		1209	9

Continued

Table A1.12. Continued.

Country	Administrative region 1	Administrative region 2	Reference ID	No. entries[a]
Bosnia and Herzegovina	Federacija Bosna i Hercegovina		1276	
Bosnia and Herzegovina	Republika Srpska		1276	
Bulgaria	Sofia	Botevgrad	1254	
Cyprus			1118	
Cyprus	Famagusta		777	4
Cyprus	Larnaca		777	4
Cyprus	Limassol		777	3
Cyprus	Nicosia		777	4
Cyprus	Paphos		777	5
Denmark	Århus		1149	
Denmark	Frederiksborg		1149	
Denmark	Fyn		1149	
Denmark	North Jutland		1149	
Denmark	Ribe		1149	
Denmark	Ringkøbing		1149	
Denmark	South Jutland		1149	
Denmark	Storstrøm		1149	
Denmark	Vejle		1149	
Denmark	Vestsjælland		1149	
Denmark	Viborg		1149	
Egypt	Al Jizah		1343	
Egypt	Qina		1343	
Egypt	Shamal Sina'		1343	
France	Centre		1164	
Germany	Baden-Württemberg		599	
Germany	Bayern		599	
Germany	Berlin		599	
Germany	Brandenburg		599	
Germany	Bremen		599	
Germany	Hamburg		599	
Germany	Hessen		599	
Germany	Mecklenburg-Vorpommern		599	
Germany	Niedersachsen		599	
Germany	Nordrhein-Westfalen		599	
Germany	Rheinland-Pfalz		599	
Germany	Saarland		599	
Germany	Sachsen		599	
Germany	Sachsen-Anhalt		599	
Germany	Schleswig-Holstein		599	
Germany	Thüringen		599	
Greece	Anatoliki Makedonia kai Thraki	Drama	1254	
Greece	Anatoliki Makedonia kai Thraki	Evros	1254	
Greece	Anatoliki Makedonia kai Thraki	Kavala	1254	
Greece	Anatoliki Makedonia kai Thraki	Rodopi	1254	
Greece	Anatoliki Makedonia kai Thraki	Xanthi	1254	

Continued

Table A1.12. Continued.

Country	Administrative region 1	Administrative region 2	Reference ID	No. entries[a]
Greece	Dytiki Makedonia	Florina	1254	
Greece	Dytiki Makedonia	Grevena	1254	
Greece	Dytiki Makedonia	Grevena	1254	
Greece	Dytiki Makedonia	Kastoria	1254	
Greece	Dytiki Makedonia	Kozani	1254	
Greece	Kentriki Makedonia	Imathia	1254	
Greece	Kentriki Makedonia	Khalkidiki	1254	
Greece	Kentriki Makedonia	Kilkis	1254	
Greece	Kentriki Makedonia	Pella	1254	
Greece	Kentriki Makedonia	Pieria	1254	
Greece	Kentriki Makedonia	Serrai	1254	
Greece	Kentriki Makedonia	Thessaloniki	1254	
Israel	Haifa		1252	
Israel	Jerusalem		1275	
Italy	Apulia		1257	3
Italy	Apulia	Bari	1247	
Italy	Basilicata		1257	2
Italy	Campania		1259	
Italy	Emilia-Romagna	Modena	1354	
Italy	Emilia-Romagna	Parma	1354	
Italy	Emilia-Romagna	Reggio Nell'Emilia	1354	
Italy	Lombardia		1197	12
Italy	Lombardia	Cremona	1300	
Italy	Lombardia	Mantua	1300	
Italy	Sardegna		1236	2
Italy	Sicily		999	
Italy	Veneto	Padua	1300	
Montenegro	Žabljak		1222	
Montenegro	Andrijevica		1222	
Montenegro	Berane		1222	
Montenegro	Cetinje		1222	
Montenegro	Danilovgrad		1222	
Montenegro	Mojkovac		1222	
Montenegro	Nikoic		1222	
Montenegro	Pljevlja		1222	
Montenegro	Plužine		1222	
Montenegro	Podgorica		1222	
Montenegro	Ulcinj		1222	
Netherlands	Gelderland		1271	
Netherlands	Noord-Brabant		1271	
Poland	Lublin		1184	
Slovakia	Košický		1193	
Spain	Castilla y León	Soria	1250	
Spain	Castilla-La Mancha	Albacete	1158	
Spain	Castilla-La Mancha	Toledo	989	3
Spain	Cataluña	Barcelona	1153	
Spain	País Vasco		1201	
Spain	País Vasco		1266	3
Spain	País Vasco		78	
Spain	País Vasco		79	
Tunisia	Sousse		1159	
Turkey	Ankara		1214	

Continued

Table A1.12. Continued.

Country	Administrative region 1	Administrative region 2	Reference ID	No. entries[a]
Turkey	Ankara		1216	
Turkey	Antalya		1160	
Turkey	Antalya		1204	
Turkey	Aydın		1215	
Turkey	Bolu		1213	
Turkey	Diyarbakır		1160	
Turkey	Kayseri		1214	
Turkey	Niğde		1214	
Turkey	Samsun		1160	
Turkey	Tokat		1203	
UK	Northern Ireland	Antrim	1239	
UK	Northern Ireland	Ballymena	1239	
UK	Northern Ireland	Banbridge	1239	
UK	Northern Ireland	Craigavon	1239	
UK	Northern Ireland	Derry	1239	
UK	Northern Ireland	Down	1239	
UK	Northern Ireland	Dungannon	1239	
UK	Northern Ireland	Fermanagh	1239	
UK	Northern Ireland	Limavady	1239	
UK	Northern Ireland	Magherafelt	1239	
UK	Northern Ireland	Moyle	1239	
UK	Northern Ireland	Omagh	1239	
UK	Northern Ireland	Strabane	1239	

[a]If more than one.

Table A1.13. Geographic distribution data for *Ornithodoros* spp. See Appendix 2 (hosted on http://www.cabi.org/vetmedresource/) for complete reference indicated by the Reference ID.

Country	Administrative region 1	Administrative region 2	Reference ID
Algeria	El Tarf	Ain El Assel	0
Armenia	Lake Sevan		0
Armenia	Lori		0
Armenia	Shirak		0
Armenia	Tavush		0
Egypt	Ad Daqahliyah		0
Egypt	Al Buhayrah		0
Egypt	Al Gharbiyah		0
Egypt	Al Iskandariyah		0
Egypt	Al Isma'iliyah		0
Egypt	Al Jizah		1101
Egypt	Al Minufiyah		0
Egypt	Al Qahirah		0
Egypt	Al Qalyubiyah		0
Egypt	Ash Sharqiyah		0
Egypt	Bur Sa'id		0
Egypt	Dumyat		0
Egypt	Kafr ash Shaykh		0
Egypt	Matruh		0
Egypt	Shamal Sina'		0

Continued

Table A1.13. Continued.

Country	Administrative region 1	Administrative region 2	Reference ID
Georgia	Abkhazia	Gagra	1398
Georgia	Ajaria	Batumi	1398
Georgia	Guria	Chokhatauri	1398
Georgia	Imereti	Bagdati	1398
Georgia	Kakheti	Akhmeta	1398
Georgia	Kvemo Kartli	Bolnisi	1398
Georgia	Mtskheta-Mtianeti	Akhalgori	1398
Georgia	Racha-Lechkhumi-Kvemo Svaneti	Ambrolauri	1398
Georgia	Samegrelo-Zemo Svaneti	Abasha	1398
Georgia	Samtskhe-Javakheti	Adigeni	1398
Georgia	Shida Kartli	Gori	1398
Georgia	Tbilisi	Tbilisi	1398
Israel	Golan		0
Israel	HaDarom		0
Israel	Haifa		0
Israel	HaMerkaz		0
Israel	HaZafon		0
Israel	Jerusalem		0
Israel	Tel Aviv		0
Jordan	Ajlun	Ajloun	0
Jordan	Amman	Amman	0
Jordan	Aqaba	Aqaba	0
Jordan	Balqa	Al-Balqa	0
Jordan	Irbid	Aghwar Shamaliyyeh	0
Jordan	Jarash	Jarash	0
Jordan	Karak	Ayy	0
Jordan	Madaba	Dhiban	0
Jordan	Tafilah	Bsaira	0
Lebanon	An Nabatiyah	Hasbaya	0
Lebanon	Beirut	Beirut	0
Lebanon	South Lebanon	Jezzine	0
Morocco	Chaouia-Ouardigha	Ben Slimane	0
Morocco	Doukkala-Abda	El Jadida	0
Morocco	Gharb-Chrarda-Béni Hssen	Kénitra	0
Morocco	Grand Casablanca	Casablanca	0
Morocco	Guelmim-Es-Semara	Assa-Zag	0
Morocco	Marrakech-Tensift-Al Haouz	Al Haouz	0
Morocco	Oriental	Berkane Taourirt	0
Morocco	Rabat-Salé-Zemmour-Zaer	Khémisset	0
Morocco	Souss-Massa-Draâ	Agadir-Ida ou Tanane	0
Morocco	Tadla-Azilal	Azilal	0
Morocco	Tanger-Tétouan	Chefchaouen	0
Palestine	Gaza	Deir Al-Balah	0
Palestine	West Bank	Bethlehem	0
Portugal	Beja	Aljustrel	0
Portugal	Castelo Branco	Belmonte	0
Portugal	Coimbra	Arganil	0
Portugal	Faro	Albufeira	0
Portugal	Leiria	Alcobaça	0
Portugal	Lisboa	Alenquer	0
Portugal	Portalegre	Alter do Chão	0
Portugal	Santarém	Mação	0

Continued

Table A1.13. Continued.

Country	Administrative region 1	Administrative region 2	Reference ID
Portugal	Setúbal	Alcácer do Sal	0
Portugal	Évora	Alandroal	0
Russia	Chechnya	Achkhoy-Martanovskiy rayon	0
Russia	Ingush	Malgobekskiy rayon	0
Russia	Kabardin-Balkar	Baksanskiy rayon	0
Russia	North Ossetia	Alagirskiy rayon	0
Spain	Andalucía	Almería	0
Spain	Extremadura	Badajoz	0
Tunisia	Ariana	Ariana Médina	0
Tunisia	Ben Arous (Tunis Sud)	Ben Arous	0
Tunisia	Bizerte	Bizerte Nord	0
Tunisia	Béja	Amdoun	0
Tunisia	Gabès	Gabès Médina	0
Tunisia	Jendouba	Aïn Draham	0
Tunisia	Manubah	Borj El Amri	0
Tunisia	Médenine	Ben Guerdane	0
Tunisia	Nabeul	Beni Khalled	0
Tunisia	Tunis	Bab Bhar	0
Tunisia	Zaghouan	Bir Mchergua	0
Turkey	Artvin		0
Turkey	Rize		0
Turkey	Trabzon		0

Table A1.14. Geographic distribution data for *Ixodes ricinus*. See Appendix 2 (hosted on http://www.cabi.org/vetmedresource/) for complete reference indicated by the Reference ID.

Country	Administrative region 1	Administrative region 2	Reference ID	No. entries[a]
Algeria	Alger	Ain Benian	0	3
Algeria	Constantine	Didouche Mourad	0	
Algeria	Guelma	Hammam Debagh	0	
Algeria	Guelma	Hammam Debagh	0	
Algeria	Tlemcen	Tlemcen	0	5
Austria	Burgenland	Oberpullendorf	0	
Austria	Burgenland	Oberwart	0	
Austria	Kärnten	Villach	0	
Austria	Kärnten	Wolfsberg	0	
Austria	Niederösterreich	Mistelbach	0	
Austria	Oberösterreich	Braunau am Inn	0	
Austria	Oberösterreich	Urfahr Umgebung	0	
Austria	Salzburg	Hallein	0	
Austria	Salzburg	Sankt Johann im Pongau	0	
Austria	Steiermark	Deutschlandsberg	0	
Austria	Steiermark	Graz Umgebung	0	
Austria	Steiermark	Mürzzuschlag	0	
Austria	Steiermark	Voitsberg	0	
Austria	Steiermark	Weiz	0	
Austria	Tirol	Imst	0	
Austria	Tirol	Schwaz	0	
Austria	Vorarlberg	Bludenz	0	1

Continued

Table A1.14. Continued.

Country	Administrative region 1	Administrative region 2	Reference ID	No. entries[a]
Austria	Vorarlberg	Feldkirch	0	
Belgium	Antwerp	Turnhout	0	
Belgium	Liège	Liège	0	
Belgium	West Flanders	Brugge	0	
Bulgaria	Blagoevgrad	Gotse Delchev	0	4
Bulgaria	Blagoevgrad	Sandanski	0	2
Bulgaria	Burgas	Burgas	0	12
Bulgaria	Burgas	Kameno	0	2
Bulgaria	Burgas	Malko Tarnovo	0	
Bulgaria	Burgas	Tsarevo	0	2
Bulgaria	Dobrich	Shabla	0	2
Bulgaria	Grad Sofiya	Stolichna	0	10
Bulgaria	Kardzhali	Kirkovo	0	2
Bulgaria	Kardzhali	Momchilgrad	0	2
Bulgaria	Kyustendil	Rila	0	2
Bulgaria	Lovech	Letnitsa	0	2
Bulgaria	Montana	Montana	0	2
Bulgaria	Pazardzhik	Pazardzhik	0	2
Bulgaria	Pazardzhik	Strelcha	0	2
Bulgaria	Plovdiv	Krichim	0	4
Bulgaria	Plovdiv	Plovdiv	0	2
Bulgaria	Sliven	Sliven	0	8
Bulgaria	Smolyan	Zlatograd	0	2
Bulgaria	Sofia	Svoge	0	2
Bulgaria	Sofia	Zlatitsa	0	6
Bulgaria	Stara Zagora	Chirpan	0	2
Bulgaria	Stara Zagora	Haskovo	69	
Bulgaria	Stara Zagora	Pavel Banya	0	2
Bulgaria	Varna	Valchi Dol	0	
Bulgaria	Vidin	Belogradchik	0	2
Bulgaria	Vratsa	Vratsa	0	2
Bulgaria	Yambol	Bolyarovo	0	2
Croatia	Medimurska		0	4
Croatia	Primorsko-Goranska		0	4
Croatia	Varaždinska		0	
Czech Republic	Jihočeský		216	
Czech Republic	Jihočeský	Ceské Budejovice	0	10
Czech Republic	Jihočeský	Jindrichuv Hradec	0	4
Czech Republic	Jihočeský	Písek	0	4
Czech Republic	Jihočeský	Tábor	0	4
Czech Republic	Jihomoravský		1082	
Czech Republic	Jihomoravský		427	4
Czech Republic	Jihomoravský		490	2
Czech Republic	Jihomoravský		742	
Czech Republic	Jihomoravský	Blansko	0	6
Czech Republic	Jihomoravský	Breclav	0	12
Czech Republic	Jihomoravský	Brno	0	4
Czech Republic	Jihomoravský	Brno	472	
Czech Republic	Jihomoravský	Brno-Venkov	0	6
Czech Republic	Jihomoravský	Vyškov	0	3
Czech Republic	Jihomoravsky	Znojmo	0	13
Czech Republic	Kraj Vysočina	Žd'ár nad Sázavou	0	

Continued

Table A1.14. Continued.

Country	Administrative region 1	Administrative region 2	Reference ID	No. entries[a]
Czech Republic	Kraj Vysočina	Trebíc	0	
Czech Republic	Královéhradecký	Náchod	0	3
Czech Republic	Královéhradecký	Trutnov	0	2
Czech Republic	Královéhradecký	Trutnov	620	
Czech Republic	Královéhradecký	Trutnov	621	
Czech Republic	Liberecký		271	24
Czech Republic	Liberecký	Liberec	0	
Czech Republic	Liberecký	Liberec	472	
Czech Republic	Moravskoslezský		220	
Czech Republic	Moravskoslezský	Bruntál	0	5
Czech Republic	Moravskoslezský	Opava	0	4
Czech Republic	Olomoucký	Jeseník	0	
Czech Republic	Olomoucký	Olomouc	0	
Czech Republic	Olomoucký	Prostejov	0	
Czech Republic	Pardubický	Ústí nad Orlicí	0	
Czech Republic	Pardubický	Ústí nad Orlicí	472	
Czech Republic	Plzeňský	Domaûlice	0	
Czech Republic	Plzeňský	Klatovy	0	4
Czech Republic	Plzeňský	Plzen	0	2
Czech Republic	Plzeňský	Plzeň-Jih	0	6
Czech Republic	Plzeňský	Plzeň-Sever	0	7
Czech Republic	Plzeňský	Rokycany	0	2
Czech Republic	Plzeňský	Tachov	0	
Czech Republic	Prague		216	
Czech Republic	Prague		427	2
Czech Republic	Středočeský	Benešov	0	5
Czech Republic	Středočeský	Beroun	0	3
Czech Republic	Středočeský	Kolín	0	2
Czech Republic	Středočeský	Praha-v chod	0	2
Czech Republic	Středočeský	Praha-západ	0	2
Czech Republic	Středočeský	Príbram	0	4
Czech Republic	Zlínský	Kromerí	0	2
Denmark	Århus		888	
Denmark	Århus	Ebeltoft	0	
Denmark	Århus	Hammel	0	
Denmark	Århus	Skanderborg	0	
Denmark	Bornholm		888	
Denmark	Copenhagen		478	2
Denmark	Copenhagen		479	
Denmark	Copenhagen		888	
Denmark	Frederiksborg		888	
Denmark	Frederiksborg	Helsinge	0	
Denmark	Fyn		888	
Denmark	North Jutland		888	
Denmark	Ribe		888	
Denmark	Ringkøbing		888	
Denmark	Roskilde		888	
Denmark	South Jutland		888	
Denmark	Storstrøm		888	
Denmark	Vejle		888	
Denmark	Vestsjælland		888	
Denmark	Vestsjælland	Korsør	973	

Continued

Table A1.14. Continued.

Country	Administrative region 1	Administrative region 2	Reference ID	No. entries[a]
Denmark	Viborg		888	
Estonia	Harju		0	27
Estonia	Hiiu		0	13
Estonia	Järva		0	18
Estonia	Jıgeva		0	13
Estonia	Lääne		0	17
Estonia	Lääne-Viru		0	26
Estonia	Pärnu		0	30
Estonia	Põlva		0	8
Estonia	Rapla		0	23
Estonia	Saare		0	24
Estonia	Tartu		0	11
Estonia	Valga		0	12
Estonia	Viljandi		0	17
Estonia	Vıru		0	11
Faroe Islands			471	5
Finland	Southern Finland	Uusimaa	0	2
France	Alsace	Bas-Rhin	0	60
France	Alsace	Haut-Rhin	0	28
France	Alsace	Haut-Rhin	331	
France	Aquitaine	Dordogne	0	39
France	Aquitaine	Gironde	0	21
France	Aquitaine	Landes	0	24
France	Aquitaine	Lot-et-Garonne	0	7
France	Aquitaine	Pyrénées-Atlantiques	0	18
France	Auvergne	Allier	0	33
France	Auvergne	Cantal	0	31
France	Auvergne	Haute-Loire	0	
France	Auvergne	Puy-de-Dôme	0	7
France	Auvergne	Puy-de-Dôme	132	7
France	Basse-Normandie	Calvados	0	22
France	Basse-Normandie	Manche	0	24
France	Basse-Normandie	Orne	0	14
France	Bourgogne	Côte-d'Or	0	11
France	Bourgogne	Nièvre	0	9
France	Bourgogne	Saône-et-Loire	0	45
France	Bourgogne	Yonne	0	30
France	Bretagne	Côtes-d'Armor	0	88
France	Bretagne	Finistère	0	37
France	Bretagne	Ille-et-Vilaine	0	43
France	Bretagne	Morbihan	0	50
France	Centre	Cher	0	30
France	Centre	Eure-et-Loir	0	10
France	Centre	Indre	0	21
France	Centre	Indre-et-Loire	0	29
France	Centre	Loir-et-Cher	0	12
France	Centre	Loiret	0	10
France	Champagne-Ardenne	Ardennes	0	3
France	Champagne-Ardenne	Aube	0	38
France	Champagne-Ardenne	Haute-Marne	0	24
France	Champagne-Ardenne	Marne	0	16
France	Corse	Corse-du-Sud	0	2

Continued

Table A1.14. Continued.

Country	Administrative region 1	Administrative region 2	Reference ID	No. entries[a]
France	Franche-Comté	Doubs	0	7
France	Franche-Comté	Haute-Saône	0	8
France	Franche-Comté	Jura	0	15
France	Haute-Normandie	Eure	0	20
France	Haute-Normandie	Seine-Maritime	0	4
France	Île-de-France	Essonne	0	18
France	Île-de-France	Seine-et-Marne	0	31
France	Île-de-France	Val-d'Oise	0	10
France	Île-de-France	Val-de-Marne	0	8
France	Île-de-France	Ville de Paris	0	8
France	Île-de-France	Yvelines	0	40
France	Île-de-France	Yvelines	1024	
France	Languedoc-Roussillon	Aude	0	8
France	Languedoc-Roussillon	Gard	0	2
France	Languedoc-Roussillon	Hérault	0	
France	Languedoc-Roussillon	Pyrénées-Orientales	0	10
France	Limousin	Corrèze	0	14
France	Limousin	Creuse	0	21
France	Limousin	Haute-Vienne	0	4
France	Lorraine	Meurthe-et-Moselle	0	24
France	Lorraine	Meuse	0	50
France	Lorraine	Vosges	0	19
France	Midi-Pyrénées	Ariège	0	17
France	Midi-Pyrénées	Aveyron	0	20
France	Midi-Pyrénées	Gers	0	11
France	Midi-Pyrénées	Haute-Garonne	0	13
France	Midi-Pyrénées	Hautes-Pyrénées	0	14
France	Midi-Pyrénées	Lot	0	2
France	Midi-Pyrénées	Tarn	0	8
France	Midi-Pyrénées	Tarn-et-Garonne	0	4
France	Nord-Pas-de-Calais	Nord	0	6
France	Nord-Pas-de-Calais	Pas-de-Calais	0	2
France	Pays de la Loire	Loire-Atlantique	0	23
France	Pays de la Loire	Maine-et-Loire	0	16
France	Pays de la Loire	Mayenne	0	14
France	Pays de la Loire	Sarthe	0	24
France	Pays de la Loire	Vendée	0	7
France	Picardie	Aisne	0	18
France	Picardie	Oise	0	26
France	Picardie	Somme	0	2
France	Poitou-Charentes	Charente	0	11
France	Poitou-Charentes	Charente-Maritime	0	9
France	Poitou-Charentes	Deux-Sèvres	0	8
France	Poitou-Charentes	Vienne	0	10
France	Provence-Alpes-Côte-d'Azur	Alpes-De-Haute-Provence	0	6
France	Provence-Alpes-Côte-d'Azur	Alpes-Maritimes	0	4
France	Provence-Alpes-Côte-d'Azur	Bouches-du-Rhône	0	14
France	Provence-Alpes-Côte-d'Azur	Hautes-Alpes	0	8
France	Provence-Alpes-Côte-d'Azur	Var	0	10
France	Provence-Alpes-Côte-d'Azur	Vaucluse	0	2
France	Rhône-Alpes	Ain	0	208

Continued

Table A1.14. Continued.

Country	Administrative region 1	Administrative region 2	Reference ID	No. entries[a]
France	Rhône-Alpes	Ardèche	0	6
France	Rhône-Alpes	Drôme	0	27
France	Rhône-Alpes	Haute-Savoie	0	87
France	Rhône-Alpes	Isère	0	159
France	Rhône-Alpes	Loire	0	4
France	Rhône-Alpes	Rhône	0	4
France	Rhône-Alpes	Savoie	0	83
Germany	Baden-Württemberg	Freiburg	0	36
Germany	Baden-Württemberg	Karlsruhe	0	18
Germany	Baden-Württemberg	Stuttgart	0	16
Germany	Baden-Württemberg	Tübingen	0	17
Germany	Bayern		332	2
Germany	Bayern		333	5
Germany	Bayern		564	
Germany	Bayern	Mittelfranken	0	6
Germany	Bayern	Niederbayern	0	52
Germany	Bayern	Oberbayern	0	29
Germany	Bayern	Oberfranken	0	2
Germany	Bayern	Oberpfalz	0	14
Germany	Bayern	Schwaben	0	4
Germany	Bayern	Unterfranken	0	4
Germany	Berlin		236	
Germany	Brandenburg		236	
Germany	Brandenburg	Brandenburg	0	294
Germany	Hessen	Darmstadt	496	5
Germany	Hessen	Gieflen	0	5
Germany	Hessen	Kassel	0	
Germany	Mecklenburg-Vorpommern	Mecklenburg-Vorpommern	0	114
Germany	Niedersachsen	Braunschweig	0	
Germany	Nordrhein-Westfalen		600	
Germany	Nordrhein-Westfalen		857	
Germany	Nordrhein-Westfalen	Köln	0	3
Germany	Nordrhein-Westfalen	Munster	0	
Germany	Rheinland-Pfalz	Koblenz	0	
Germany	Rheinland-Pfalz	Rheinhessen-Pfalz	0	
Germany	Saarland	Saarland	0	
Germany	Sachsen	Chemnitz	0	2
Germany	Sachsen	Leipzig	0	
Germany	Sachsen-Anhalt	Dessau	0	15
Germany	Sachsen-Anhalt	Halle	0	
Germany	Sachsen-Anhalt	Magdeburg	0	12
Germany	Thüringen		268	
Germany	Thüringen	Thüringen	0	121
Greece	Anatoliki Makedonia kai Thraki		735	2
Greece	Anatoliki Makedonia kai Thraki	Drama	0	15
Greece	Anatoliki Makedonia kai Thraki	Evros	0	9
Greece	Anatoliki Makedonia kai Thraki	Kavala	0	12
Greece	Anatoliki Makedonia kai Thraki	Rodopi	0	
Greece	Dytiki Makedonia		735	2
Greece	Dytiki Makedonia	Florina	0	3
Greece	Dytiki Makedonia	Grevena	0	12

Continued

Table A1.14. Continued.

Country	Administrative region 1	Administrative region 2	Reference ID	No. entries[a]
Greece	Dytiki Makedonia	Kozani	0	8
Greece	Ipeiros	Preveza	0	2
Greece	Kentriki Makedonia		735	3
Greece	Kentriki Makedonia	Khalkidiki	0	6
Greece	Kentriki Makedonia	Khalkidiki	1303	3
Greece	Kentriki Makedonia	Kilkis	0	16
Greece	Kentriki Makedonia	Pieria	0	8
Greece	Kentriki Makedonia	Serrai	0	24
Greece	Kentriki Makedonia	Thessaloniki	0	7
Hungary	Bács-Kiskun		0	2
Hungary	Bács-Kiskun		322	2
Hungary	Bács-Kiskun		341	
Hungary	Baranya		0	
Hungary	Baranya		322	2
Hungary	Baranya		341	
Hungary	Békés		341	
Hungary	Borsod-Abaúj-Zemplén		0	
Hungary	Borsod-Abaúj-Zemplén		322	5
Hungary	Borsod-Abaúj-Zemplén		341	
Hungary	Borsod-Abaúj-Zemplén		443	
Hungary	Budapest		322	8
Hungary	Csongrád		341	
Hungary	Fejér		0	2
Hungary	Fejér		322	7
Hungary	Gyor-Moson-Sopron		0	
Hungary	Gyor-Moson-Sopron		322	
Hungary	Gyor-Moson-Sopron		341	
Hungary	Hajdú-Bihar		0	
Hungary	Hajdú-Bihar		322	
Hungary	Heves		0	
Hungary	Jász-Nagykun-Szolnok		0	
Hungary	Komárom-Esztergom		341	
Hungary	Pest		0	4
Hungary	Pest		344	3
Hungary	Somogy		341	
Hungary	Szabolcs-Szatmár-Bereg		341	
Hungary	Vas		341	
Hungary	Veszprém		0	3
Hungary	Veszprém		341	
Hungary	Zala		0	
Ireland	Cavan		0	3
Ireland	Clare		0	2
Ireland	Cork		0	5
Ireland	Donegal		0	2
Ireland	Dublin		0	3
Ireland	Galway		0	10
Ireland	Galway		380	
Ireland	Kerry		0	8
Ireland	Kerry		419	
Ireland	Kerry		420	
Ireland	Kerry		754	
Ireland	Kilkenny		0	

Continued

Table A1.14. Continued.

Country	Administrative region 1	Administrative region 2	Reference ID	No. entries[a]
Ireland	Limerick		0	2
Ireland	Longford		0	2
Ireland	Mayo		0	6
Ireland	Meath		0	
Ireland	Monaghan		0	2
Ireland	Offaly		0	
Ireland	Roscommon		0	2
Ireland	Sligo		0	
Ireland	Waterford		0	
Ireland	Wexford		0	
Ireland	Wicklow		0	2
Italy	Abruzzo	L'Aquila	0	32
Italy	Abruzzo	Teramo	0	4
Italy	Apulia	Bari	0	2
Italy	Apulia	Lecce	0	2
Italy	Basilicata	Matera	0	6
Italy	Basilicata	Potenza	0	8
Italy	Calabria	Catanzaro	0	2
Italy	Calabria	Reggio di Calabria	0	2
Italy	Campania		802	
Italy	Campania	Avellino	0	
Italy	Emilia-Romagna	Bologna	0	8
Italy	Emilia-Romagna	Ferrara	0	12
Italy	Emilia-Romagna	Forlì-Cesena	0	3
Italy	Emilia-Romagna	Parma	0	12
Italy	Emilia-Romagna	Piacenza	0	9
Italy	Emilia-Romagna	Ravenna	0	6
Italy	Emilia-Romagna	Reggio nell'Emilia	0	2
Italy	Emilia-Romagna	Rimini	0	2
Italy	Friuli-Venezia Giulia		215	
Italy	Friuli-Venezia Giulia	Pordenone	0	10
Italy	Friuli-Venezia Giulia	Trieste	0	7
Italy	Friuli-Venezia Giulia	Udine	0	9
Italy	Lazio	Frosinone	0	5
Italy	Lazio	Latina	0	18
Italy	Lazio	Rieti	0	10
Italy	Lazio	Roma	0	23
Italy	Lazio	Viterbo	0	13
Italy	Liguria	Genova	0	8
Italy	Liguria	Imperia	0	2
Italy	Liguria	Savona	0	12
Italy	Lombardia		847	
Italy	Lombardia	Brescia	0	
Italy	Lombardia	Como	0	4
Italy	Lombardia	Lecco	0	2
Italy	Lombardia	Monza and Brianza	0	2
Italy	Lombardia	Pavia	0	4
Italy	Lombardia	Sondrio	0	4
Italy	Lombardia	Varese	0	2
Italy	Piemonte	Alessandria	0	2
Italy	Piemonte	Biella	0	2
Italy	Piemonte	Cuneo	0	6

Continued

Table A1.14. Continued.

Country	Administrative region 1	Administrative region 2	Reference ID	No. entries[a]
Italy	Piemonte	Novara	0	12
Italy	Piemonte	Torino	0	8
Italy	Piemonte	Vercelli	0	2
Italy	Sardegna	Oristano	0	2
Italy	Sicily	Palermo	0	6
Italy	Sicily	Palermo	242	
Italy	Toscana		1331	
Italy	Toscana	Florence	0	3
Italy	Toscana	Grosseto	0	6
Italy	Toscana	Livorno	0	2
Italy	Toscana	Lucca	0	
Italy	Toscana	Pisa	0	6
Italy	Toscana	Pistoia	0	2
Italy	Toscana	Siena	0	2
Italy	Trentino-Alto Adige	Bolzano	0	28
Italy	Trentino-Alto Adige	Trento	0	150
Italy	Trentino-Alto Adige	Trento	160	
Italy	Trentino-Alto Adige	Trento	609	11
Italy	Trentino-Alto Adige	Trento	741	
Italy	Umbria	Perugia	0	6
Italy	Umbria	Terni	0	2
Italy	Veneto	Belluno	0	9
Italy	Veneto	Padua	0	2
Italy	Veneto	Treviso	0	2
Italy	Veneto	Verona	0	
Italy	Veneto	Verona	0	9
Italy	Veneto	Vicenza	0	14
Kosovo	Gnjilane	Kosovska Kamenica	0	
Kosovo	Kosovska Mitrovica	Kosovska Mitrovica	0	
Kosovo	Kosovska Mitrovica	Kosovska Mitrovica	346	
Kosovo	Prizren	Dragaö	0	
Kosovo	Prizren	Prizren	0	4
Latvia	Kurzeme	Kuldiga	0	8
Latvia	Kurzeme	Kuldiga	126	
Latvia	Kurzeme	Liepaja	0	
Latvia	Kurzeme	Liepaja	126	
Latvia	Kurzeme	Saldus	0	7
Latvia	Kurzeme	Saldus	126	
Latvia	Kurzeme	Talsi	0	4
Latvia	Kurzeme	Talsi	126	
Latvia	Kurzeme	Ventspils	0	
Latvia	Kurzeme	Ventspils	126	
Latvia	Latgale	Daugavpils	126	
Latvia	Latgale	Kraslava	126	
Latvia	Riga	Limbaži	0	3
Latvia	Riga	Limbaži	126	
Latvia	Riga	Ogre	0	5
Latvia	Riga	Ogre	126	
Latvia	Riga	Riga	0	5
Latvia	Riga	Riga	126	
Latvia	Riga	Tukums	0	
Latvia	Riga	Tukums	126	

Continued

Table A1.14. Continued.

Country	Administrative region 1	Administrative region 2	Reference ID	No. entries[a]
Latvia	Vidzeme	Cesis	0	
Latvia	Vidzeme	Cesis	0	2
Latvia	Vidzeme	Cesis	126	
Latvia	Vidzeme	Valka	0	4
Latvia	Vidzeme	Valka	126	
Latvia	Vidzeme	Valmiera	0	4
Latvia	Vidzeme	Valmiera	126	
Latvia	Zemgale	Aizkraukle	0	3
Latvia	Zemgale	Aizkraukle	126	
Latvia	Zemgale	Bauska	0	2
Latvia	Zemgale	Bauska	126	
Latvia	Zemgale	Dobele	0	4
Latvia	Zemgale	Dobele	126	
Latvia	Zemgale	Jelgava	0	2
Latvia	Zemgale	Jelgava	126	
Libya	Al Hizam Al Akhdar		0	2
Lithuania	Alytaus	Alytaus	0	
Lithuania	Alytaus	Varenos	0	
Lithuania	Alytaus	Varenos	732	
Lithuania	Kauno	Kauno	0	2
Lithuania	Kauno	Kauno	732	
Lithuania	Kauno	Prienu	0	
Lithuania	Kauno	Prienu	732	
Lithuania	Klaipėdos		733	
Lithuania	Klaipėdos	Klaipėdos	0	
Lithuania	Klaipėdos	Klaipėdos	732	
Lithuania	Klaipėdos	Neringos	0	
Lithuania	Klaipėdos	Neringos	732	
Lithuania	Klaipėdos	Šilutes	0	3
Lithuania	Klaipėdos	Šilutes	732	
Lithuania	Marijampoles	Marijampoles	0	
Lithuania	Marijampoles	Marijampoles	732	
Lithuania	Panevezio		733	
Lithuania	Panevezio	Paneveûio	0	
Lithuania	Panevezio	Paneveûio	732	
Lithuania	Panevezio	Rokiökio	0	
Lithuania	Panevezio	Rokiökio	732	
Lithuania	Šiauliai	Joniökio	0	
Lithuania	Šiauliai	Joniökio	732	
Lithuania	Šiauliai	Kelmes	0	
Lithuania	Šiauliai	Kelmes	732	
Lithuania	Šiauliai	Pakruojo	0	
Lithuania	Šiauliai	Šiauliu	0	2
Lithuania	Šiauliai	Šiauliu	732	
Lithuania	Utenos	Ignalinos	0	
Lithuania	Utenos	Utenos	0	2
Lithuania	Utenos	Utenos	732	2
Lithuania	Utenos	Zarasu	0	
Lithuania	Vilniaus		733	
Lithuania	Vilniaus	Ukmerges	0	
Lithuania	Vilniaus	Ukmerges	732	
Lithuania	Vilniaus	Vilniaus	0	

Continued

Table A1.14. Continued.

Country	Administrative region 1	Administrative region 2	Reference ID	No. entries[a]
Lithuania	Vilniaus	Vilniaus	732	
Moldova	Anenii Noi		667	
Moldova	Balti		667	
Moldova	Basarabeasca		667	
Moldova	Bender		667	
Moldova	Briceni		667	
Moldova	Cahul		667	
Moldova	Calarasi		667	
Moldova	Cantemir		667	
Moldova	Causeni		667	
Moldova	Chisinau		667	
Moldova	Cimislia		667	
Moldova	Criuleni		667	
Moldova	Donduseni		667	
Moldova	Drochia		667	
Moldova	Dubasari		667	
Moldova	Edinet		667	
Moldova	Falesti		667	
Moldova	Floresti		667	
Moldova	Gagauzia		667	
Moldova	Glodeni		667	
Moldova	Hîncesti		667	
Moldova	Ialoveni		667	
Moldova	Leova		667	
Moldova	Nisporeni		667	
Moldova	Ocnita		667	
Moldova	Orhei		667	
Moldova	Rezina		667	
Moldova	Rîscani		667	
Moldova	Sîngerei		667	
Moldova	Soldanesti		667	
Moldova	Soroca		667	
Moldova	Stefan Voda		667	
Moldova	Straseni		667	
Moldova	Taraclia		667	
Moldova	Telenesti		667	
Moldova	Transnistria		667	
Moldova	Ungheni		667	
Morocco	Chaouia-Ouardigha	Ben Slimane	0	13
Morocco	Gharb-Chrarda-Béni Hssen	Kénitra	0	2
Morocco	Meknès-Tafilalet	Ifrane	0	7
Morocco	Rabat-Salé-Zemmour-Zaer	Khémisset	0	13
Morocco	Tadla-Azilal	Azilal	0	3
Morocco	Tanger-Tétouan	Chefchaouen	0	3
Netherlands	Drenthe	Westerveld	0	
Netherlands	Friesland	Ameland	0	2
Netherlands	Friesland	Opsterland	0	
Netherlands	Friesland	Terschelling	0	
Netherlands	Gelderland	Brummen	0	
Netherlands	Gelderland	Buren	0	
Netherlands	Gelderland	Epe	0	
Netherlands	Gelderland	Nijkerk	0	2

Continued

Table A1.14. Continued.

Country	Administrative region 1	Administrative region 2	Reference ID	No. entries[a]
Netherlands	Gelderland	Nunspeet	0	3
Netherlands	Noord-Brabant	Halderberge	0	
Netherlands	Noord-Brabant	Moerdijk	0	
Netherlands	Noord-Brabant	Oisterwijk	0	
Netherlands	Noord-Holland	Amsterdam	0	
Netherlands	Noord-Holland	Heemskerk	0	
Netherlands	Noord-Holland	Hilversum	0	
Netherlands	Overijssel	Deventer	0	
Netherlands	Overijssel	Raalte	0	
Netherlands	Overijssel	Zwolle	0	
Netherlands	Utrecht	Bunnik	0	4
Netherlands	Utrecht	De Bilt	0	2
Netherlands	Utrecht	Doorn	0	
Netherlands	Utrecht	Leersum	0	
Netherlands	Utrecht	Maarn	0	
Netherlands	Utrecht	Soest	0	
Netherlands	Utrecht	Utrecht	0	4
Netherlands	Utrecht	Zeist	0	2
Netherlands	Zeeland	Middelburg	0	
Netherlands	Zeeland	Schouwen-Duiveland	0	
Netherlands	Zeeland	Sluis	0	
Netherlands	Zuid-Holland	Pijnacker-Nootdorp	0	
Netherlands	Zuid-Holland	Waddinxveen	0	
Netherlands	Zuid-Holland	Warmond	0	
Norway	Akershus	Aurskog-Høland	0	
Norway	Akershus	Sørum	0	
Norway	Ãstfold	Aremark	0	
Norway	Ãstfold	Halden	0	
Norway	Ãstfold	Marker	0	
Norway	Ãstfold	Sarpsborg	0	
Norway	Ãstfold	Spydeberg	0	
Norway	Aust-Agder		733	
Norway	Aust-Agder	Arendal	0	
Norway	Aust-Agder	Birkenes	0	2
Norway	Aust-Agder	Birkenes	732	
Norway	Aust-Agder	Froland	0	
Norway	Aust-Agder	Grimstad	0	
Norway	Aust-Agder	Lillesand	0	
Norway	Aust-Agder	Lillesand	732	
Norway	Aust-Agder	Tvedestrand	0	
Norway	Aust-Agder	Vegårshei	0	
Norway	Buskerud	Flesberg	0	
Norway	Hordaland	Bergen	0	
Norway	Hordaland	Eidfjord	0	
Norway	Hordaland	Etne	0	
Norway	Hordaland	Kvinnherad	0	4
Norway	Hordaland	Kvinnherad	732	
Norway	Hordaland	Lindås	0	
Norway	Hordaland	Masfjorden	0	
Norway	Hordaland	Masfjorden	0	
Norway	Hordaland	Modalen	0	
Norway	Hordaland	Odda	0	

Continued

Table A1.14. Continued.

Country	Administrative region 1	Administrative region 2	Reference ID	No. entries[a]
Norway	Hordaland	Samnanger	0	
Norway	Hordaland	Ullensvang	0	
Norway	Hordaland	Vaksdal	0	
Norway	Hordaland	Vaksdal	0	
Norway	Hordaland	Voss	0	4
Norway	Møre og Romsdal		1116	8
Norway	Møre og Romsdal	Norddal	0	
Norway	Møre og Romsdal	Ørsta	0	
Norway	Møre og Romsdal	Rauma	0	
Norway	Møre og Romsdal	Sunndal	0	
Norway	Møre og Romsdal	Surnadal	0	
Norway	Møre og Romsdal	Ulstein	0	
Norway	Møre og Romsdal	Volda	0	
Norway	Rogaland	Bjerkreim	0	2
Norway	Rogaland	Gjesdal	0	
Norway	Rogaland	Hjelmeland	0	2
Norway	Rogaland	Lund	0	
Norway	Rogaland	Sandnes	0	
Norway	Rogaland	Sauda	0	2
Norway	Rogaland	Suldal	0	
Norway	Sogn og Fjordane	Askvoll	0	
Norway	Sogn og Fjordane	Aurland	0	
Norway	Sogn og Fjordane	Førde	0	
Norway	Sogn og Fjordane	Gaular	0	3
Norway	Sogn og Fjordane	Gloppen	0	3
Norway	Sogn og Fjordane	Gulen	0	
Norway	Sogn og Fjordane	Gulen	732	
Norway	Sogn og Fjordane	Hyllestad	0	
Norway	Sogn og Fjordane	Hyllestad	732	
Norway	Sogn og Fjordane	Stryn	0	2
Norway	Sør-Trøndelag	Hitra	1323	
Norway	Telemark		1116	8
Norway	Telemark	Drangedal	0	
Norway	Telemark	Drangedal	732	
Norway	Telemark	Kragerø	0	3
Norway	Telemark	Kragerø	732	
Norway	Telemark	Skien	0	2
Norway	Vest-Agder		1116	8
Norway	Vest-Agder		733	2
Norway	Vest-Agder	Audnedal	0	
Norway	Vest-Agder	Flekkefjord	0	
Norway	Vest-Agder	Hægebostad	0	
Norway	Vest-Agder	Kristiansand	0	
Norway	Vest-Agder	Kristiansand	732	
Norway	Vest-Agder	Kvinesdal	0	
Norway	Vest-Agder	Marnardal	0	2
Norway	Vest-Agder	Songdalen	0	2
Norway	Vest-Agder	Vennesla	0	
Norway	Vestfold		1116	7
Norway	Vestfold		810	
Norway	Vestfold	Larvik	0	2
Norway	Vestfold	Vaale	0	

Continued

Table A1.14. Continued.

Country	Administrative region 1	Administrative region 2	Reference ID	No. entries[a]
Poland	Greater Poland		603	
Poland	Greater Poland		645	2
Poland	Greater Poland	Chodziez	0	2
Poland	Greater Poland	Czarnków-Trzcianka	0	4
Poland	Greater Poland	Gniezno	0	
Poland	Greater Poland	Gostyn	0	
Poland	Greater Poland	Gostyn	0	
Poland	Greater Poland	Grodzisk Wielkopolski	0	2
Poland	Greater Poland	Kalisz	0	
Poland	Greater Poland	Kalisz City	0	
Poland	Greater Poland	Kepno	0	3
Poland	Greater Poland	Koscian	0	2
Poland	Greater Poland	Krotoszyn	0	
Poland	Greater Poland	Leszno	0	2
Poland	Greater Poland	Miedzychód	0	2
Poland	Greater Poland	Nowy Tomyśl	0	2
Poland	Greater Poland	Oborniki	0	
Poland	Greater Poland	Ostrów Wielkopolski	0	2
Poland	Greater Poland	Ostrzeszów	0	3
Poland	Greater Poland	Pleszew	0	
Poland	Greater Poland	Poznan	0	13
Poland	Greater Poland	Poznan	644	
Poland	Greater Poland	Poznan	645	3
Poland	Greater Poland	Poznan City	0	3
Poland	Greater Poland	Rawicz	0	
Poland	Greater Poland	Słupca	0	
Poland	Greater Poland	Śrem	0	4
Poland	Greater Poland	Środa Wielkopolska	0	3
Poland	Greater Poland	Turek	0	
Poland	Greater Poland	Wągrowiec	0	3
Poland	Greater Poland	Wolsztyn	0	
Poland	Greater Poland	Zlotów	0	
Poland	Kuyavian-Pomeranian	Bydgoszcz City	0	
Poland	Kuyavian-Pomeranian	Chelmno	0	2
Poland	Kuyavian-Pomeranian	Grudziadz	0	
Poland	Kuyavian-Pomeranian	Grudziadz City	0	
Poland	Kuyavian-Pomeranian	Inowrocław	0	2
Poland	Kuyavian-Pomeranian	Lipno	0	
Poland	Kuyavian-Pomeranian	Nakło	0	2
Poland	Kuyavian-Pomeranian	Rypin	0	3
Poland	Kuyavian-Pomeranian	Świecie	0	
Poland	Kuyavian-Pomeranian	Torun City	0	2
Poland	Kuyavian-Pomeranian	Tuchola	0	4
Poland	Lesser Poland	Bochnia	0	
Poland	Lesser Poland	Brzesko Brzeg	0	
Poland	Lesser Poland	Chrzanów	0	4
Poland	Lesser Poland	Gorlice	0	
Poland	Lesser Poland	Kraków	0	12
Poland	Lesser Poland	Kraków City	0	3
Poland	Lesser Poland	Limanowa	0	2
Poland	Lesser Poland	Miechów	0	5
Poland	Lesser Poland	Myślenice	0	2

Continued

Table A1.14. Continued.

Country	Administrative region 1	Administrative region 2	Reference ID	No. entries[a]
Poland	Lesser Poland	Nowy Sącz	0	3
Poland	Lesser Poland	Nowy Targ	0	6
Poland	Lesser Poland	Olkusz	0	5
Poland	Lesser Poland	Oświęcim	0	4
Poland	Lesser Poland	Proszowice	0	
Poland	Lesser Poland	Sucha	0	3
Poland	Lesser Poland	Tarnów	0	
Poland	Lesser Poland	Tarnów City	0	2
Poland	Lesser Poland	Tatra	0	
Poland	Lesser Poland	Wieliczka	0	4
Poland	Lódz	Kutno	0	
Poland	Lódz	Lowicz	0	2
Poland	Lódz	LÛdz	0	
Poland	Lódz	Pabianice	0	3
Poland	Lódz	Sieradz	0	3
Poland	Lódz	TomaszÛw Mazowiecki	0	
Poland	Lódz	WieruszÛw	0	
Poland	Lódz	Zgierz	0	
Poland	Lower Silesian	Bolesławiec	0	
Poland	Lower Silesian	Dzierzoniów	0	4
Poland	Lower Silesian	Jelenia Góra	0	4
Poland	Lower Silesian	Kłodzko	0	8
Poland	Lower Silesian	Legnica	0	
Poland	Lower Silesian	Lubin	0	
Poland	Lower Silesian	Lwówek Śląski	0	
Poland	Lower Silesian	Milicz	0	2
Poland	Lower Silesian	Oleśnica	0	
Poland	Lower Silesian	Środa Śląska	0	4
Poland	Lower Silesian	Strzelin	0	
Poland	Lower Silesian	Świdnica	0	
Poland	Lower Silesian	Trzebnica	0	2
Poland	Lower Silesian	Wołów	0	
Poland	Lower Silesian	Wrocław	0	6
Poland	Lower Silesian	Wrocław	0	4
Poland	Lower Silesian	Wrocław	513	
Poland	Lower Silesian	Ząbkowice	0	3
Poland	Lower Silesian	Zgorzelec	0	
Poland	Lublin		102	
Poland	Lublin		104	3
Poland	Lublin		105	2
Poland	Lublin		953	4
Poland	Lublin	Bilgoraj	0	2
Poland	Lublin	Chelm	0	2
Poland	Lublin	Hrubieszów	0	
Poland	Lublin	Janów	0	
Poland	Lublin	Krasnik	0	4
Poland	Lublin	Lubartów	0	4
Poland	Lublin	Lublin	0	2
Poland	Lublin	Lublin City	0	2
Poland	Lublin	Puławy	0	3
Poland	Lublin	Radzyn	0	
Poland	Lublin	Tomaszów	0	3

Continued

Table A1.14. Continued.

Country	Administrative region 1	Administrative region 2	Reference ID	No. entries[a]
Poland	Lublin	Włodawa	0	5
Poland	Lublin	Zamosc	0	6
Poland	Lubusz	Gorzów	0	3
Poland	Lubusz	Miedzyrzecz	0	3
Poland	Lubusz	Slubice	0	
Poland	Lubusz	Strzelce-Drezdenko	0	
Poland	Lubusz	Sulęcin	0	3
Poland	Lubusz	Świebodzin	0	
Poland	Lubusz	Zielona Góra	0	
Poland	Masovian		964	
Poland	Masovian	Ciechanów	0	2
Poland	Masovian	Garwolin	0	
Poland	Masovian	Legionowo	0	
Poland	Masovian	Losice	0	
Poland	Masovian	Minsk	0	
Poland	Masovian	Mława	0	2
Poland	Masovian	Nowy Dwór Mazowiecki	0	
Poland	Masovian	Ostrołęka	0	2
Poland	Masovian	Otwock	0	
Poland	Masovian	Piaseczno	0	3
Poland	Masovian	Płock	0	
Poland	Masovian	Przysucha	0	
Poland	Masovian	Radom	0	2
Poland	Masovian	Warsaw	0	3
Poland	Masovian	Warsaw	1099	
Poland	Masovian	Warsaw West	0	4
Poland	Masovian	Wegrów	0	
Poland	Masovian	Wyszków	0	
Poland	Masovian	Zuromin	0	
Poland	Opole	Glubczyce	0	
Poland	Opole	Kędzierzyn-Koźle	0	
Poland	Opole	Namysłów	0	3
Poland	Opole	Nysa	0	4
Poland	Opole	Opole	0	2
Poland	Opole	Prudnik	0	
Poland	Opole	Strzelce	0	3
Poland	Podlachian		1340	
Poland	Podlachian		464	
Poland	Podlachian	Augustów	0	9
Poland	Podlachian	Białystok	0	4
Poland	Podlachian	Bielsk	0	3
Poland	Podlachian	Grajewo	0	2
Poland	Podlachian	Hajnówka	0	8
Poland	Podlachian	Hajnówka	385	
Poland	Podlachian	Lomza	0	
Poland	Podlachian	Monki	0	2
Poland	Podlachian	Siemiatycze	0	5
Poland	Podlachian	Sokółka	0	2
Poland	Podlachian	Suwałki	0	2
Poland	Podlachian	Zambrów	0	
Poland	Pomeranian		843	
Poland	Pomeranian	Bytów	0	4

Continued

Table A1.14. Continued.

Country	Administrative region 1	Administrative region 2	Reference ID	No. entries[a]
Poland	Pomeranian	Chojnice	0	5
Poland	Pomeranian	Czluchów	0	3
Poland	Pomeranian	Gdansk	0	6
Poland	Pomeranian	Gdansk City	0	3
Poland	Pomeranian	Gdynia	0	5
Poland	Pomeranian	Kartuzy	0	9
Poland	Pomeranian	Koscierzyna	0	5
Poland	Pomeranian	Kwidzyn	0	4
Poland	Pomeranian	Lebork	0	4
Poland	Pomeranian	Malbork	0	2
Poland	Pomeranian	Nowy Dwór Gdanski	0	6
Poland	Pomeranian	Puck	0	7
Poland	Pomeranian	Słupsk	0	5
Poland	Pomeranian	Słupsk City	0	
Poland	Pomeranian	Sopot	0	
Poland	Pomeranian	Starogard	0	3
Poland	Pomeranian	Sztum	0	2
Poland	Pomeranian	Tczew	0	2
Poland	Pomeranian	Wejherowo	0	9
Poland	Silesian	Bytom	0	
Poland	Silesian	Chorzów	0	
Poland	Silesian	Cieszyn	0	
Poland	Silesian	Czestochowa	0	4
Poland	Silesian	Gliwice	0	
Poland	Silesian	Katowice City	0	3
Poland	Silesian	Kłobuck	0	
Poland	Silesian	Lubliniec	0	2
Poland	Silesian	Mikolów	0	2
Poland	Silesian	Myslowice	0	
Poland	Silesian	Myszków	0	3
Poland	Silesian	Rybnik	0	
Poland	Silesian	Siemianowice Slaskie	0	
Poland	Silesian	Zabrze	0	
Poland	Silesian	Zawiercie	0	7
Poland	Silesian	Zywiec	0	2
Poland	Subcarpathian	Bieszczady	0	3
Poland	Subcarpathian	Jasło	0	
Poland	Subcarpathian	Kolbuszowa	0	
Poland	Subcarpathian	Krosno	0	
Poland	Subcarpathian	Lesko	0	3
Poland	Subcarpathian	Lezajsk	0	2
Poland	Subcarpathian	Lubaczów	0	2
Poland	Subcarpathian	Nisko	0	
Poland	Subcarpathian	Przemyśl	0	
Poland	Subcarpathian	Przeworsk	0	
Poland	Subcarpathian	Rzeszów City	0	2
Poland	Subcarpathian	Sanok	0	
Poland	Świętokrzyskie	Busko	0	
Poland	Świętokrzyskie	Kielce	0	3
Poland	Świętokrzyskie	Kielce City	0	2
Poland	Świętokrzyskie	Sandomierz	0	
Poland	Świętokrzyskie	Skarzysko	0	2

Continued

Table A1.14. Continued.

Country	Administrative region 1	Administrative region 2	Reference ID	No. entries[a]
Poland	Świętokrzyskie	StaszÛw	0	
Poland	Warmian-Masurian		738	
Poland	Warmian-Masurian		739	2
Poland	Warmian-Masurian	Bartoszyce	0	3
Poland	Warmian-Masurian	Braniewo	0	2
Poland	Warmian-Masurian	Działdowo	0	
Poland	Warmian-Masurian	Elbląg	0	3
Poland	Warmian-Masurian	Ełk	0	
Poland	Warmian-Masurian	Gizycko	0	3
Poland	Warmian-Masurian	Iława	0	3
Poland	Warmian-Masurian	Ketrzyn	0	
Poland	Warmian-Masurian	Lidzbark	0	
Poland	Warmian-Masurian	Mragowo	0	3
Poland	Warmian-Masurian	Nidzica	0	
Poland	Warmian-Masurian	Olecko	0	
Poland	Warmian-Masurian	Olsztyn	0	18
Poland	Warmian-Masurian	Olsztyn	546	
Poland	Warmian-Masurian	Olsztyn City	0	
Poland	Warmian-Masurian	Ostróda	0	4
Poland	Warmian-Masurian	Pisz	0	5
Poland	Warmian-Masurian	Szczytno	0	2
Poland	Warmian-Masurian	Wegorzewo	0	2
Poland	West Pomeranian	Białogard	0	2
Poland	West Pomeranian	Choszczno	0	2
Poland	West Pomeranian	Drawsko	0	3
Poland	West Pomeranian	Goleniów	0	
Poland	West Pomeranian	Gryfice	0	3
Poland	West Pomeranian	Gryfino	0	3
Poland	West Pomeranian	Kamien	0	8
Poland	West Pomeranian	Kołobrzeg	0	
Poland	West Pomeranian	Koszalin	0	3
Poland	West Pomeranian	Koszalin City	0	2
Poland	West Pomeranian	Lobez	0	2
Poland	West Pomeranian	Myślibórz	0	
Poland	West Pomeranian	Police	0	2
Poland	West Pomeranian	Pyrzyce	0	
Poland	West Pomeranian	Sławno	0	8
Poland	West Pomeranian	Stargard	0	5
Poland	West Pomeranian	Świdwin	0	
Poland	West Pomeranian	Szczecin	0	2
Poland	West Pomeranian	Szczecin	138	
Poland	West Pomeranian	Szczecin	893	8
Poland	West Pomeranian	Szczecin	894	
Poland	West Pomeranian	Szczecinek	0	3
Poland	West Pomeranian	Wałcz	0	3
Portugal	Braga		1143	
Portugal	Évora	Montemor-o-Novo	0	2
Portugal	Évora	Viana do Alentejo	0	2
Portugal	Faro	Monchique	0	2
Portugal	Faro	Vila do Bispo	0	2
Portugal	Lisboa		1143	
Portugal	Lisboa	Mafra	0	2

Continued

Table A1.14. Continued.

Country	Administrative region 1	Administrative region 2	Reference ID	No. entries[a]
Portugal	Madeira		1143	
Portugal	Portalegre		1143	
Portugal	Portalegre	Avis	0	2
Portugal	Santarém	Alpiarça	0	2
Portugal	Setúbal		1143	
Portugal	Setúbal	Alcácer do Sal	0	8
Portugal	Setúbal	Barreiro	0	2
Portugal	Setúbal	Grândola	0	6
Portugal	Setúbal	Montijo	0	2
Portugal	Setúbal	Palmela	0	10
Portugal	Viana do Castelo	Arcos de Valdevez	0	2
Romania	Alba		769	
Romania	Arad		769	
Romania	Arges		769	
Romania	Bihor		769	
Romania	Botosani		769	
Romania	Braila		769	
Romania	Buzau		769	
Romania	Calarasi		769	
Romania	Caras-Severin		769	
Romania	Cluj		769	
Romania	Constanta		769	
Romania	Covasna		0	
Romania	Dâmbovita		460	
Romania	Dolj		769	
Romania	Galati		769	
Romania	Giurgiu		769	
Romania	Gorj		769	
Romania	Hunedoara		0	
Romania	Iasi		0	
Romania	Mehedinti		769	
Romania	Mures		769	
Romania	Satu Mare		769	
Romania	Suceava		0	6
Romania	Suceava		460	
Romania	Suceava		462	
Romania	Suceava		769	
Romania	Teleorman		460	
Romania	Teleorman		462	
Romania	Timis		0	
Romania	Timis		177	4
Romania	Tulcea		0	4
Romania	Tulcea		198	2
Romania	Tulcea		460	
Romania	Tulcea		603	
Romania	Tulcea		769	
Romania	Vâlcea		0	2
Romania	Vrancea		769	
Russia	City of St Petersburg		35	
Russia	Kaliningrad		35	
Russia	Moskva		522	
Serbia	Grad Beograd		654	3

Continued

Table A1.14. Continued.

Country	Administrative region 1	Administrative region 2	Refe-ence ID	No. entries[a]
Serbia	Grad Beograd	Grocka	0	
Serbia	Grad Beograd	Mladenovac	0	
Serbia	Grad Beograd	Rakovica	785	
Serbia	Grad Beograd	Stari Grad	785	
Serbia	Grad Beograd	Voûdovac	0	
Serbia	Moravicki	Cacak	0	
Serbia	Moravicki	Gornji Milanovac	0	
Serbia	Ni'avski	Aleksinac	0	
Serbia	Ni'avski	Svrljig	0	
Serbia	Pirotski	Bela Palanka	0	
Serbia	Ra'ki	Kraljevo	0	
Serbia	Ra'ki	Novi Pazar	0	
Serbia	Šumadijski	Arandelovac	0	
Serbia	Šumadijski	Kragujevac	0	
Serbia	Šumadijski	Topola	0	
Serbia	Zajecarski	Boljevac	0	
Serbia	Zajecarski	Knjaževac	0	2
Serbia	Zajecarski	Zajecar	0	2
Serbia	Zapadno-Backi	Sombor	0	
Serbia	Zlatiborski	Bajina Baöta	0	
Serbia	Zlatiborski	Priboj	0	
Serbia	Zlatiborski	Prijepolje	0	2
Slovakia	Banskobystrický		0	
Slovakia	Banskobystrický		140	6
Slovakia	Banskobystrický		449	
Slovakia	Banskobystrický	Žarnovica	0	3
Slovakia	Banskobystrický	Žiar nad Hronom	0	2
Slovakia	Bratislavský	Bratislava II	0	2
Slovakia	Bratislavský	Malacky	0	10
Slovakia	Bratislavský	Senec	0	
Slovakia	Bratislavský		1349	
Slovakia	Košický		140	14
Slovakia	Košický		602	
Slovakia	Košický		914	16
Slovakia	Košický	Koöice I	0	5
Slovakia	Košický	Koöice III	0	
Slovakia	Košický	Koöice IV	0	
Slovakia	Košický	Koöice-okolie	0	5
Slovakia	Nitrianský		1349	
Slovakia	Nitrianský	Levice	0	9
Slovakia	Nitrianský	Nitra	0	2
Slovakia	Nitrianský	Nové Zámky	0	3
Slovakia	Nitrianský	Topolcany	0	
Slovakia	Nitrianský	Zlaté Moravce	0	5
Slovakia	Prešovský	Kežmarok	0	
Slovakia	Prešovský	Stará Lubovna	0	2
Slovakia	Trenčianský		140	2
Slovakia	Trenčianský		603	
Slovakia	Trenčianský	Ilava	0	4
Slovakia	Trenčianský	Myjava	0	4
Slovakia	Trenčianský	Nové Mesto nad Váhom	0	4
Slovakia	Trenčianský	Partizánske	0	

Continued

Table A1.14. Continued.

Country	Administrative region 1	Administrative region 2	Reference ID	No. entries[a]
Slovakia	Trenčianský	Považská Bystrica	0	4
Slovakia	Trenčianský	Prievidza	0	
Slovakia	Trenčianský	Púchov	0	3
Slovakia	Trenčianský	Trencín	0	3
Slovakia	Trnavský		140	3
Slovakia	Trnavský	Dunajská Streda	0	3
Slovakia	Trnavský	Piešt'any	0	3
Slovakia	Trnavský	Senica	0	8
Slovakia	Trnavský	Skalica	0	
Slovakia	Trnavský	Trnava	0	
Slovakia	Žilinský		140	2
Slovakia	Žilinský	Bytca	0	2
Slovakia	Žilinský	Cadca	0	2
Slovakia	Žilinský	Martin	0	
Slovakia	Žilinský	Námestovo	0	
Slovakia	Žilinský	Tvrdošín	0	
Slovenia	Gorenjska	Bled	0	
Slovenia	Gorenjska	Cerklje na Gorenjskem	0	
Slovenia	Gorenjska	Cerklje na Gorenjskem	527	
Slovenia	Gorenjska	Železniki	0	
Slovenia	Gorenjska	Gorenja Vas-Poljane	0	
Slovenia	Gorenjska	Preddvor	0	
Slovenia	Goriška	Brda	0	2
Slovenia	Goriška	Cerkno	0	
Slovenia	Jugovzhodna Slovenija	Kocevje	0	2
Slovenia	Jugovzhodna Slovenija	Sodrazica	527	
Slovenia	Notranjsko-kraška	Ilirska Bistrica	0	
Slovenia	Notranjsko-kraška	Pivka	0	
Slovenia	Obalno-kraška		527	
Slovenia	Obalno-kraška	Divaca	0	
Slovenia	Obalno-kraška	Hrpelje-Kozina	0	
Slovenia	Obalno-kraška	Koper	0	
Slovenia	Osrednjeslovenska	Grosuplje	0	
Slovenia	Osrednjeslovenska	Kamnik	527	
Slovenia	Osrednjeslovenska	Ljubljana	0	3
Slovenia	Osrednjeslovenska	Medvode	0	2
Slovenia	Osrednjeslovenska	Medvode	527	2
Slovenia	Osrednjeslovenska	Velike Lašč	0	
Slovenia	Savinjska	Mozirje	527	
Slovenia	Savinjska	Tabor	0	
Slovenia	Zasavska	Zagorje ob Savi	0	
Spain	Andalucía	Cádiz	0	8
Spain	Andalucía	Córdoba	0	4
Spain	Andalucía	Granada	0	12
Spain	Andalucía	Huelva	0	12
Spain	Andalucía	Jaén	0	8
Spain	Andalucía	Jaén	611	3
Spain	Andalucía	Málaga	0	2
Spain	Aragón	Huesca	0	6
Spain	Aragón	Zaragoza	0	2
Spain	Cantabria	Cantabria	0	9
Spain	Castilla y León	Burgos	0	9

Continued

Table A1.14. Continued.

Country	Administrative region 1	Administrative region 2	Reference ID	No. entries[a]
Spain	Castilla y León	León	0	3
Spain	Castilla y León	Toledo	821	2
Spain	Castilla-La Mancha	Ciudad Real	0	2
Spain	Castilla-La Mancha	Cuenca	821	
Spain	Cataluña	Barcelona	0	2
Spain	Cataluña	Girona	0	4
Spain	Comunidad Foral de Navarra	Navarra	0	4
Spain	Extremadura	Badajoz	0	14
Spain	Extremadura	Cáceres	0	51
Spain	Galicia	Lugo	0	2
Spain	Islas Baleares	Baleares	345	
Spain	La Rioja	La Rioja	0	21
Spain	La Rioja	La Rioja	315	17
Spain	País Vasco		77	
Spain	País Vasco	Álava	0	38
Spain	País Vasco	Guipúzcoa	0	9
Spain	País Vasco	Vizcaya	0	41
Spain	Principado de Asturias		821	
Spain	Región de Murcia	Murcia	616	
Sweden	Blekinge		1339	2
Sweden	Blekinge		350	
Sweden	Dalarna		1339	2
Sweden	Gävleborg		1339	3
Sweden	Gävleborg		350	
Sweden	Gotland		1339	2
Sweden	Gotland	Gotland	470	2
Sweden	Halland		1339	2
Sweden	Halland	Kungsbacka	470	
Sweden	Halland	Varberg	470	
Sweden	Jämtland		1339	
Sweden	Jönköping		1339	2
Sweden	Kalmar		1339	2
Sweden	Kalmar	Kalmar	350	
Sweden	Kronoberg		1339	2
Sweden	Norrbotten		1339	
Sweden	Orebro		1339	
Sweden	Östergötland		1339	2
Sweden	Skåne		1339	2
Sweden	Skåne		350	
Sweden	Skåne	Båstad	470	
Sweden	Skåne	Simrishamn	584	
Sweden	Södermanland		1339	
Sweden	Stockholm		1339	2
Sweden	Stockholm	Norrtälje	470	
Sweden	Stockholm	Nynäshamn	470	2
Sweden	Stockholm	Stockholm	350	
Sweden	Uppsala		1339	2
Sweden	Uppsala	Älvkarleby	470	
Sweden	Uppsala	Tierp	470	
Sweden	Uppsala	Uppsala	470	
Sweden	Värmland		1339	2
Sweden	Västerbotten		1339	

Continued

Table A1.14. Continued.

Country	Administrative region 1	Administrative region 2	Reference ID	No. entries[a]
Sweden	Västerbotten	Umeå	1322	
Sweden	Västerbotten	Umeå	470	
Sweden	Västernorrland		1339	2
Sweden	Västmanland		1339	
Sweden	Västra Götaland		1339	2
Switzerland	Aargau		0	7
Switzerland	Basel-Landschaft		524	
Switzerland	Bern		0	7
Switzerland	Fribourg		0	
Switzerland	Graubünden		435	5
Switzerland	Jura		0	
Switzerland	Lucerne		0	2
Switzerland	Neuchâtel		0	2
Switzerland	Neuchâtel		450	2
Switzerland	Neuchâtel		487	
Switzerland	Sankt Gallen		0	3
Switzerland	Schaffhausen		0	6
Switzerland	Ticino		0	8
Switzerland	Ticino		435	2
Switzerland	Valais		0	
Switzerland	Valais		1284	4
Switzerland	Valais		1321	
Switzerland	Vaud		0	
Switzerland	Zürich		0	19
Tunisia	Béja	Béja Nord	0	2
Tunisia	Béja	Mejez El Bab	0	
Tunisia	Béja	Nefza	0	6
Tunisia	Béja	Testour	0	2
Tunisia	Bizerte	Sejnane	0	7
Tunisia	Jendouba	Aïn Draham	0	10
Tunisia	Jendouba	Balta Bou Aouane	0	
Tunisia	Jendouba	Ghardimaou	0	3
Tunisia	Jendouba	Tabarka	0	3
Tunisia	Kairouan	Chebika	0	
Tunisia	Le Kef	Nebeur	0	3
Tunisia	Siliana	Bouarada	0	
Tunisia	Siliana	Kesra	0	2
Tunisia	Sousse	Kondar	0	
Tunisia	Tataouine	Ghomrassen	0	
Tunisia	Zaghouan	Zriba	0	3
Turkey	Artvin		0	
Turkey	Burdur		0	2
Turkey	Burdur		1008	
Turkey	Giresun		0	5
Turkey	Istanbul		0	3
Turkey	Istanbul		145	
Turkey	Rize		0	4
Turkey	Samsun		0	7
Turkey	Sinop		0	12
Turkey	Trabzon		0	5
UK	England	Bedfordshire	0	3
UK	England	Berkshire	0	4

Continued

Table A1.14. Continued.

Country	Administrative region 1	Administrative region 2	Reference ID	No. entries[a]
UK	England	Buckinghamshire	0	2
UK	England	Cambridgeshire	0	15
UK	England	Cheshire	0	
UK	England	Cornwall	0	17
UK	England	Croydon	0	
UK	England	Cumbria	0	17
UK	England	Derbyshire	0	
UK	England	Devon	0	13
UK	England	Dorset	0	41
UK	England	Durham	0	
UK	England	East Sussex	0	2
UK	England	Essex	0	12
UK	England	Gloucestershire	0	3
UK	England	Hampshire	0	28
UK	England	Herefordshire	0	7
UK	England	Hounslow	0	
UK	England	Isle of Wight	0	2
UK	England	Kensington and Chelsea	0	
UK	England	Kent	0	7
UK	England	Kingston upon Thames	0	
UK	England	Lambeth	0	
UK	England	Lancashire	0	
UK	England	Lancashire	559	
UK	England	Leicester	0	
UK	England	Leicestershire	0	4
UK	England	Lincolnshire	0	2
UK	England	Luton	0	
UK	England	Merseyside	0	2
UK	England	Merton	0	
UK	England	Milton Keynes	0	
UK	England	Norfolk	0	53
UK	England	North Somerset	0	
UK	England	North Yorkshire	0	5
UK	England	North Yorkshire	1297	3
UK	England	Northamptonshire	0	
UK	England	Northumberland	0	20
UK	England	Northumberland	131	
UK	England	Nottinghamshire	0	2
UK	England	Oxfordshire	0	3
UK	England	Peterborough	0	
UK	England	Plymouth	0	2
UK	England	Poole	0	
UK	England	Portsmouth	0	2
UK	England	Richmond upon Thames	0	
UK	England	Somerset	0	4
UK	England	Staffordshire	0	4
UK	England	Suffolk	0	17
UK	England	Surrey	0	8
UK	England	Sutton	0	
UK	England	Thurrock	0	
UK	England	Torbay	0	
UK	England	Tyne and Wear	0	

Continued

Table A1.14. Continued.

Country	Administrative region 1	Administrative region 2	Reference ID	No. entries[a]
UK	England	Waltham Forest	0	
UK	England	Warwickshire	0	2
UK	England	West Sussex	0	12
UK	England	Wiltshire	0	8
UK	England	Worcestershire	0	
UK	Northern Ireland	Down	0	
UK	Northern Ireland	Dungannon	0	
UK	Northern Ireland	Fermanagh	0	4
UK	Northern Ireland	Newry and Mourne	0	
UK	Northern Ireland	Omagh	0	
UK	Scotland		1114	
UK	Scotland	Aberdeen	302	
UK	Scotland	Aberdeenshire	0	19
UK	Scotland	Angus	0	8
UK	Scotland	Argyll and Bute	0	40
UK	Scotland	Dumfries and Galloway	0	17
UK	Scotland	East Ayrshire	0	2
UK	Scotland	Eilean Siar	0	5
UK	Scotland	Fife	0	
UK	Scotland	Highland	0	104
UK	Scotland	Moray	0	5
UK	Scotland	North Ayshire	0	6
UK	Scotland	Perthshire and Kinross	0	17
UK	Scotland	Renfrewshire	0	
UK	Scotland	Scottish Borders	0	11
UK	Scotland	Shetland Islands	0	3
UK	Scotland	South Ayrshire	0	4
UK	Scotland	South Lanarkshire	0	2
UK	Scotland	Stirling	0	8
UK	Scotland	West Dunbartonshire	0	2
UK	Wales	Anglesey	0	6
UK	Wales	Bridgend	0	
UK	Wales	Cardiff	0	
UK	Wales	Carmarthenshire	0	14
UK	Wales	Ceredigion	0	19
UK	Wales	Conwy	0	5
UK	Wales	Denbighshire	0	
UK	Wales	Gwynedd	0	30
UK	Wales	Monmouthshire	0	2
UK	Wales	Neath Port Talbot	0	2
UK	Wales	Pembrokeshire	0	9
UK	Wales	Powys	0	16
UK	Wales	Swansea	631	
UK	Wales	Torfaen	0	
UK	Wales	Vale of Glamorgan	0	2

(*) If more than one.

Table A1.15. Geographic distribution data for *Haemaphysalis punctata*. See Appendix 2 (hosted on http://www.cabi.org/vetmedresource/) for complete reference indicated by the Refe-ence ID.

Country	Administrative region 1	Administrative region 2	Reference ID	No. entries[a]
Albania	Elbasan	Elbasanit	0	
Albania	Gjirokastër	Gjirokastrës	0	
Albania	Korçë	Pogradecit	0	
Albania	Vlorë	Vlorës	0	
Algeria	Alger	Bouzareah	0	2
Algeria	Bouira	Ain el Hadjar	0	2
Algeria	Bouira	Ain Turk	0	
Algeria	Bouira	Lakhdaria	0	
Algeria	Médéa	Medea	0	
Algeria	Relizane	Oued Rhiou	0	
Algeria	Sétif	Ain Arnat	0	2
Algeria	Tiaret	Sougueur	0	2
Algeria	Tizi Ouzou	Boghni	0	2
Algeria	Tizi Ouzou	Tizi-Ghenif	0	2
Algeria	Tlemcen	Tlemcen	0	2
Bulgaria	Montana	Montana	0	
Bulgaria	Vratsa	Vratsa	0	
Croatia	Karlovacka		0	
Croatia	Splitsko-Dalmatinska		0	2
Croatia	Splitsko-Dalmatinska		277	
Croatia	Splitsko-Dalmatinska		781	
Croatia	Zadarska		0	
France	Aquitaine	Dordogne	0	3
France	Aquitaine	Dordogne	1280	
France	Aquitaine	Gironde	0	
France	Aquitaine	Gironde	1280	
France	Aquitaine	Landes	0	7
France	Aquitaine	Landes	1280	
France	Aquitaine	Pyrénées-Atlantiques	0	5
France	Aquitaine	Pyrénées-Atlantiques	1280	
France	Auvergne	Puy-de-Dôme	0	2
France	Auvergne	Puy-de-Dôme	0	
France	Basse-Normandie	Manche	1280	
France	Bourgogne	Côte-d'Or	0	
France	Bourgogne	Côte-d'Or	1280	
France	Centre	Eure-et-Loir	0	
France	Centre	Indre	0	
France	Centre	Indre	1280	
France	Centre	Indre-et-Loire	1280	
France	Corse		1280	
France	Corse	Corse-du-Sud	0	7
France	Corse	Haute-Corse	0	
France	Île-de-France	Seine-et-Marne	0	
France	Île-de-France	Seine-et-Marne	1280	
France	Île-de-France	Seine-Maritime	1280	
France	Île-de-France	Val-de-Marne	0	
France	Île-de-France	Ville de Paris	1280	
France	Languedoc-Roussillon	Aude	0	
France	Languedoc-Roussillon	Hérault	0	
France	Languedoc-Roussillon	Pyrénées-Orientales	0	10
France	Languedoc-Roussillon	Pyrénées-Orientales	1280	
France	Limousin	Haute-Vienne	0	3

Continued

Table A1.15. Continued.

Country	Administrative region 1	Administrative region 2	Reference ID	No. entries[a]
France	Midi-Pyrénées	Aveyron	0	5
France	Midi-Pyrénées	Aveyron	1280	
France	Midi-Pyrénées	Lot	0	4
France	Midi-Pyrénées	Lot	1280	
France	Midi-Pyrénées	Tarn	0	
France	Pays de la Loire	Vendée	0	7
France	Poitou-Charentes	Vienne	1280	
France	Provence-Alpes-Côte-d'Azur	Alpes-de-Haute-Provence	0	9
France	Provence-Alpes-Côte-d'Azur	Alpes-de-Haute-Provence	1280	
France	Provence-Alpes-Côte-d'Azur	Alpes-Maritimes	0	
France	Provence-Alpes-Côte-d'Azur	Bouches-du-Rhône	0	7
France	Provence-Alpes-Côte-d'Azur	Hautes-Alpes	0	3
France	Provence-Alpes-Côte-d'Azur	Var	0	3
France	Rhône-Alpes	Ain	0	2
France	Rhône-Alpes	Ain	1280	
France	Rhône-Alpes	Ardèche	0	8
France	Rhône-Alpes	Drôme	0	14
France	Rhône-Alpes	Isère	0	
France	Rhône-Alpes	Savoie	0	3
Greece	Anatoliki Makedonia kai Thraki	Drama	0	6
Greece	Anatoliki Makedonia kai Thraki	Evros	0	6
Greece	Anatoliki Makedonia kai Thraki	Kavala	0	6
Greece	Attiki	Attica	0	2
Greece	Dytiki Makedonia	Grevena	0	10
Greece	Dytiki Makedonia	Kozani	0	5
Greece	Ipeiros	Ioannina	0	2
Greece	Ipeiros	Preveza	0	2
Greece	Kentriki Makedonia	Khalkidiki	0	2
Greece	Kentriki Makedonia	Kilkis	0	3
Greece	Kentriki Makedonia	Pella	0	3
Greece	Kentriki Makedonia	Pieria	0	
Greece	Kentriki Makedonia	Serrai	0	8
Greece	Kentriki Makedonia	Thessaloniki	0	3
Greece	Stereá Elláda	Boeotia	0	
Greece	Stereá Elláda	Evritania	0	
Italy	Abruzzo	Chieti	0	2
Italy	Abruzzo	L'Aquila	0	26
Italy	Abruzzo	Teramo	0	5
Italy	Apulia	Foggia	0	5
Italy	Apulia	Lecce	0	3
Italy	Campania		802	
Italy	Campania	Avellino	0	4
Italy	Campania	Caserta	0	
Italy	Emilia-Romagna	Ferrara	0	2
Italy	Emilia-Romagna	Forlì-Cesena	0	2
Italy	Emilia-Romagna	Modena	0	
Italy	Emilia-Romagna	Parma	0	2
Italy	Emilia-Romagna	Piacenza	0	
Italy	Lazio	Latina	0	3
Italy	Lazio	Rieti	0	3
Italy	Lazio	Roma	0	10
Italy	Lazio	Viterbo	0	3
Italy	Liguria	Genova	0	2

Continued

Table A1.15. Continued.

Country	Administrative region 1	Administrative region 2	Reference ID	No. entries[a]
Italy	Lombardia	Varese	0	
Italy	Marche	Ancona	0	2
Italy	Molise	Isernia	0	
Italy	Piemonte	Alessandria	0	
Italy	Sardegna	Cagliari	0	
Italy	Sardegna	Carbonia-Iglesias	0	
Italy	Sardegna	Nuoro	0	3
Italy	Sardegna	Olbia-Tempio	0	
Italy	Sardegna	Oristano	0	4
Italy	Sardegna	Sassari	0	11
Italy	Sicily	Palermo	0	3
Italy	Sicily	Palermo	242	
Italy	Sicily	Palermo	366	
Italy	Sicily	Palermo	998	
Italy	Sicily	Trapani	0	
Italy	Toscana		1331	
Italy	Toscana	Florence	0	2
Italy	Toscana	Pisa	0	3
Italy	Toscana	Pistoia	0	
Italy	Trentino-Alto Adige	Bolzano	0	2
Italy	Trentino-Alto Adige	Trento	0	
Italy	Umbria	Perugia	0	2
Italy	Umbria	Terni	0	
Italy	Veneto	Verona	0	
Kosovo	Kosovska Mitrovica	Kosovska Mitrovica	0	
Kosovo	Pristina	Priötina	0	
Kosovo	Uroöevac	Uroöevac	0	
Libya	Ajdabiya		0	
Libya	Al Jabal al Akhdar		0	2
Libya	An Nuqat al Khams		0	
Libya	Tarhunah-Masallatah		0	
Macedonia	Pelagonia	Bitola	0	
Macedonia	Polog	Tetovo	0	2
Macedonia	Skopje	Centar	0	2
Macedonia	Southwestern	Ohrid	0	
Moldova	Cahul		667	
Moldova	Cantemir		667	
Moldova	Causeni		667	
Moldova	Cimislia		667	
Moldova	Leova		667	
Moldova	Taraclia		667	
Montenegro	Kolacin		0	
Morocco	Gharb-Chrarda-Béni Hssen	Kénitra	0	
Morocco	Grand Casablanca	Mohammedia	0	
Morocco	Meknès-Tafilalet	Ifrane	0	
Morocco	Meknès-Tafilalet	Khénifra	0	
Morocco	Meknès-Tafilalet	Meknès	0	
Portugal	Évora	Montemor-o-Novo	0	
Portugal	Lisboa		1143	
Portugal	Santarém	Coruche	0	
Portugal	Setúbal		1143	
Portugal	Setúbal	Alcácer do Sal	0	2
Portugal	Setúbal	Grândola	0	

Continued

Table A1.15. Continued.

Country	Administrative region 1	Administrative region 2	Reference ID	No. entries[a]
Romania	Timis		177	5
Serbia	Macvanski	Ljubovija	0	
Serbia	Pcinjski	Surdulica	0	
Serbia	Toplicki	Prokuplje	0	
Slovakia	Košický		140	
Spain	Andalucía	Cádiz	0	5
Spain	Andalucía	Córdoba	0	
Spain	Andalucía	Granada	0	
Spain	Andalucía	Huelva	0	3
Spain	Andalucía	Jaén	0	2
Spain	Andalucía	Jaén	611	
Spain	Aragón	Huesca	0	7
Spain	Aragón	Teruel	0	
Spain	Aragón	Zaragoza	0	3
Spain	Cantabria	Cantabria	0	2
Spain	Castilla y León	Burgos	0	2
Spain	Castilla y León	Burgos	821	2
Spain	Castilla y León	Soria	0	
Spain	Cataluña	Girona	0	3
Spain	Cataluña	Lleida	0	3
Spain	Cataluña	Tarragona	0	
Spain	Comunidad Foral de Navarra	Navarra	0	6
Spain	Extremadura	Badajoz	0	
Spain	Extremadura	Cáceres	0	3
Spain	Islas Baleares	Baleares	1345	2
Spain	La Rioja	La Rioja	0	4
Spain	País Vasco		77	
Spain	País Vasco	Álava	0	25
Spain	País Vasco	Guipúzcoa	0	14
Spain	País Vasco	Vizcaya	0	15
Spain	Principado de Asturias		821	
Sweden	Blekinge		1339	
Sweden	Gotland		1339	
Sweden	Kalmar		1339	
Sweden	Västra Götaland		1339	
Switzerland	Ticino		435	
Switzerland	Ticino		435	
Tunisia	Ariana	Kalaat El Andalous	0	
Tunisia	Béja	Béja Nord	0	
Tunisia	Béja	Nefza	0	2
Tunisia	Béja	Téboursouk	0	2
Tunisia	Bizerte	Bizerte Sud	0	
Tunisia	Bizerte	Mateur	0	2
Tunisia	Bizerte	Sejnane	0	3
Tunisia	Jendouba	Aïn Draham	0	
Tunisia	Jendouba	Ghardimaou	0	
Tunisia	Le Kef	Nebeur	0	
Tunisia	Nabeul	Haouaria	0	
Tunisia	Nabeul	Menzel Temime	0	
Tunisia	Zaghouan	Zriba	0	2
Turkey	Aksaray		1353	
Turkey	Ankara		0	8
Turkey	Ankara		1353	
Turkey	Elâziğ		0	

[a]If more than one.

Table A1.16. Geographic distribution data for *Haemaphysalis concinna*. See Appendix 2 (hosted on http://www.cabi.org/vetmedresource/) for complete reference indicated by the Reference ID.

Country	Administrative region 1	Administrative region 2	Reference ID
Czech Republic	Jihomoravský		490
France	Aquitaine	Landes	1280
France	Aquitaine	Pyrénées-Atlantiques	1280
France	Centre	Indre	1280
France	Centre	Indre-et-Loire	1280
France	Île-de-France	Seine-et-Marne	1280
France	Île-de-France	Seine-Maritime	1280
France	Limousin	Haute-Vienne	1280
France	Midi-Pyrénées	Ariège	1280
France	Picardie	Oise	1280
France	Poitou-Charentes	Vienne	1280
Hungary	Bacs-Kiskun		1417
Hungary	Baranya		1417
Hungary	Csongrad		1417
Hungary	Fejer		1417
Hungary	Nograd		1417
Hungary	Somogy		1417
Hungary	Zala		1417
Romania	Timis		177
Slovakia	Banskobystrický		140
Slovakia	Bratislavský		1349
Slovakia	Bratislavský		1349
Slovakia	Nitrianský		1349
Slovakia	Trenčianský		1349
Slovakia	Trnavský		900
Spain	País Vasco		77
Spain	Principado de Asturias		821

Table A1.17. Geographic distribution data for *Haemaphysalis inermis*. See Appendix 2 (hosted on http://www.cabi.org/vetmedresource/) for complete reference indicated by the Reference ID.

Country	Administrative region 1	Administrative region 2	Reference ID
Bulgaria	Smolyan		884
Bulgaria	Yuzhen tsentralen		
France	Centre	Indre	1280
France	Centre	Indre-et-Loire	1280
France	Île-de-France	Seine-et-Marne	1280
France	Languedoc-Roussillon	Pyrénées-Orientales	1280
France	Limousin	Haute-Vienne	1280
France	Poitou-Charentes	Vienne	1280
Hungary	Borsod-Abauj-Zemplen		1417
Hungary	Fejer		1417
Hungary	Heves		1417
Italy	Campania		802
Portugal	Lisboa		1143
Slovakia	Banskobystrický		140
Slovakia	Bratislavský		1349
Slovakia	Košický		1349
Slovakia	Trenčianský		1349
Slovakia	Zilinský		1349
Spain	País Vasco		77

Table A1.18. Geographic distribution data for *Hyalomma marginatum*. See Appendix 2 (hosted on http://www.cabi.org/vetmedresource/) for complete reference indicated by the Reference ID.

Country	Administrative region 1	Administrative region 2	Reference ID	No. entries[a]
Albania	Elbasan	Elbasanit	0	
Albania	Elbasan	Librazhdit	0	
Albania	Korçë	Korçës	0	
Albania	Korçë	Pogradecit	0	2
Albania	Lezhë	Lezhës	0	
Albania	Vlorë	Sarandës	0	
Albania	Vlorë	Vlorës	0	3
Algeria	Aïn Defla	Ain Benian	0	
Algeria	Aïn Defla	Ain Lechiakh	0	
Algeria	Aïn Defla	Arib	0	
Algeria	Aïn Témouchent	Ain Kihel	0	
Algeria	Alger	Ain Benian	0	
Algeria	Alger	Ouled Chebel	0	
Algeria	Annaba	Annaba	0	
Algeria	Batna	Ain Touta	0	
Algeria	Batna	Batna	0	
Algeria	Batna	Chemora	0	
Algeria	Batna	Ksar Bellezma	0	
Algeria	Batna	Ras El Aioun	0	
Algeria	Batna	Seriana	0	
Algeria	Beja	Bejaia	0	
Algeria	Biskra	Besbes	0	
Algeria	Biskra	El Ouitaya	0	
Algeria	Biskra	Ouled Djellal	0	
Algeria	Biskra	Tolga	0	
Algeria	Blida	Bouaarfa	0	
Algeria	Blida	Bouinan	0	
Algeria	Blida	Soumaa	0	
Algeria	Bouira	Ain El Hadjar	0	
Algeria	Bouira	Ain Turk	0	
Algeria	Bouira	Djebahia	0	
Algeria	Bouira	El Asnam	0	
Algeria	Bouira	Lakhdaria	0	2
Algeria	Bouira	Sour El Ghouzlane	0	3
Algeria	Boumerdès	Boudouaou	0	
Algeria	Boumerdès	Corso	0	
Algeria	Boumerdès	Isser	0	
Algeria	Chlef	Bouzeghaia	0	
Algeria	Chlef	Chlef	0	2
Algeria	Chlef	Sendjas	0	
Algeria	Chlef	Sidi Akkacha	0	
Algeria	Chlef	Taougrit	0	
Algeria	Chlef	Tenes	0	
Algeria	Constantine	Constantine	0	
Algeria	Constantine	El Khroub	0	
Algeria	Djelfa	M'Liliha	0	
Algeria	El Tarf	El Aioun	0	
Algeria	Guelma	Ain Rekada	0	
Algeria	Guelma	Hammam Debagh	0	2
Algeria	Jijel	Chekfa	0	
Algeria	Jijel	Jijel	0	

Continued

Table A1.18. Continued.

Country	Administrative region 1	Administrative region 2	Reference ID	No. entries[a]
Algeria	Jijel	Jijel	110	
Algeria	Jijel	Jijel	110	
Algeria	Laghouat	Laghouat	0	
Algeria	M'Sila	Ain Errich	0	
Algeria	Mascara	Ain Ferah	0	
Algeria	Médéa	Berrouaghia	0	
Algeria	Médéa	Boughezoul	0	
Algeria	Médéa	Medea	0	
Algeria	Mila	Chelghoum Laid	0	
Algeria	Oran	Oran	0	
Algeria	Oran	Oued Tlelat	0	
Algeria	Oum el Bouaghi	Meskiana	0	
Algeria	Relizane	Relizane	0	
Algeria	Relizane	Sidi M'Hamed Benaouda	0	
Algeria	Sétif		110	
Algeria	Sétif		110	
Algeria	Sétif	Ain Arnat	0	5
Algeria	Sétif	Ain Lahdjar	0	
Algeria	Sétif	El Eulma	0	
Algeria	Sétif	Guellal	0	
Algeria	Sétif	Mezloug	0	
Algeria	Sétif	Ouled Saber	0	
Algeria	Sétif	Setif	0	
Algeria	Sidi Bel Abbès	Dhaya	0	
Algeria	Skikda	Ouled Attia	0	
Algeria	Tébessa	Bir El Ater	0	
Algeria	Tébessa	Boulhaf Dyr	0	
Algeria	Tébessa	Tebessa	0	4
Algeria	Tiaret		129	
Algeria	Tipaza	Hadjout	0	
Algeria	Tissemsilt	Lardjem	0	
Algeria	Tissemsilt	Theniet El Had	0	
Algeria	Tizi Ouzou	Boghni	0	
Algeria	Tizi Ouzou	Draa-Ben-Khedda	0	
Algeria	Tizi Ouzou	Tadmait	0	
Algeria	Tizi Ouzou	Tizi Ouzou	0	2
Algeria	Tizi Ouzou	Tizi-Ghenif	0	
Algeria	Tizi Ouzou	Zekri	0	
Bosnia and Herzegovina	Federacija Bosna i Hercegovina	Tuzla	0	
Bulgaria	Blagoevgrad	Bansko	0	2
Bulgaria	Burgas	Karnobat	0	
Bulgaria	Burgas	Malko Tarnovo	0	
Bulgaria	Burgas	Tsarevo	0	
Bulgaria	Haskovo	Svilengrad	0	
Bulgaria	Kardzhali	Ardino	0	
Bulgaria	Kardzhali	Kardzhali	0	2
Bulgaria	Kardzhali	Momchilgrad	0	2
Bulgaria	Kyustendil	Rila	0	
Bulgaria	Lovech	Troyan	0	
Bulgaria	Montana	Montana	0	

Continued

Table A1.18. Continued.

Country	Administrative region 1	Administrative region 2	Reference ID	No. entries[a]
Bulgaria	Pazardzhik	Pazardzhik	0	
Bulgaria	Pernik	Breznik	0	
Bulgaria	Pleven	Pleven	0	
Bulgaria	Plovdiv	Plovdiv	0	
Bulgaria	Plovdiv	Saedinenie	0	
Bulgaria	Shumen	Preslav	0	
Bulgaria	Sliven	Sliven	0	
Bulgaria	Sofia	Samokov	0	
Bulgaria	Sofia	Zlatitsa	0	
Bulgaria	Stara Zagora	Chirpan	0	
Bulgaria	Stara Zagora	Haskovo	69	
Bulgaria	Stara Zagora	Radnevo	0	
Bulgaria	Varna	Dolni Chiflik	0	
Bulgaria	Varna	Provadiya	0	
Bulgaria	Varna	Varna	0	
Bulgaria	Vidin	Dimovo	0	3
Bulgaria	Vratsa	Vratsa	0	
Bulgaria	Yambol	Tundzha	0	3
Bulgaria	Yambol	Yambol	0	2
Croatia	Grad Zagreb		0	2
Croatia	Istarska		0	
Croatia	Karlovacka		0	
Croatia	Primorsko-Goranska		0	3
Croatia	Šibensko-Kninska		0	
Croatia	Splitsko-Dalmatinska		0	3
Croatia	Splitsko-Dalmatinska		781	
Croatia	Zadarska		0	6
Cyprus	Famagusta		0	
Cyprus	Nicosia		0	2
Cyprus	Paphos		0	2
France	Aquitaine	Pyrénées-Atlantiques	0	
France	Aquitaine	Pyrénées-Atlantiques	1280	
France	Centre	Indre-et-Loire	0	2
France	Corse		1280	
France	Corse	Corse-du-Sud	0	14
France	Corse	Haute-Corse	0	7
France	Languedoc-Roussillon	Lozère	0	
France	Languedoc-Roussillon	Pyrénées-Orientales	1280	
France	Provence-Alpes-Côte-d'Azur	Alpes-Maritimes	0	
France	Provence-Alpes-Côte-d'Azur	Bouches-du-Rhône	0	7
France	Provence-Alpes-Côte-d'Azur	Var	0	
France	Provence-Alpes-Côte-d'Azur	Var	1280	
Germany	Baden-Württemberg	Freiburg	1302	
Greece	Anatoliki Makedonia kai Thraki		735	
Greece	Anatoliki Makedonia kai Thraki		735	

Continued

Table A1.18. Continued.

Country	Administrative region 1	Administrative region 2	Reference ID	No. entries[a]
Greece	Anatoliki Makedonia kai Thraki	Drama	0	4
Greece	Anatoliki Makedonia kai Thraki	Evros	0	5
Greece	Anatoliki Makedonia kai Thraki	Kavala	0	4
Greece	Dytiki Makedonia		735	
Greece	Dytiki Makedonia		735	
Greece	Dytiki Makedonia	Grevena	0	
Greece	Dytiki Makedonia	Kozani	0	
Greece	Ipeiros	Ioannina	0	3
Greece	Ipeiros	Preveza	0	
Greece	Kentriki Makedonia		735	
Greece	Kentriki Makedonia	Khalkidiki	0	5
Greece	Kentriki Makedonia	Khalkidiki	1303	
Greece	Kentriki Makedonia	Khalkidiki	1303	
Greece	Kentriki Makedonia	Khalkidiki	1303	
Greece	Kentriki Makedonia	Kilkis	0	4
Greece	Kentriki Makedonia	Pella	0	
Greece	Kentriki Makedonia	Serrai	0	2
Greece	Kentriki Makedonia	Thessaloniki	0	3
Greece	Kriti	Heraklion	0	
Greece	Peloponnisos	Arcadia	0	
Greece	Peloponnisos	Messinia	0	3
Greece	Stereá Elláda	Boeotia	0	
Greece	Stereá Elláda	Euboea	0	
Greece	Stereá Elláda	Fthiotis	0	
Greece	Thessalia	Larisa	0	
Italy	Abruzzo	L'Aquila	0	2
Italy	Abruzzo	Teramo	0	
Italy	Apulia	Bari	0	11
Italy	Apulia	Barletta-Andria-Trani	0	2
Italy	Apulia	Brindisi	0	2
Italy	Apulia	Foggia	0	19
Italy	Apulia	Lecce	0	2
Italy	Apulia	Taranto	0	2
Italy	Basilicata	Matera	0	7
Italy	Calabria	Cosenza	0	
Italy	Calabria	Reggio di Calabria	0	3
Italy	Campania	Napoli	0	
Italy	Emilia-Romagna	Forlì-Cesena	0	
Italy	Emilia-Romagna	Parma	0	3
Italy	Emilia-Romagna	Ravenna	0	4
Italy	Emilia-Romagna	Rimini	0	
Italy	Friuli-Venezia Giulia	Udine	0	
Italy	Lazio	Latina	0	3
Italy	Lazio	Rieti	0	8
Italy	Lazio	Roma	0	14
Italy	Lazio	Viterbo	0	5
Italy	Liguria	Genova	0	2
Italy	Liguria	Savona	0	
Italy	Lombardia	Brescia	0	

Continued

Table A1.18. Continued.

Country	Administrative region 1	Administrative region 2	Reference ID	No. entries[a]
Italy	Marche	Macerata	0	
Italy	Molise	Campobasso	0	6
Italy	Piemonte	Alessandria	0	
Italy	Piemonte	Torino	0	
Italy	Piemonte	Verbano-Cusio-Ossola	0	3
Italy	Sardegna	Cagliari	0	
Italy	Sardegna	Carbonia-Iglesias	0	
Italy	Sardegna	Ogliastra	0	2
Italy	Sardegna	Oristano	0	2
Italy	Sicily	Agrigento	0	2
Italy	Sicily	Enna	0	
Italy	Sicily	Messina	0	
Italy	Sicily	Palermo	0	2
Italy	Sicily	Palermo	242	
Italy	Sicily	Palermo	366	
Italy	Sicily	Palermo	366	
Italy	Sicily	Palermo	366	
Italy	Sicily	Palermo	366	
Italy	Sicily	Palermo	998	
Italy	Sicily	Ragusa	366	
Italy	Sicily	Trapani	0	2
Italy	Toscana		1331	
Italy	Toscana	Florence	0	2
Italy	Toscana	Grosseto	0	
Italy	Toscana	Pisa	0	2
Italy	Trentino-Alto Adige	Trento	0	
Italy	Umbria	Perugia	0	2
Italy	Umbria	Terni	0	
Italy	Veneto	Venezia	0	
Italy	Veneto	Verona	0	
Kosovo	Đakovica	Dečani	0	3
Kosovo	Gnjilane	Gnjilane	0	
Kosovo	Gnjilane	Vitina	0	
Kosovo	Kosovska Mitrovica	Kosovska Mitrovica	0	
Kosovo	Kosovska Mitrovica	Kosovska Mitrovica	346	
Kosovo	Kosovska Mitrovica	Kosovska Mitrovica	346	
Kosovo	Pristina	Lipljan	0	
Kosovo	Pristina	Priötina	0	
Libya	Al Marqab		0	
Macedonia	Eastern	Pehcevo	0	
Macedonia	Pelagonia	Bitola	0	2
Macedonia	Polog	Tetovo	0	2
Macedonia	Skopje	Centar	0	5
Macedonia	Southeastern	Lake Dojran	0	2
Macedonia	Southwestern	Ohrid	0	2
Macedonia	Vardar	Kavadartsi	0	
Morocco	Chaouia-Ouardigha	Ben Slimane	0	3
Morocco	Chaouia-Ouardigha	Settat	0	3
Morocco	Doukkala-Abda	El Jadida	0	
Morocco	Doukkala-Abda	Safi	0	3
Morocco	Fès-Boulemane	Zouagha-Moulay Yacoub	0	

Continued

Table A1.18. Continued.

Country	Administrative region 1	Administrative region 2	Reference ID	No. entries[a]
Morocco	Gharb-Chrarda-Béni Hssen	Kénitra	0	3
Morocco	Grand Casablanca	Mohammedia	0	
Morocco	Marrakech-Tensift-Al Haouz	Al Haouz	0	3
Morocco	Marrakech-Tensift-Al Haouz	El Kelaâ des Sraghna	0	6
Morocco	Marrakech-Tensift-Al Haouz	Marrakech	0	5
Morocco	Marrakech-Tensift-Al Haouz	Marrakech	0	
Morocco	Meknès-Tafilalet	Meknès	0	2
Morocco	Rabat-Salé-Zemmour-Zaer	Khémisset	0	7
Morocco	Rabat-Salé-Zemmour-Zaer	Skhirate-Témara	0	4
Morocco	Souss-Massa-Draâ	Taroudannt	0	
Morocco	Tanger-Tétouan	Chefchaouen	0	3
Morocco	Taza-Al Hoceima-Taounate	Al Hoceïma	0	
Morocco	Taza-Al Hoceima-Taounate	Taza	0	2
Portugal	Aveiro	Oliveira de Azeméis	0	
Portugal	Coimbra	Coimbra	0	
Portugal	Évora	Montemor-o-Novo	0	2
Portugal	Évora	Portel	0	4
Portugal	Guarda	Guarda	0	
Portugal	Leiria	Peniche	0	
Portugal	Lisboa	Alenquer	0	
Portugal	Portalegre	Marvão	0	
Portugal	Porto	Marco de Canaveses	0	
Portugal	Santarém	Benavente	0	
Portugal	Santarém	Coruche	0	
Portugal	Setúbal	Alcácer do Sal	0	15
Portugal	Setúbal	Grândola	0	
Portugal	Setúbal	Palmela	0	
Portugal	Setúbal	Setúbal	0	
Portugal	Viana do Castelo	Paredes de Coura	0	
Romania	Alba		0	
Romania	Bucharest		0	
Romania	Constanta		0	8
Romania	Dolj		0	3
Romania	Gorj		769	
Romania	Mehedinti		769	
Romania	Mures		769	
Romania	Timis		0	3
Romania	Tulcea		0	
Romania	Tulcea		460	
Romania	Tulcea		769	
Russia	Stavropol		1144	
Russia	Stavropol		1144	
Serbia	Borski	Kladovo	0	2
Serbia	Macvanski	Ljubovija	0	2
Serbia	Pcinjski	Surdulica	0	
Serbia	Pirotski	Pirot	0	
Serbia	Ra'ki	Novi Pazar	0	2
Serbia	Ra'ki	Vrnjacka Banja	0	
Serbia	Toplicki	Kuröumlija	0	
Spain	Andalucía	Cádiz	0	6
Spain	Andalucía	Cádiz	241	

Continued

Table A1.18. Continued.

Country	Administrative region 1	Administrative region 2	Reference ID	No. entries[a]
Spain	Andalucía	Cádiz	241	
Spain	Andalucía	Cádiz	821	
Spain	Andalucía	Córdoba	0	3
Spain	Andalucía	Granada	0	3
Spain	Andalucía	Jaén	0	5
Spain	Andalucía	Sevilla	0	8
Spain	Aragón	Zaragoza	0	8
Spain	Castilla y León	Salamanca	0	
Spain	Castilla y León	Toledo	821	
Spain	Castilla y León	Toledo	821	
Spain	Castilla-La Mancha	Albacete	821	
Spain	Castilla-La Mancha	Ciudad Real	246	
Spain	Castilla-La Mancha	Cuenca	821	
Spain	Castilla-La Mancha	Toledo	0	
Spain	Castilla-La Mancha	Toledo	989	
Spain	Comunidad Foral de Navarra	Navarra	0	
Spain	Extremadura	Badajoz	0	14
Spain	Extremadura	Cáceres	0	50
Sweden	Gotland		1339	
Sweden	Kalmar		1339	
Sweden	Skåne		1339	
Switzerland	Ticino		1284	
Switzerland	Ticino		1284	
Tunisia	Ariana	Ettadhamen	0	
Tunisia	Ariana	Kalaat El Andalous	0	
Tunisia	Ariana	Sidi Thabet	0	3
Tunisia	Béja	Amdoun	0	
Tunisia	Béja	Béja Nord	0	
Tunisia	Béja	Béja Sud	0	
Tunisia	Béja	Nefza	0	3
Tunisia	Ben Arous (Tunis Sud)	Ezzahra	0	
Tunisia	Bizerte	Bizerte Nord	0	2
Tunisia	Bizerte	Bizerte Sud	0	
Tunisia	Bizerte	Ghar El Melh	0	
Tunisia	Bizerte	Ghazala	0	
Tunisia	Gabès	Hamma	0	
Tunisia	Gabès	Matmata	0	
Tunisia	Gafsa	Ksar	0	
Tunisia	Jendouba	Aïn Draham	0	2
Tunisia	Jendouba	Fernana	0	
Tunisia	Jendouba	Tabarka	0	
Tunisia	Kairouan	Kairouan Sud	0	
Tunisia	Kairouan	Sbikha	0	
Tunisia	Kebili	Kebili Nord	0	
Tunisia	Le Kef	Sakiet Sidi Youssef	0	2
Tunisia	Mahdia	Chebba	0	
Tunisia	Manubah	El Battan	0	
Tunisia	Manubah	Oued Ellil	0	
Tunisia	Manubah	Tebourba	0	
Tunisia	Médenine	Houmt Souk	0	
Tunisia	Monastir	Jammel	0	

Continued

Table A1.18. Continued.

Country	Administrative region 1	Administrative region 2	Reference ID	No. entries[a]
Tunisia	Nabeul	Haouaria	0	
Tunisia	Nabeul	Korba	0	2
Tunisia	Nabeul	Soliman	0	
Tunisia	Nabeul	Takelsa	0	
Tunisia	Sidi Bou Zid	Meknassi	0	
Tunisia	Tunis	Cité El Khadra	0	
Tunisia	Tunis	La Goulette	0	
Tunisia	Tunis	Sidi El Béchir	0	3
Tunisia	Zaghouan	Fahs	0	2
Tunisia	Zaghouan	Zriba	0	2
Turkey	Aksaray		1353	
Turkey	Ankara		0	25
Turkey	Ankara		1353	
Turkey	Ankara		1353	
Turkey	Ankara		844	
Turkey	Artvin		0	2
Turkey	Balıkesir		0	
Turkey	Burdur		0	
Turkey	Çorum		0	6
Turkey	Elâziğ		0	3
Turkey	Erzurum		0	2
Turkey	Giresun		0	3
Turkey	Gümüshane		0	
Turkey	İzmir		0	
Turkey	Kayseri		0	8
Turkey	Kırklareli		0	
Turkey	Ordu		0	
Turkey	Samsun		0	2
Turkey	Sivas		0	12
Turkey	Tokat		0	45
Turkey	Yozgat		0	6

[a]If more than one.

Table A1.19. Geographic distribution data for *Rhipicephalus sanguineus* group (*Rh. sanguineus* and *Rh. turanicus*). See Appendix 2 (hosted on http://www.cabi.org/vetmedresource/) for complete reference indicated by the Reference ID.

Species	Country	Administrative region 1	Administrative region 2	Reference ID	No. entries[a]
Rhipicephalus sanguineus	Algeria	Alger		110	
Rhipicephalus sanguineus	Algeria	El Tarf	El Tarf	110	
Rhipicephalus sanguineus	Bulgaria	Stara Zagora	Haskovo	69	2
Rhipicephalus sanguineus	Egypt	Al Jizah		1101	
Rhipicephalus sanguineus	Egypt	Al Wadi al Jadid		595	

Continued

Table A1.19. Continued.

Species	Country	Administrative region 1	Administrative region 2	Reference ID	No. entries[a]
Rhipicephalus sanguineus	France	Aquitaine	Dordogne	1280	
Rhipicephalus sanguineus	France	Aquitaine	Gironde	1280	
Rhipicephalus sanguineus	France	Aquitaine	Landes	1280	
Rhipicephalus sanguineus	France	Aquitaine	Lot-et-Garonne	1280	
Rhipicephalus sanguineus	France	Auvergne	Allier	1280	
Rhipicephalus sanguineus	France	Bourgogne	Côte-d'Or	1280	
Rhipicephalus sanguineus	France	Corse		1280	
Rhipicephalus sanguineus	France	Corse	Corse-du-Sud	1306	4
Rhipicephalus sanguineus	France	Île-de-France	Seine-Saint-Denis	1280	
Rhipicephalus sanguineus	France	Île-de-France	Ville de Paris	1280	
Rhipicephalus sanguineus	France	Provence-Alpes-Côte-d'Azur	Alpes-Maritimes	1280	
Rhipicephalus sanguineus	France	Provence-Alpes-Côte-d'Azur	Bouches-du-Rhône	1280	
Rhipicephalus sanguineus	France	Provence-Alpes-Côte-d'Azur	Var	1280	
Rhipicephalus sanguineus	France	Rhône-Alpes	Isère	1280	
Rhipicephalus sanguineus	Greece	Dytiki Makedonia		735	2
Rhipicephalus sanguineus	Greece	Kentriki Makedonia		735	
Rhipicephalus sanguineus	Greece	Kentriki Makedonia		735	
Rhipicephalus sanguineus	Greece	Kentriki Makedonia	Khalkidiki	1303	
Rhipicephalus sanguineus	Greece	Station Stereá Elláda	Fokis	778	
Rhipicephalus sanguineus	Hungary	Nógrád		445	
Rhipicephalus sanguineus	Hungary	Nógrád		829	
Rhipicephalus sanguineus	Italy	Sicily	Palermo	242	
Rhipicephalus sanguineus	Italy	Sicily	Palermo	998	
Rhipicephalus sanguineus	Italy	Toscana		1331	
Rhipicephalus sanguineus	Kosovo	Kosovska Mitrovica	Kosovska Mitrovica	346	
Rhipicephalus sanguineus	Kosovo	Kosovska Mitrovica	Kosovska Mitrovica	346	

Continued

Table A1.19. Continued.

Species	Country	Administrative region 1	Administrative region 2	Reference ID	No. entries[a]
Rhipicephalus sanguineus	Romania	Timis		177	
Rhipicephalus sanguineus	Spain	Andalucía	Jaén	611	
Rhipicephalus sanguineus	Spain	Andalucía	Sevilla	614	
Rhipicephalus sanguineus	Spain	Castilla-La Mancha	Toledo	989	2
Rhipicephalus sanguineus	Spain	Islas Baleares	Baleares	1345	2
Rhipicephalus sanguineus	Spain	País Vasco		77	
Rhipicephalus sanguineus	Spain	Región de Murcia	Murcia	616	
Rhipicephalus sanguineus	Sweden	Östergötland		1339	
Rhipicephalus sanguineus	Sweden	Södermanland		1339	
Rhipicephalus sanguineus	Sweden	Stockholm		1339	
Rhipicephalus sanguineus	Sweden	Uppsala		1339	
Rhipicephalus sanguineus	Sweden	Västernorrland		1339	
Rhipicephalus sanguineus	Sweden	Västmanland		1339	
Rhipicephalus sanguineus	Turkey	Afyon		186	
Rhipicephalus sanguineus	Turkey	Elâziğ		22	
Rhipicephalus sanguineus	Turkey	Erzurum		32	
Rhipicephalus sanguineus	Turkey	Istanbul		145	
Rhipicephalus sanguineus	Turkey	Şanlıurfa		303	
Rhipicephalus sanguineus group	Albania	Durrës	Durrësit	0	
Rhipicephalus sanguineus group	Albania	Elbasan	Librazhdit	0	2
Rhipicephalus sanguineus group	Albania	Gjirokastër	Përmetit	0	
Rhipicephalus sanguineus group	Albania	Korçë	Korçës	0	
Rhipicephalus sanguineus group	Albania	Lezhë	Lezhës	0	2

Continued

Table A1.19. Continued.

Species	Country	Administrative region 1	Administrative region 2	Reference ID	No. entries[a]
Rhipicephalus sanguineus group	Albania	Shkodër	Shkodrës	0	
Rhipicephalus sanguineus group	Albania	Tiranë	Tiranës	0	2
Rhipicephalus sanguineus group	Albania	Vlorë	Sarandës	0	2
Rhipicephalus sanguineus group	Albania	Vlorë	Vlorës	0	2
Rhipicephalus sanguineus group	Algeria	Aïn Témouchent	Oued Sebbah	0	
Rhipicephalus sanguineus group	Algeria	Alger	Ain Benian	0	
Rhipicephalus sanguineus group	Algeria	Alger	Bouzareah	0	
Rhipicephalus sanguineus group	Algeria	Alger	Ouled Chebel	0	
Rhipicephalus sanguineus group	Algeria	Batna	Ain Touta	0	
Rhipicephalus sanguineus group	Algeria	Batna	Chemora	0	
Rhipicephalus sanguineus group	Algeria	Batna	Ras El Aioun	0	
Rhipicephalus sanguineus group	Algeria	Béchar	Beni Abbes	0	2
Rhipicephalus sanguineus group	Algeria	Blida	Blida	0	
Rhipicephalus sanguineus group	Algeria	Blida	Boufarik	0	
Rhipicephalus sanguineus group	Algeria	Bouira	Ain Turk	0	
Rhipicephalus sanguineus group	Algeria	Bouira	Guerrouma	0	2
Rhipicephalus sanguineus group	Algeria	Bouira	Lakhdaria	0	

Continued

Table A1.19. Continued.

Species	Country	Administrative region 1	Administrative region 2	Reference ID	No. entries[a]
Rhipicephalus sanguineus group	Algeria	Bouira	Sour El Ghouzlane	0	
Rhipicephalus sanguineus group	Algeria	Chlef	Chlef	0	
Rhipicephalus sanguineus group	Algeria	El Bayadh	Boualem	0	
Rhipicephalus sanguineus group	Algeria	El Bayadh	Bougtoub	0	
Rhipicephalus sanguineus group	Algeria	El Bayadh	El Bayadh	0	3
Rhipicephalus sanguineus group	Algeria	El Bayadh	Stitten	0	
Rhipicephalus sanguineus group	Algeria	Guelma	Hammam Debagh	0	
Rhipicephalus sanguineus group	Algeria	Jijel	Chekfa	0	
Rhipicephalus sanguineus group	Algeria	Laghouat	Hassi R'Mel	0	
Rhipicephalus sanguineus group	Algeria	Laghouat	Laghouat	0	
Rhipicephalus sanguineus group	Algeria	M'Sila	Sidi M'Hamed	0	2
Rhipicephalus sanguineus group	Algeria	Mascara	Mascara	0	
Rhipicephalus sanguineus group	Algeria	Mostaganem	Mostaganem	0	
Rhipicephalus sanguineus group	Algeria	Oran	Oran	0	2
Rhipicephalus sanguineus group	Algeria	Relizane	Sidi M'Hamed Benaouda	0	
Rhipicephalus sanguineus group	Algeria	Sétif	Ain Arnat	0	
Rhipicephalus sanguineus group	Algeria	Sétif	Setif	0	

Continued

Table A1.19. Continued.

Species	Country	Administrative region 1	Administrative region 2	Reference ID	No. entries[a]
Rhipicephalus sanguineus group	Algeria	Sidi Bel Abbès	Dhaya	0	
Rhipicephalus sanguineus group	Algeria	Souk Ahras	Ouled Moumen	0	
Rhipicephalus sanguineus group	Algeria	Tiaret	Meghila	0	
Rhipicephalus sanguineus group	Algeria	Tindouf	Tindouf	0	3
Rhipicephalus sanguineus group	Algeria	Tissemsilt	Theniet El Had	0	
Rhipicephalus sanguineus group	Algeria	Tizi Ouzou	Ain El Hammam	0	
Rhipicephalus sanguineus group	Bulgaria	Burgas	Burgas	0	6
Rhipicephalus sanguineus group	Bulgaria	Burgas	Malko Tarnovo	0	3
Rhipicephalus sanguineus group	Bulgaria	Burgas	Primorsko	0	
Rhipicephalus sanguineus group	Bulgaria	Kardzhali	Ardino	0	
Rhipicephalus sanguineus group	Bulgaria	Kardzhali	Momchilgrad	0	4
Rhipicephalus sanguineus group	Bulgaria	Shumen	Shumen	0	3
Rhipicephalus sanguineus group	Bulgaria	Smolyan	Dospat	0	
Rhipicephalus sanguineus group	Bulgaria	Sofia	Ihtiman	0	
Rhipicephalus sanguineus group	Bulgaria	Sofia	Zlatitsa	0	2
Rhipicephalus sanguineus group	Croatia	Dubrovacko-Neretvanska		0	3
Rhipicephalus sanguineus group	Croatia	Istarska		0	

Continued

Table A1.19. Continued.

Species	Country	Administrative region 1	Administrative region 2	Reference ID	No. entries[a]
Rhipicephalus sanguineus group	Croatia	Splitsko-Dalmatinska		0	2
Rhipicephalus sanguineus group	Croatia	Zadarska		0	4
Rhipicephalus sanguineus group	Cyprus	Famagusta		0	4
Rhipicephalus sanguineus group	Cyprus	Larnaca		0	
Rhipicephalus sanguineus group	Cyprus	Limassol		0	
Rhipicephalus sanguineus group	Cyprus	Nicosia		0	2
Rhipicephalus sanguineus group	Cyprus	Paphos		0	3
Rhipicephalus sanguineus group	Egypt	Al Bahr al Ahmar		0	3
Rhipicephalus sanguineus group	Egypt	Al Iskandariyah		0	
Rhipicephalus sanguineus group	Egypt	Al Jizah		0	4
Rhipicephalus sanguineus group	Egypt	Al Minufiyah		0	9
Rhipicephalus sanguineus group	Egypt	Al Minya		0	
Rhipicephalus sanguineus group	Egypt	Aswan		0	3
Rhipicephalus sanguineus group	Egypt	Bani Suwayf		0	2
Rhipicephalus sanguineus group	Egypt	Janub Sina'		0	6
Rhipicephalus sanguineus group	France	Alsace	Bas-Rhin	0	4
Rhipicephalus sanguineus group	France	Aquitaine	Dordogne	0	

Continued

Table A1.19. Continued.

Species	Country	Administrative region 1	Administrative region 2	Reference ID	No. entries[a]
Rhipicephalus sanguineus group	France	Aquitaine	Gironde	0	2
Rhipicephalus sanguineus group	France	Aquitaine	Lot-et-Garonne	0	2
Rhipicephalus sanguineus group	France	Aquitaine	Pyrénées-Atlantiques	0	
Rhipicephalus sanguineus group	France	Auvergne	Allier	0	2
Rhipicephalus sanguineus group	France	Auvergne	Cantal	0	
Rhipicephalus sanguineus group	France	Auvergne	Puy-de-Dôme	0	
Rhipicephalus sanguineus group	France	Basse-Normandie	Calvados	0	
Rhipicephalus sanguineus group	France	Basse-Normandie	Orne	0	
Rhipicephalus sanguineus group	France	Centre	Indre	0	
Rhipicephalus sanguineus group	France	Centre	Indre-et-Loire	0	
Rhipicephalus sanguineus group	France	Champagne-Ardenne	Ardennes	0	
Rhipicephalus sanguineus group	France	Champagne-Ardenne	Marne	0	2
Rhipicephalus sanguineus group	France	Corse	Corse-du-Sud	0	8
Rhipicephalus sanguineus group	France	Corse	Haute-Corse	0	4
Rhipicephalus sanguineus group	France	Franche-Comté	Doubs	0	
Rhipicephalus sanguineus group	France	Île-de-France	Hauts-de-Seine	0	
Rhipicephalus sanguineus group	France	Île-de-France	Seine-et-Marne	0	2

Continued

Table A1.19. Continued.

Species	Country	Administrative region 1	Administrative region 2	Reference ID	No. entries[a]
Rhipicephalus sanguineus group	France	Île-de-France	Seine-Saint-Denis	0	
Rhipicephalus sanguineus group	France	Île-de-France	Val-de-Marne	0	5
Rhipicephalus sanguineus group	France	Île-de-France	Ville de Paris	0	3
Rhipicephalus sanguineus group	France	Île-de-France	Yvelines	0	2
Rhipicephalus sanguineus group	France	Languedoc-Roussillon	Aude	0	2
Rhipicephalus sanguineus group	France	Languedoc-Roussillon	Gard	0	8
Rhipicephalus sanguineus group	France	Languedoc-Roussillon	Hérault	0	3
Rhipicephalus sanguineus group	France	Languedoc-Roussillon	Pyrénées-Orientales	0	10
Rhipicephalus sanguineus group	France	Lorraine	Meurthe-et-Moselle	0	
Rhipicephalus sanguineus group	France	Midi-Pyrénées	Ariège	0	2
Rhipicephalus sanguineus group	France	Midi-Pyrénées	Aveyron	0	2
Rhipicephalus sanguineus group	France	Midi-Pyrénées	Haute-Garonne	0	
Rhipicephalus sanguineus group	France	Midi-Pyrénées	Lot	0	2
Rhipicephalus sanguineus group	France	Midi-Pyrénées	Tarn	0	
Rhipicephalus sanguineus group	France	Midi-Pyrénées	Tarn-et-Garonne	0	
Rhipicephalus sanguineus group	France	Pays de la Loire	Vendée	0	
Rhipicephalus sanguineus group	France	Picardie	Somme	0	

Continued

Table A1.19. Continued.

Species	Country	Administrative region 1	Administrative region 2	Reference ID	No. entries[a]
Rhipicephalus sanguineus group	France	Provence-Alpes-Côte-d'Azur	Alpes-de-Haute-Provence	0	8
Rhipicephalus sanguineus group	France	Provence-Alpes-Côte-d'Azur	Alpes-Maritimes	0	11
Rhipicephalus sanguineus group	France	Provence-Alpes-Côte-d'Azur	Bouches-du-Rhône	0	35
Rhipicephalus sanguineus group	France	Provence-Alpes-Côte-d'Azur	Var	0	26
Rhipicephalus sanguineus group	France	Provence-Alpes-Côte-d'Azur	Vaucluse	0	24
Rhipicephalus sanguineus group	France	Rhône-Alpes	Ain	0	
Rhipicephalus sanguineus group	France	Rhône-Alpes	Ardèche	0	11
Rhipicephalus sanguineus group	France	Rhône-Alpes	Drôme	0	5
Rhipicephalus sanguineus group	France	Rhône-Alpes	Isère	0	2
Rhipicephalus sanguineus group	France	Rhône-Alpes	Loire	0	
Rhipicephalus sanguineus group	France	Rhône-Alpes	Rhône	0	
Rhipicephalus sanguineus group	Greece	Anatoliki Makedonia kai Thraki	Drama	0	2
Rhipicephalus sanguineus group	Greece	Anatoliki Makedonia kai Thraki	Evros	0	4
Rhipicephalus sanguineus group	Greece	Anatoliki Makedonia kai Thraki	Kavala	0	5
Rhipicephalus sanguineus group	Greece	Attiki	Attica	0	
Rhipicephalus sanguineus group	Greece	Dytiki Makedonia	Grevena	0	
Rhipicephalus sanguineus group	Greece	Dytiki Makedonia	Kastoria	0	3

Continued

Table A1.19. Continued.

Species	Country	Administrative region 1	Administrative region 2	Reference ID	No. entries[a]
Rhipicephalus sanguineus group	Greece	Ionioi Nisoi	Corfu	0	
Rhipicephalus sanguineus group	Greece	Ionioi Nisoi	Kefallinia	0	2
Rhipicephalus sanguineus group	Greece	Ipeiros	Ioannina	0	2
Rhipicephalus sanguineus group	Greece	Kentriki Makedonia	Imathia	0	
Rhipicephalus sanguineus group	Greece	Kentriki Makedonia	Khalkidiki	0	10
Rhipicephalus sanguineus group	Greece	Kentriki Makedonia	Kilkis	0	5
Rhipicephalus sanguineus group	Greece	Kentriki Makedonia	Pella	0	
Rhipicephalus sanguineus group	Greece	Kentriki Makedonia	Pieria	0	8
Rhipicephalus sanguineus group	Greece	Kentriki Makedonia	Serrai	0	2
Rhipicephalus sanguineus group	Greece	Kentriki Makedonia	Thessaloniki	0	13
Rhipicephalus sanguineus group	Greece	Notio Aigaio	Cyclades	0	
Rhipicephalus sanguineus group	Greece	Peloponnisos	Arcadia	0	
Rhipicephalus sanguineus group	Greece	Peloponnisos	Messinia	0	
Rhipicephalus sanguineus group	Greece	Stereá Elláda	Boeotia	0	
Rhipicephalus sanguineus group	Greece	Thessalia	Larisa	0	
Rhipicephalus sanguineus group	Greece	Thessalia	Magnesia	0	
Rhipicephalus sanguineus group	Greece	Voreio Aigaio	Lesvos	0	4

Continued

Table A1.19. Continued.

Species	Country	Administrative region 1	Administrative region 2	Reference ID	No. entries[a]
Rhipicephalus sanguineus group	Italy	Abruzzo	L'Aquila	0	
Rhipicephalus sanguineus group	Italy	Abruzzo	Teramo	0	6
Rhipicephalus sanguineus group	Italy	Apulia	Bari	0	29
Rhipicephalus sanguineus group	Italy	Apulia	Barletta-Andria-Trani	0	4
Rhipicephalus sanguineus group	Italy	Apulia	Brindisi	0	20
Rhipicephalus sanguineus group	Italy	Apulia	Foggia	0	49
Rhipicephalus sanguineus group	Italy	Apulia	Lecce	0	11
Rhipicephalus sanguineus group	Italy	Apulia	Taranto	0	8
Rhipicephalus sanguineus group	Italy	Basilicata	Matera	0	18
Rhipicephalus sanguineus group	Italy	Basilicata	Potenza	0	
Rhipicephalus sanguineus group	Italy	Calabria	Cosenza	0	12
Rhipicephalus sanguineus group	Italy	Calabria	Reggio di Calabria	0	10
Rhipicephalus sanguineus group	Italy	Campania		802	
Rhipicephalus sanguineus group	Italy	Campania	Avellino	0	2
Rhipicephalus sanguineus group	Italy	Campania	Caserta	0	2
Rhipicephalus sanguineus group	Italy	Emilia-Romagna	Forlè-Cesena	0	2
Rhipicephalus sanguineus group	Italy	Emilia-Romagna	Modena	0	

Continued

Table A1.19. Continued.

Species	Country	Administrative region 1	Administrative region 2	Reference ID	No. entries[a]
Rhipicephalus sanguineus group	Italy	Emilia-Romagna	Parma	0	
Rhipicephalus sanguineus group	Italy	Emilia-Romagna	Ravenna	0	3
Rhipicephalus sanguineus group	Italy	Emilia-Romagna	Reggio nell'Emilia	0	
Rhipicephalus sanguineus group	Italy	Friuli-Venezia Giulia	Udine	0	
Rhipicephalus sanguineus group	Italy	Lazio	Latina	0	4
Rhipicephalus sanguineus group	Italy	Lazio	Rieti	0	2
Rhipicephalus sanguineus group	Italy	Lazio	Roma	0	14
Rhipicephalus sanguineus group	Italy	Lazio	Viterbo	0	3
Rhipicephalus sanguineus group	Italy	Liguria	Genova	0	3
Rhipicephalus sanguineus group	Italy	Liguria	Imperia	0	2
Rhipicephalus sanguineus group	Italy	Marche	Ascoli Piceno	0	2
Rhipicephalus sanguineus group	Italy	Molise	Campobasso	0	15
Rhipicephalus sanguineus group	Italy	Piemonte	Cuneo	0	
Rhipicephalus sanguineus group	Italy	Sardegna	Cagliari	0	
Rhipicephalus sanguineus group	Italy	Sardegna	Carbonia-Iglesias	0	2
Rhipicephalus sanguineus group	Italy	Sardegna	Nuoro	0	
Rhipicephalus sanguineus group	Italy	Sardegna	Ogliastra	0	

Continued

Table A1.19. Continued.

Species	Country	Administrative region 1	Administrative region 2	Reference ID	No. entries[a]
Rhipicephalus sanguineus group	Italy	Sardegna	Olbia-Tempio	0	3
Rhipicephalus sanguineus group	Italy	Sardegna	Oristano	0	
Rhipicephalus sanguineus group	Italy	Sardegna	Sassari	0	
Rhipicephalus sanguineus group	Italy	Sicily	Agrigento	0	8
Rhipicephalus sanguineus group	Italy	Sicily	Catania	0	
Rhipicephalus sanguineus group	Italy	Sicily	Enna	0	
Rhipicephalus sanguineus group	Italy	Sicily	Palermo	0	17
Rhipicephalus sanguineus group	Italy	Sicily	Trapani	0	6
Rhipicephalus sanguineus group	Italy	Toscana	Florence	0	2
Rhipicephalus sanguineus group	Italy	Toscana	Grosseto	0	5
Rhipicephalus sanguineus group	Italy	Toscana	Livorno	0	
Rhipicephalus sanguineus group	Italy	Toscana	Pisa	0	
Rhipicephalus sanguineus group	Italy	Umbria	Perugia	0	
Rhipicephalus sanguineus group	Italy	Umbria	Terni	0	2
Rhipicephalus sanguineus group	Italy	Veneto	Belluno	0	
Rhipicephalus sanguineus group	Italy	Veneto	Verona	0	2
Rhipicephalus sanguineus group	Libya	Ajdabiya		0	

Continued

Table A1.19. Continued.

Species	Country	Administrative region 1	Administrative region 2	Reference ID	No. entries[a]
Rhipicephalus sanguineus group	Libya	Al Butnan		0	3
Rhipicephalus sanguineus group	Libya	Al Hizam al Akhdar		0	13
Rhipicephalus sanguineus group	Libya	Al Jabal al Akhdar		0	
Rhipicephalus sanguineus group	Libya	Al Jabal al Akhdar		0	3
Rhipicephalus sanguineus group	Libya	Al Kufrah		0	2
Rhipicephalus sanguineus group	Libya	Al Marj		0	3
Rhipicephalus sanguineus group	Libya	Al Marqab		0	
Rhipicephalus sanguineus group	Libya	Al Qubbah		0	3
Rhipicephalus sanguineus group	Libya	Al Wahah		0	2
Rhipicephalus sanguineus group	Libya	An Nuqat al Khams		0	
Rhipicephalus sanguineus group	Libya	Ghadamis		0	2
Rhipicephalus sanguineus group	Libya	Gharyan		0	
Rhipicephalus sanguineus group	Libya	Misratah		0	5
Rhipicephalus sanguineus group	Libya	Nalut		0	3
Rhipicephalus sanguineus group	Libya	Surt		0	2
Rhipicephalus sanguineus group	Libya	Tajura' wa an Nawahi al Arba		0	
Rhipicephalus sanguineus group	Libya	Tarabulus		0	2

Continued

Table A1.19. Continued.

Species	Country	Administrative region 1	Administrative region 2	Reference ID	No. entries[a]
Rhipicephalus sanguineus group	Libya	Tarhunah-Masallatah		0	
Rhipicephalus sanguineus group	Libya	Yafran-Jadu		0	2
Rhipicephalus sanguineus group	Monaco	Monaco		0	
Rhipicephalus sanguineus group	Morocco	Chaouia-Ouardigha	Ben Slimane	0	3
Rhipicephalus sanguineus group	Morocco	Chaouia-Ouardigha	Settat	0	4
Rhipicephalus sanguineus group	Morocco	Fès-Boulemane	Zouagha-Moulay Yacoub	0	2
Rhipicephalus sanguineus group	Morocco	Gharb-Chrarda-Béni Hssen	Kénitra	0	4
Rhipicephalus sanguineus group	Morocco	Gharb-Chrarda-Béni Hssen	Sidi Kacem	0	
Rhipicephalus sanguineus group	Morocco	Grand Casablanca	Mohammedia	0	2
Rhipicephalus sanguineus group	Morocco	Guelmim-Es-Semara	Assa-Zag	0	2
Rhipicephalus sanguineus group	Morocco	Guelmim-Es-Semara	Guelmim	0	4
Rhipicephalus sanguineus group	Morocco	Marrakech-Tensift-Al Haouz	Al Haouz	0	2
Rhipicephalus sanguineus group	Morocco	Marrakech-Tensift-Al Haouz	El Kelaâ Des Sraghna	0	4
Rhipicephalus sanguineus group	Morocco	Marrakech-Tensift-Al Haouz	Essaouira	0	
Rhipicephalus sanguineus group	Morocco	Marrakech-Tensift-Al Haouz	Marrakech	0	5
Rhipicephalus sanguineus group	Morocco	Meknès-Tafilalet	Ifrane	0	2
Rhipicephalus sanguineus group	Morocco	Meknès-Tafilalet	Khénifra	0	

Continued

Table A1.19. Continued.

Species	Country	Administrative region 1	Administrative region 2	Reference ID	No. entries[a]
Rhipicephalus sanguineus group	Morocco	Meknès-Tafilalet	Meknès	0	4
Rhipicephalus sanguineus group	Morocco	Oriental	Oujda Angad	0	
Rhipicephalus sanguineus group	Morocco	Rabat-Salé-Zemmour-Zaer	Khémisset	0	4
Rhipicephalus sanguineus group	Morocco	Rabat-Salé-Zemmour-Zaer	Skhirate-Témara	0	3
Rhipicephalus sanguineus group	Morocco	Souss-Massa-Draâ	Agadir-Ida Ou Tanane	0	4
Rhipicephalus sanguineus group	Morocco	Souss-Massa-Draâ	Chtouka-Aït Baha	0	
Rhipicephalus sanguineus group	Morocco	Souss-Massa-Draâ	Ouarzazate	0	
Rhipicephalus sanguineus group	Morocco	Souss-Massa-Draâ	Taroudannt	0	
Rhipicephalus sanguineus group	Morocco	Souss-Massa-Draâ	Zagora	0	
Rhipicephalus sanguineus group	Morocco	Tadla-Azilal	Azilal	0	3
Rhipicephalus sanguineus group	Morocco	Tadla-Azilal	Béni Mellal	0	
Rhipicephalus sanguineus group	Morocco	Tanger-Tétouan	Chefchaouen	0	2
Rhipicephalus sanguineus group	Morocco	Tanger-Tétouan	Larache	0	
Rhipicephalus sanguineus group	Morocco	Tanger-Tétouan	Tanger-Assilah	0	
Rhipicephalus sanguineus group	Morocco	Taza-Al Hoceima-Taounate	Taounate	0	3
Rhipicephalus sanguineus group	Portugal	Beja	Beja	0	
Rhipicephalus sanguineus group	Portugal	Beja	Ferreira do Alentejo	0	2

Continued

Table A1.19. Continued.

Species	Country	Administrative region 1	Administrative region 2	Reference ID	No. entries[a]
Rhipicephalus sanguineus group	Portugal	Beja	Mértola	0	7
Rhipicephalus sanguineus group	Portugal	Beja	Odemira	0	
Rhipicephalus sanguineus group	Portugal	Braga	Cabeceiras de Basto	0	3
Rhipicephalus sanguineus group	Portugal	Bragança		1143	
Rhipicephalus sanguineus group	Portugal	Castelo Branco	Idanha-a-Nova	0	
Rhipicephalus sanguineus group	Portugal	Évora	Montemor-o-Novo	0	
Rhipicephalus sanguineus group	Portugal	Évora	Portel	0	
Rhipicephalus sanguineus group	Portugal	Faro	Lagos	0	3
Rhipicephalus sanguineus group	Portugal	Faro	Loulé	0	
Rhipicephalus sanguineus group	Portugal	Faro	Monchique	0	
Rhipicephalus sanguineus group	Portugal	Faro	Vila do Bispo	0	3
Rhipicephalus sanguineus group	Portugal	Faro	Vila Real de Santo António	0	
Rhipicephalus sanguineus group	Portugal	Guarda	Guarda	0	2
Rhipicephalus sanguineus group	Portugal	Leiria		1143	
Rhipicephalus sanguineus group	Portugal	Leiria	Ansião	0	
Rhipicephalus sanguineus group	Portugal	Leiria	Ansião	0	
Rhipicephalus sanguineus group	Portugal	Leiria	Pombal	0	2

Continued

Table A1.19. Continued.

Species	Country	Administrative region 1	Administrative region 2	Reference ID	No. entries[a]
Rhipicephalus sanguineus group	Portugal	Lisboa	Sintra	0	
Rhipicephalus sanguineus group	Portugal	Lisboa	Sobral de Monte Agraço	0	2
Rhipicephalus sanguineus group	Portugal	Lisboa	Torres Vedras	0	
Rhipicephalus sanguineus group	Portugal	Portalegre		1143	
Rhipicephalus sanguineus group	Portugal	Portalegre	Marvão	0	3
Rhipicephalus sanguineus group	Portugal	Portalegre	Ponte de Súr	0	
Rhipicephalus sanguineus group	Portugal	Porto	Matosinhos	0	
Rhipicephalus sanguineus group	Portugal	Santarém		1143	
Rhipicephalus sanguineus group	Portugal	Santarém	Benavente	0	3
Rhipicephalus sanguineus group	Portugal	Santarém	Coruche	0	
Rhipicephalus sanguineus group	Portugal	Santarém	Santarém	0	
Rhipicephalus sanguineus group	Portugal	Setúbal		1143	
Rhipicephalus sanguineus group	Portugal	Setúbal	Alcácer do Sal	0	12
Rhipicephalus sanguineus group	Portugal	Setúbal	Almada	0	2
Rhipicephalus sanguineus group	Portugal	Setúbal	Barreiro	0	2
Rhipicephalus sanguineus group	Portugal	Setúbal	Grândola	0	7
Rhipicephalus sanguineus group	Portugal	Setúbal	Moita	0	

Continued

Table A1.19. Continued.

Species	Country	Administrative region 1	Administrative region 2	Reference ID	No. entries[a]
Rhipicephalus sanguineus group	Portugal	Setúbal	Palmela	0	18
Rhipicephalus sanguineus group	Portugal	Setúbal	Seixal	0	5
Rhipicephalus sanguineus group	Portugal	Setúbal	Setúbal	0	3
Rhipicephalus sanguineus group	Portugal	Setúbal	Sines	0	
Rhipicephalus sanguineus group	Portugal	Viana do Castelo	Melgaço	0	2
Rhipicephalus sanguineus group	Portugal	Viana do Castelo	Ponte da Barca	0	
Rhipicephalus sanguineus group	Romania	Constanta		0	4
Rhipicephalus sanguineus group	Romania	Dolj		0	2
Rhipicephalus sanguineus group	Romania	Timis		0	2
Rhipicephalus sanguineus group	Slovenia	Obalno-kraška	Piran	0	
Rhipicephalus sanguineus group	Slovenia	Osrednjesloven-ska	Ljubljana	0	
Rhipicephalus sanguineus group	Spain	Andalucía	Almería	0	
Rhipicephalus sanguineus group	Spain	Andalucía	Cádiz	0	
Rhipicephalus sanguineus group	Spain	Andalucía	Granada	0	18
Rhipicephalus sanguineus group	Spain	Andalucía	Huelva	0	3
Rhipicephalus sanguineus group	Spain	Andalucía	Jaén	0	4
Rhipicephalus sanguineus group	Spain	Andalucía	Málaga	0	

Continued

Table A1.19. Continued.

Species	Country	Administrative region 1	Administrative region 2	Reference ID	No. entries[a]
Rhipicephalus sanguineus group	Spain	Andalucía	Sevilla	0	3
Rhipicephalus sanguineus group	Spain	Aragón	Zaragoza	0	2
Rhipicephalus sanguineus group	Spain	Castilla y León	Soria	0	
Rhipicephalus sanguineus group	Spain	Castilla y León	Valladolid	0	
Rhipicephalus sanguineus group	Spain	Castilla y León	Zamora	0	
Rhipicephalus sanguineus group	Spain	Castilla-La Mancha	Guadalajara	0	
Rhipicephalus sanguineus group	Spain	Cataluña	Barcelona	0	5
Rhipicephalus sanguineus group	Spain	Cataluña	Girona	0	12
Rhipicephalus sanguineus group	Spain	Cataluña	Lleida	0	2
Rhipicephalus sanguineus group	Spain	Cataluña	Lleida	0	
Rhipicephalus sanguineus group	Spain	Cataluña	Tarragona	0	8
Rhipicephalus sanguineus group	Spain	Extremadura	Badajoz	0	8
Rhipicephalus sanguineus group	Spain	Extremadura	Cáceres	0	46
Rhipicephalus sanguineus group	Spain	País Vasco	Álava	0	3
Rhipicephalus sanguineus group	Spain	País Vasco	Vizcaya	0	3
Rhipicephalus sanguineus group	Tunisia	Ariana	Sidi Thabet	0	2
Rhipicephalus sanguineus group	Tunisia	Béja	Amdoun	0	

Continued

Table A1.19. Continued.

Species	Country	Administrative region 1	Administrative region 2	Reference ID	No. entries[a]
Rhipicephalus sanguineus group	Tunisia	Béja	Béja Nord	0	
Rhipicephalus sanguineus group	Tunisia	Béja	Mejez El Bab	0	
Rhipicephalus sanguineus group	Tunisia	Ben Arous (Tunis Sud)	Fouchana	0	
Rhipicephalus sanguineus group	Tunisia	Bizerte	Bizerte Sud	0	
Rhipicephalus sanguineus group	Tunisia	Bizerte	Ghar El Melh	0	2
Rhipicephalus sanguineus group	Tunisia	Bizerte	Sejnane	0	
Rhipicephalus sanguineus group	Tunisia	Gabès	Gabès Médina	0	
Rhipicephalus sanguineus group	Tunisia	Gabès	Gabès Ouest	0	
Rhipicephalus sanguineus group	Tunisia	Gabès	Hamma	0	3
Rhipicephalus sanguineus group	Tunisia	Gabès	Matmata	0	3
Rhipicephalus sanguineus group	Tunisia	Gafsa	Ksar	0	2
Rhipicephalus sanguineus group	Tunisia	Gafsa	Metlaoui	0	2
Rhipicephalus sanguineus group	Tunisia	Jendouba	Aïn Draham	0	
Rhipicephalus sanguineus group	Tunisia	Jendouba	Tabarka	0	
Rhipicephalus sanguineus group	Tunisia	Kairouan	Bouhajla	0	
Rhipicephalus sanguineus group	Tunisia	Kairouan	Kairouan Sud	0	2
Rhipicephalus sanguineus group	Tunisia	Kairouan	Sbikha	0	

Continued

Table A1.19. Continued.

Species	Country	Administrative region 1	Administrative region 2	Reference ID	No. entries[a]
Rhipicephalus sanguineus group	Tunisia	Kebili	Faouar	0	
Rhipicephalus sanguineus group	Tunisia	Kebili	Kebili Nord	0	
Rhipicephalus sanguineus group	Tunisia	Le Kef	Kef Ouest	0	
Rhipicephalus sanguineus group	Tunisia	Médenine	Houmt Souk	0	
Rhipicephalus sanguineus group	Tunisia	Nabeul	Menzel Temime	0	
Rhipicephalus sanguineus group	Tunisia	Nabeul	Soliman	0	
Rhipicephalus sanguineus group	Tunisia	Nabeul	Takelsa	0	3
Rhipicephalus sanguineus group	Tunisia	Sfax	Sfax Sud	0	2
Rhipicephalus sanguineus group	Tunisia	Sidi Bou Zid	Meknassi	0	2
Rhipicephalus sanguineus group	Tunisia	Siliana	Siliana Nord	0	
Rhipicephalus sanguineus group	Tunisia	Tataouine	Tataouine Nord	0	
Rhipicephalus sanguineus group	Tunisia	Tozeur	Tamaghza	0	
Rhipicephalus sanguineus group	Tunisia	Tozeur	Tozeur	0	2
Rhipicephalus sanguineus group	Tunisia	Tunis	Carthage	0	
Rhipicephalus sanguineus group	Tunisia	Tunis	La Goulette	0	
Rhipicephalus sanguineus group	Tunisia	Tunis	Sidi El Béchir	0	7
Rhipicephalus sanguineus group	Turkey	Ankara		0	21

Continued

Table A1.19. Continued.

Species	Country	Administrative region 1	Administrative region 2	Reference ID	No. entries[a]
Rhipicephalus sanguineus group	Turkey	Burdur		0	14
Rhipicephalus sanguineus group	Turkey	Çorum		0	2
Rhipicephalus sanguineus group	Turkey	Edirne		0	3
Rhipicephalus sanguineus group	Turkey	Elâziğ		0	14
Rhipicephalus sanguineus group	Turkey	Erzincan		0	
Rhipicephalus sanguineus group	Turkey	Kayseri		0	4
Rhipicephalus sanguineus group	Turkey	Kırklareli		0	3
Rhipicephalus sanguineus group	Turkey	Samsun		0	
Rhipicephalus sanguineus group	Turkey	Sivas		0	4
Rhipicephalus sanguineus group	Turkey	Tokat		0	4
Rhipicephalus sanguineus group	Turkey	Van		0	8
Rhipicephalus sanguineus group	Turkey	Yozgat		0	
Rhipicephalus turanicus	Algeria	Tizi Ouzou	Tizi Ouzou	110	
Rhipicephalus turanicus	Bulgaria	Stara Zagora	Haskovo	69	2
Rhipicephalus turanicus	Croatia	Splitsko-Dalmatinska		277	
Rhipicephalus turanicus	Egypt	Al Isma'iliyah		595	
Rhipicephalus turanicus	France	Provence-Alpes-Côte-d'Azur	Bouches-du-Rhône	1280	
Rhipicephalus turanicus	Greece	Anatoliki Makedonia kai Thraki		735	2
Rhipicephalus turanicus	Greece	Kentriki Makedonia		735	

Continued

Table A1.19. Continued.

Species	Country	Administrative region 1	Administrative region 2	Reference ID	No. entries[a]
Rhipicephalus turanicus	Greece	Kentriki Makedonia	Khalkidiki	1303	3
Rhipicephalus turanicus	Greece	Stereá Elláda	Fokis	778	
Rhipicephalus turanicus	Israel	Jerusalem		1071	
Rhipicephalus turanicus	Italy	Sicily	Palermo	242	2
Rhipicephalus turanicus	Spain	Andalucía	Jaén	611	2
Rhipicephalus turanicus	Spain	Andalucía	Jaén	611	
Rhipicephalus turanicus	Spain	Islas Baleares	Baleares	1345	2
Rhipicephalus turanicus	Turkey	Afyon		186	
Rhipicephalus turanicus	Turkey	Aksaray		1353	
Rhipicephalus turanicus	Turkey	Ankara		1353	2
Rhipicephalus turanicus	Turkey	Şanlıurfa		303	

[a]If more than one.

Table A1.20. Geographic distribution data for *Rhipicephalus bursa*. See Appendix 2 (hosted on http://www.cabi.org/vetmedresource/) for complete reference indicated by the Reference ID.

Country	Administrative region 1	Administrative region 2	Reference ID	No. entries[a]
Albania	Durrës	Durrësit	0	
Albania	Elbasan	Librazhdit	0	2
Albania	Korçë	Korçës	0	3
Albania	Lezhë	Lezhës	0	
Albania	Tiranë	Tiranës	0	2
Albania	Vlorë	Sarandës	0	
Albania	Vlorë	Vlorës	0	4
Algeria	Aïn Defla	Ain Benian	0	
Algeria	Aïn Defla	Ain Lechiakh	0	
Algeria	Aïn Defla	Arib	0	
Algeria	Aïn Defla	Ben Allal	0	
Algeria	Aïn Defla	Rouina	0	
Algeria	Aïn Témouchent	Ain Kihel	0	
Algeria	Aïn Témouchent	Terga	0	
Algeria	Alger	Ain Benian	0	
Algeria	Alger	Ain Benian	0	
Algeria	Alger	Bordj El Kiffan	0	
Algeria	Alger	Bouzareah	0	
Algeria	Alger	Ouled Chebel	0	
Algeria	Batna	Ain Touta	0	
Algeria	Biskra	El Ouitaya	0	

Continued

Table A1.20. Continued.

Country	Administrative region 1	Administrative region 2	Reference ID	No. entries[a]
Algeria	Blida	Boufarik	0	
Algeria	Bouira	Ain Turk	0	
Algeria	Bouira	Lakhdaria	0	
Algeria	Bouira	Sour El Ghouzlane	0	2
Algeria	Boumerdès	Boudouaou	0	
Algeria	Boumerdès	Corso	0	
Algeria	Boumerdès	Dellys	0	
Algeria	Boumerdès	Isser	0	
Algeria	Boumerdès	Sidi Daoud	0	
Algeria	Chlef	Abou El Hassen	0	
Algeria	Chlef	Boukadir	0	
Algeria	Chlef	Bouzeghaia	0	
Algeria	Chlef	Chlef	0	2
Algeria	Chlef	Ouled Fares	0	
Algeria	Chlef	Sidi Akkacha	0	
Algeria	Chlef	Tenes	0	
Algeria	Constantine	Constantine	0	
Algeria	Djelfa	Benhar	0	
Algeria	El Tarf	El Aioun	0	
Algeria	Guelma	Hammam Debagh	0	
Algeria	Jijel	Jijel	0	
Algeria	Jijel	Taher	0	
Algeria	M'Sila	Ain Errich	0	
Algeria	M'Sila	Sidi M'Hamed	0	
Algeria	Mascara	Ain Ferah	0	
Algeria	Médéa	Medea	0	
Algeria	Mila	Mila	0	
Algeria	Oran	Misserghin	0	2
Algeria	Oran	Oran	0	
Algeria	Relizane	Relizane	0	
Algeria	Relizane	Sidi M'Hamed Benaouda	0	
Algeria	Sétif	Ain Arnat	0	3
Algeria	Sétif	Setif	0	
Algeria	Sidi Bel Abbès	Dhaya	0	
Algeria	Sidi Bel Abbès	Sidi Bel Abbes	0	
Algeria	Tébessa	Bir El Ater	0	
Algeria	Tébessa	Tebessa	0	
Algeria	Tiaret		129	
Algeria	Tiaret	Meghila	0	
Algeria	Tiaret	Sougueur	0	2
Algeria	Tipaza	Douaouda	0	
Algeria	Tissemsilt	Theniet El Had	0	
Algeria	Tizi Ouzou		110	
Algeria	Tizi Ouzou	Ain-El-Hammam	0	
Algeria	Tizi Ouzou	Boghni	0	
Algeria	Tizi Ouzou	Draa-Ben-Khedda	0	
Algeria	Tizi Ouzou	Larbaa-Nath-Irathen	0	
Algeria	Tizi Ouzou	Tadmait	0	
Algeria	Tizi Ouzou	Tizi Ouzou	0	2
Algeria	Tizi Ouzou	Tizi-Rached	0	
Algeria	Tizi Ouzou	Zekri	0	
Algeria	Tlemcen	Tlemcen	0	2

Continued

Table A1.20. Continued.

Country	Administrative region 1	Administrative region 2	Reference ID	No. entries[a]
Bulgaria	Blagoevgrad	Petrich	0	
Bulgaria	Blagoevgrad	Sandanski	0	2
Bulgaria	Burgas	Burgas	0	2
Bulgaria	Burgas	Kameno	0	
Bulgaria	Burgas	Malko Tarnovo	0	2
Bulgaria	Kardzhali	Ardino	0	2
Bulgaria	Montana	Montana	0	
Bulgaria	Pazardzhik	Septemvri	0	
Bulgaria	Plovdiv	Asenovgrad	0	
Bulgaria	Shumen	Preslav	0	
Bulgaria	Shumen	Shumen	0	2
Bulgaria	Stara Zagora	Haskovo	69	2
Bulgaria	Stara Zagora	Pavel Banya	0	
Bulgaria	Veliko Tarnovo	Veliko Tarnovo	0	
Bulgaria	Vratsa	Vratsa	0	
Bulgaria	Yambol	Tundzha	0	2
Bulgaria	Yambol	Yambol	0	
Croatia	Bjelovarska-Bilogorska		0	5
Croatia	Dubrovacko-Neretvanska		0	
Croatia	Grad Zagreb		0	2
Croatia	Istarska		0	
Croatia	Primorsko-Goranska		0	2
Croatia	Splitsko-Dalmatinska		0	5
Croatia	Splitsko-Dalmatinska		277	
Croatia	Splitsko-Dalmatinska		781	
Croatia	Varaždinska		0	
Croatia	Zadarska		0	6
Cyprus	Famagusta		0	2
Egypt	Al Wadi al Jadid		0	
France	Aquitaine	Dordogne	0	2
France	Aquitaine	Landes	0	2
France	Aquitaine	Pyrénées-Atlantiques	0	5
France	Aquitaine	Pyrénées-Atlantiques	1280	
France	Bourgogne	Saône-et-Loire	0	
France	Corse		1280	
France	Corse	Corse-du-Sud	0	12
France	Corse	Haute-Corse	0	9
France	Languedoc-Roussillon	Aude	0	3
France	Languedoc-Roussillon	Pyrénées-Orientales	0	6
France	Midi-Pyrénées	Aveyron	0	
France	Midi-Pyrénées	Lot	0	2
France	Provence-Alpes-Côte-d'Azur	Alpes-de-Haute-Provence	0	3
France	Provence-Alpes-Côte-d'Azur	Bouches-du-Rhône	0	4

Continued

Table A1.20. Continued.

Country	Administrative region 1	Administrative region 2	Reference ID	No. entries[a]
France	Provence-Alpes-Côte-d'Azur	Bouches-du-Rhône	1280	
France	Provence-Alpes-Côte-d'Azur	Var	0	3
France	Provence-Alpes-Côte-d'Azur	Vaucluse	0	
France	Rhône-Alpes	Ardèche	0	2
France	Rhône-Alpes	Drôme	0	4
Greece	Anatoliki Makedonia kai Thraki		735	2
Greece	Anatoliki Makedonia kai Thraki	Drama	0	10
Greece	Anatoliki Makedonia kai Thraki	Evros	0	9
Greece	Anatoliki Makedonia kai Thraki	Kavala	0	10
Greece	Attiki	Attica	0	8
Greece	Dytiki Ellada	Achaea	0	2
Greece	Dytiki Makedonia		735	
Greece	Dytiki Makedonia		735	
Greece	Dytiki Makedonia	Florina	0	
Greece	Dytiki Makedonia	Grevena	0	10
Greece	Dytiki Makedonia	Kastoria	0	2
Greece	Dytiki Makedonia	Kozani	0	8
Greece	Ipeiros	Ioannina	0	
Greece	Ipeiros	Preveza	0	3
Greece	Kentriki Makedonia		735	
Greece	Kentriki Makedonia	Khalkidiki	0	13
Greece	Kentriki Makedonia	Khalkidiki	1303	
Greece	Kentriki Makedonia	Khalkidiki	1303	
Greece	Kentriki Makedonia	Khalkidiki	1303	
Greece	Kentriki Makedonia	Kilkis	0	8
Greece	Kentriki Makedonia	Pella	0	4
Greece	Kentriki Makedonia	Pieria	0	5
Greece	Kentriki Makedonia	Serrai	0	14
Greece	Kentriki Makedonia	Thessaloniki	0	12
Greece	Kriti	Rethymnon	0	7
Greece	Stereá Elláda	Boeotia	0	7
Greece	Stereá Elláda	Euboea	0	7
Greece	Stereá Elláda	Fokis	778	
Greece	Thessalia	Karditsa	0	7
Greece	Thessalia	Larisa	0	
Israel	Jerusalem		1071	
Italy	Abruzzo	Chieti	0	2
Italy	Abruzzo	L'Aquila	0	8
Italy	Abruzzo	Teramo	0	2
Italy	Apulia	Bari	0	17
Italy	Apulia	Barletta-Andria-Trani	0	3
Italy	Apulia	Brindisi	0	
Italy	Apulia	Foggia	0	26
Italy	Apulia	Lecce	0	2
Italy	Apulia	Taranto	0	7

Continued

Table A1.20. Continued.

Country	Administrative region 1	Administrative region 2	Reference ID	No. entries[a]
Italy	Basilicata	Matera	0	12
Italy	Calabria	Cosenza	0	
Italy	Calabria	Reggio di Calabria	0	3
Italy	Campania		802	
Italy	Campania	Avellino	0	4
Italy	Campania	Caserta	0	3
Italy	Emilia-Romagna	Ravenna	0	4
Italy	Lazio	Latina	0	14
Italy	Lazio	Rieti	0	
Italy	Lazio	Roma	0	7
Italy	Liguria	Genova	0	
Italy	Liguria	Savona	0	
Italy	Marche	Ascoli Piceno	0	
Italy	Molise	Campobasso	0	6
Italy	Molise	Isernia	0	
Italy	Sardegna	Cagliari	0	
Italy	Sardegna	Carbonia-Iglesias	0	
Italy	Sardegna	Oristano	0	2
Italy	Sardegna	Sassari	0	4
Italy	Sicily	Agrigento	0	2
Italy	Sicily	Enna	0	
Italy	Sicily	Palermo	0	10
Italy	Sicily	Palermo	242	
Italy	Sicily	Palermo	366	
Italy	Sicily	Palermo	366	
Italy	Sicily	Palermo	366	
Italy	Sicily	Palermo	366	
Italy	Sicily	Palermo	998	
Italy	Sicily	Trapani	0	
Italy	Toscana	Grosseto	0	
Italy	Toscana	Livorno	0	3
Italy	Toscana	Pisa	0	
Italy	Umbria	Perugia	0	2
Italy	Veneto	Treviso	0	
Italy	Veneto	Verona	0	2
Kosovo	Đakovica	Deçani	0	3
Kosovo	Kosovska Mitrovica	Kosovska Mitrovica	0	
Kosovo	Pristina	Obilic	0	
Kosovo	Uroöevac	Uroöevac	0	
Lebanon	Mount Lebanon	Matn	0	
Libya	Al Jabal al Akhdar		0	
Macedonia	Pelagonia	Bitola	0	3
Macedonia	Polog	Tetovo	0	2
Macedonia	Skopje	Centar	0	4
Macedonia	Skopje	Zelenikovo	0	2
Macedonia	Southeastern	Lake Dojran	0	
Macedonia	Southwestern	Ohrid	0	2
Macedonia	Vardar	Kavadartsi	0	
Montenegro	Rožaje		0	
Morocco	Doukkala-Abda	Safi	0	2
Morocco	Fès-Boulemane	Zouagha-Moulay Yacoub	0	2
Morocco	Gharb-Chrarda-Béni Hssen	Kénitra	0	6

Continued

Table A1.20. Continued.

Country	Administrative region 1	Administrative region 2	Reference ID	No. entries[a]
Morocco	Marrakech-Tensift-Al Haouz	Al Haouz	0	3
Morocco	Marrakech-Tensift-Al Haouz	El Kelaâ des Sraghna	0	5
Morocco	Marrakech-Tensift-Al Haouz	Essaouira	0	
Morocco	Marrakech-Tensift-Al Haouz	Marrakech	0	2
Morocco	Meknès-Tafilalet	Meknès	0	
Morocco	Tadla-Azilal	Béni Mellal	0	
Morocco	Tanger-Tétouan	Chefchaouen	0	2
Morocco	Tanger-Tétouan	Fahs Anjra	0	
Morocco	Tanger-Tétouan	Tanger-Assilah	0	2
Morocco	Tanger-Tétouan	Tétouan	0	
Morocco	Taza-Al Hoceima-Taounate	Taounate	0	2
Morocco	Taza-Al Hoceima-Taounate	Taza	0	2
Portugal	Braga		1143	
Portugal	Castelo Branco	Castelo Branco	0	3
Portugal	Évora	Montemor-o-Novo	0	
Portugal	Évora	Portel	0	
Portugal	Portalegre		1143	
Portugal	Portalegre	Marvão	0	3
Portugal	Porto	Marco de Canaveses	0	
Portugal	Santarém	Coruche	0	
Portugal	Setúbal	Alcácer do Sal	0	7
Portugal	Setúbal	Palmela	0	
Portugal	Viana do Castelo	Arcos de Valdevez	0	
Romania	Constanta		0	5
Romania	Dâmbovita		460	
Romania	Dolj		0	4
Romania	Timis		0	4
Romania	Tulcea		769	
Serbia	Borski	Kladovo	0	
Serbia	Juûno-Backi	Bac	0	
Serbia	Moravicki	Cacak	0	
Serbia	Ni'avski	Doljevac	0	
Serbia	Pcinjski	Surdulica	0	2
Serbia	Pirotski	Pirot	0	2
Serbia	Ra'ki	Kraljevo	0	
Serbia	Ra'ki	Novi Pazar	0	3
Serbia	Ra'ki	Vrnjacka Banja	0	
Serbia	Rasinski	Brus	0	
Serbia	Zlatiborski	Uûice	0	
Spain	Andalucía	Cádiz	0	21
Spain	Andalucía	Cádiz	821	
Spain	Andalucía	Córdoba	0	14
Spain	Andalucía	Granada	0	9
Spain	Andalucía	Huelva	0	
Spain	Andalucía	Jaén	0	12
Spain	Andalucía	Jaén	611	

Continued

Table A1.20. Continued.

Country	Administrative region 1	Administrative region 2	Reference ID	No. entries[a]
Spain	Andalucía	Jaén	611	
Spain	Andalucía	Jaén	611	
Spain	Andalucía	Jaén	611	
Spain	Andalucía	Sevilla	0	12
Spain	Aragón	Huesca	0	2
Spain	Aragón	Teruel	0	
Spain	Aragón	Zaragoza	0	
Spain	Castilla y León	Salamanca	0	3
Spain	Castilla y León	Toledo	821	
Spain	Castilla y León	Toledo	821	
Spain	Castilla y León	Ávila	0	
Spain	Castilla y León	Zamora	0	
Spain	Castilla-La Mancha	Albacete	0	2
Spain	Castilla-La Mancha	Albacete	821	
Spain	Castilla-La Mancha	Ciudad Real	0	3
Spain	Castilla-La Mancha	Guadalajara	0	
Spain	Castilla-La Mancha	Toledo	0	
Spain	Cataluña	Girona	0	3
Spain	Cataluña	Tarragona	0	5
Spain	Ceuta y Melilla	Ceuta	0	
Spain	Comunidad Foral de Navarra	Navarra	0	3
Spain	Extremadura	Badajoz	0	17
Spain	Extremadura	Cáceres	0	110
Spain	La Rioja	La Rioja	0	
Spain	País Vasco		77	
Spain	País Vasco	Álava	0	10
Spain	País Vasco	Guipúzcoa	0	8
Spain	País Vasco	Vizcaya	0	5
Spain	Principado de Asturias		821	
Tunisia	Ariana	Kalaat El Andalous	0	
Tunisia	Béja	Béja Sud	0	
Tunisia	Béja	Nefza	0	
Tunisia	Bizerte	Bizerte Sud	0	
Tunisia	Bizerte	Ghazala	0	
Tunisia	Bizerte	Sejnane	0	
Tunisia	Jendouba	Aïn Draham	0	
Tunisia	Jendouba	Ghardimaou	0	
Tunisia	Jendouba	Jendouba Nord	0	
Tunisia	Jendouba	Tabarka	0	
Tunisia	Le Kef	Nebeur	0	
Tunisia	Le Kef	Sakiet Sidi Youssef	0	
Tunisia	Nabeul	Soliman	0	
Tunisia	Nabeul	Takelsa	0	
Tunisia	Sidi Bou Zid	Meknassi	0	
Tunisia	Tunis	Carthage	0	
Tunisia	Tunis	Sidi El Béchir	0	
Tunisia	Zaghouan	Fahs	0	
Turkey	Afyon		186	
Turkey	Aksaray		1353	
Turkey	Ankara		0	20
Turkey	Ankara		1353	

Continued

Table A1.20. Continued.

Country	Administrative region 1	Administrative region 2	Reference ID	No. entries[a]
Turkey	Ankara		1353	
Turkey	Artvin		0	
Turkey	Burdur		0	
Turkey	Burdur		1008	
Turkey	Çorum		0	
Turkey	Elazig		0	16
Turkey	Giresun		0	2
Turkey	Gümüshane		0	
Turkey	Istanbul		145	
Turkey	Kayseri		0	2
Turkey	Mus		22	
Turkey	Ordu		0	
Turkey	Samsun		0	
Turkey	Sivas		0	4
Turkey	Tokat		0	18
Turkey	Van		0	8
Turkey	Yozgat		0	5

[a]If more than one.

Table A1.21. Geographic distribution data for *Rhipicephalus* (*Boophilus*) *annulatus*. See Appendix 2 (hosted on http://www.cabi.org/vetmedresource/) for complete reference indicated by the Reference ID.

Country	Administrative region 1	Administrative region 2	Reference ID	No. entries[a]
Albania	Tiranë	Tiranës	0	
Algeria	Alger	Ain Benian	0	
Algeria	Guelma	Hammam Debagh	0	
Algeria	Mascara	Mascara	0	
Algeria	Sidi Bel Abbès	Sidi Bel Abbes	0	
Algeria	Tiaret		129	
Bulgaria	Montana	Montana	0	
Bulgaria	Vratsa	Vratsa	0	
Egypt	Al Iskandariyah		0	
Egypt	Al Iskandariyah		595	
Egypt	Al Isma'iliyah		1039	2
Egypt	Al Isma'iliyah		595	
Egypt	Al Jizah		0	
Egypt	Al Jizah		1101	
Egypt	Al Minufiyah		0	3
Egypt	Al Minya		0	
Egypt	Bur Sa'id		595	
France	Corse		1280	
Greece	Anatoliki Makedonia kai Thraki	Drama	0	3
Greece	Anatoliki Makedonia kai Thraki	Kavala	0	4
Greece	Kentriki Makedonia		735	2
Greece	Kentriki Makedonia	Khalkidiki	0	
Greece	Kentriki Makedonia	Khalkidiki	1303	2
Greece	Kentriki Makedonia	Kilkis	0	2
Greece	Kentriki Makedonia	Serrai	0	2
Greece	Kentriki Makedonia	Thessaloniki	0	

Continued

Table A1.21. Continued.

Country	Administrative region 1	Administrative region 2	Reference ID	No. entries[a]
Greece	Thessalia	Larisa	0	
Italy	Sicily	Palermo	366	2
Libya	Al Butnan		0	
Libya	Al Hizam Al Akhdar		0	2
Libya	Al Jifarah		0	
Libya	Al Wahah		0	
Morocco	Chaouia-Ouardigha	Ben Slimane	0	2
Morocco	Gharb-Chrard-Béni Hssen	Kénitra	0	
Morocco	Grand Casablanca	Mohammedia	0	2
Morocco	Marrakech-Tensift-Al Haouz	El Kelaâ des Sraghna	0	3
Morocco	Marrakech-Tensift-Al Haouz	Marrakech	0	
Morocco	Meknès-Tafilalet	Meknès	0	3
Morocco	Rabat-Salé-Zemmour-Zaer	Khémisset	0	
Morocco	Tanger-Tétouan	Larache	0	
Morocco	Tanger-Tétouan	Tanger-Assilah	0	
Morocco	Taza-Al Hoceima-Taounate	Taounate	0	2
Morocco	Taza-Al Hoceima-Taounate	Taza	0	
Romania	Tulcea		769	
Spain	Andalucía	Cádiz	0	9
Spain	Andalucía	Cádiz	241	
Spain	Andalucía	Cádiz	821	
Spain	Andalucía	Huelva	0	2
Spain	Andalucía	Sevilla	0	
Tunisia	Bizerte	Sejnane	0	2
Tunisia	Jendouba	Fernana	0	
Tunisia	Jendouba	Tabarka	0	
Tunisia	Le Kef	Nebeur	0	
Tunisia	Nabeul	Takelsa	0	
Tunisia	Tozeur	Tozeur	0	
Tunisia	Tunis	Sidi El Béchir	0	
Tunisia	Zaghouan	Zriba	0	
Turkey	Ankara		0	3
Turkey	Artvin		0	
Turkey	Burdur		0	3
Turkey	Çorum		0	2
Turkey	Edirne		0	
Turkey	Elâziğ		0	4
Turkey	Elâziğ		22	
Turkey	Erzurum		0	
Turkey	Erzurum		22	
Turkey	Kayseri		0	7
Turkey	Kırklareli		0	
Turkey	Mus		22	
Turkey	Samsun		0	
Turkey	Sivas		0	95
Turkey	Tokat		0	7

[a]If more than one.

Table A1.22. Geographic distribution data for *Dermacentor marginatus*. See Appendix 2 (hosted on http://www.cabi.org/vetmedresource/) for complete reference indicated by the Reference ID.

Country	Administrative region 1	Administrative region 2	Reference ID	No. entries[a]
Albania	Vlorë	Vlorës	0	
Algeria	Mostaganem	Mostaganem	0	
Algeria	Naâma	El Biod	0	
Algeria	Tlemcen	Tlemcen	0	
Bulgaria	Grad Sofiya	Stolichna	0	
Bulgaria	Montana	Montana	0	
Bulgaria	Stara Zagora	Haskovo	69	
Bulgaria	Vratsa	Vratsa	0	
Croatia	Primorsko-Goranska		0	2
Croatia	Splitsko-Dalmatinska		0	
Croatia	Splitsko-Dalmatinska		781	
Croatia	Zadarska		0	2
France	Alsace	Bas-Rhin	0	3
France	Aquitaine	Dordogne	0	4
France	Aquitaine	Dordogne	1280	
France	Aquitaine	Gironde	0	11
France	Aquitaine	Gironde	1280	
France	Aquitaine	Landes	0	17
France	Aquitaine	Landes	1280	
France	Aquitaine	Lot	1280	
France	Aquitaine	Lot-et-Garonne	0	
France	Aquitaine	Lot-et-Garonne	1280	
France	Aquitaine	Pyrénées-Atlantiques	0	7
France	Aquitaine	Pyrénées-Atlantiques	1280	
France	Auvergne	Allier	0	2
France	Auvergne	Allier	1280	
France	Auvergne	Cantal	0	3
France	Auvergne	Puy-de-Dôme	0	3
France	Auvergne	Puy-de-Dôme	1280	
France	Bourgogne	Côte-d'Or	0	10
France	Bourgogne	Côte-d'Or	1280	
France	Bourgogne	Nièvre	0	2
France	Bourgogne	Nièvre	1280	
France	Bourgogne	Saône-et-Loire	0	3
France	Bourgogne	Yonne	0	9
France	Bourgogne	Yonne	1280	
France	Bretagne	Ille-et-Vilaine	0	2
France	Bretagne	Ille-et-Vilaine	1280	
France	Bretagne	Morbihan	0	
France	Centre	Cher	0	2
France	Centre	Cher	1280	
France	Centre	Eure-et-Loir	0	
France	Centre	Indre	0	2
France	Centre	Indre-et-Loire	0	2
France	Centre	Indre-et-Loire	1280	
France	Centre	Loiret	0	2
France	Champagne-Ardenne	Aube	0	

Continued

Table A1.22. Continued.

Country	Administrative region 1	Administrative region 2	Reference ID	No. entries[a]
France	Champagne-Ardenne	Aube	1280	
France	Corse		1280	
France	Corse	Corse-du-Sud	0	10
France	Corse	Haute-Corse	0	6
France	Franche-Comté	Jura	0	4
France	Île-de-France	Essonne	1280	
France	Île-de-France	Hauts-de-Seine	1280	
France	Île-de-France	Seine-et-Marne	0	5
France	Île-de-France	Seine-Saint-Denis	1280	
France	Île-de-France	Val-d'Oise	0	
France	Île-de-France	Val-d'Oise	1280	
France	Île-de-France	Val-de-Marne	0	3
France	Île-de-France	Val-de-Marne	1280	
France	Île-de-France	Ville de Paris	0	
France	Île-de-France	Ville de Paris	1280	
France	Île-de-France	Yvelines	1280	
France	Languedoc-Roussillon	Aude	0	3
France	Languedoc-Roussillon	Gard	0	9
France	Languedoc-Roussillon	Hérault	0	3
France	Languedoc-Roussillon	Hérault	1280	
France	Languedoc-Roussillon	Pyrénées-Orientales	0	20
France	Languedoc-Roussillon	Pyrénées-Orientales	1280	
France	Limousin	Creuse	0	
France	Limousin	Haute-Vienne	0	4
France	Limousin	Haute-Vienne	1280	
France	Lorraine	Moselle	1280	
France	Midi-Pyrénées	Ariège	0	2
France	Midi-Pyrénées	Aveyron	0	12
France	Midi-Pyrénées	Aveyron	1280	
France	Midi-Pyrénées	Gers	0	3
France	Midi-Pyrénées	Gers	1280	
France	Midi-Pyrénées	Haute-Garonne	0	5
France	Midi-Pyrénées	Haute-Garonne	1280	
France	Midi-Pyrénées	Hautes-Pyrénées	0	3
France	Midi-Pyrénées	Lot	0	5
France	Midi-Pyrénées	Tarn	0	2
France	Midi-Pyrénées	Tarn	1280	
France	Midi-Pyrénées	Tarn-et-Garonne	1280	
France	Pays de la Loire	Loire-Atlantique	0	
France	Pays de la Loire	Loire-Atlantique	1280	
France	Pays de la Loire	Maine-et-Loire	0	2
France	Pays de la Loire	Maine-et-Loire	1280	
France	Pays de la Loire	Mayenne	0	

Continued

Table A1.22. Continued.

Country	Administrative region 1	Administrative region 2	Reference ID	No. entries[a]
France	Pays de la Loire	Vendée	0	7
France	Picardie	Oise	0	
France	Poitou-Charentes	Charente	0	2
France	Poitou-Charentes	Charente	1280	
France	Poitou-Charentes	Charente-Maritime	0	4
France	Poitou-Charentes	Charente-Maritime	1280	
France	Poitou-Charentes	Deux-Sèvres	0	4
France	Poitou-Charentes	Vienne	0	2
France	Poitou-Charentes	Vienne	1280	
France	Provence-Alpes-Côte-d'Azur	Alpes-de-Haute-Provence	0	26
France	Provence-Alpes-Côte-d'Azur	Alpes-Maritimes	0	15
France	Provence-Alpes-Côte-d'Azur	Alpes-Maritimes	1280	
France	Provence-Alpes-Côte-d'Azur	Bouches-du-Rhône	0	20
France	Provence-Alpes-Côte-d'Azur	Bouches-du-Rhône	1280	
France	Provence-Alpes-Côte-d'Azur	Hautes-Alpes	0	14
France	Provence-Alpes-Côte-d'Azur	Var	0	21
France	Provence-Alpes-Côte-d'Azur	Var	1280	
France	Provence-Alpes-Côte-d'Azur	Vaucluse	0	8
France	Rhône-Alpes	Ain	0	24
France	Rhône-Alpes	Ain	1280	
France	Rhône-Alpes	Ardèche	0	18
France	Rhône-Alpes	Ardèche	1280	
France	Rhône-Alpes	Drôme	0	23
France	Rhône-Alpes	Drôme	1280	
France	Rhône-Alpes	Haute-Savoie	0	3
France	Rhône-Alpes	Isère	0	34
France	Rhône-Alpes	Isère	1280	
France	Rhône-Alpes	Loire	0	4
France	Rhône-Alpes	Rhône	0	2
France	Rhône-Alpes	Rhône	1280	
France	Rhône-Alpes	Savoie	0	3
Germany	Baden-Württemberg		948	
Greece	Anatoliki Makedonia kai Thraki	Drama	0	2
Greece	Anatoliki Makedonia kai Thraki	Evros	0	3
Greece	Anatoliki Makedonia kai Thraki	Kavala	0	3
Greece	Attiki	Attica	0	2
Greece	Dytiki Makedonia	Florina	0	
Greece	Dytiki Makedonia	Grevena	0	9

Continued

Table A1.22. Continued.

Country	Administrative region 1	Administrative region 2	Reference ID	No. entries[a]
Greece	Dytiki Makedonia	Kastoria	0	
Greece	Dytiki Makedonia	Kozani	0	3
Greece	Ipeiros	Preveza	0	3
Greece	Kentriki Makedonia	Khalkidiki	0	3
Greece	Kentriki Makedonia	Kilkis	0	
Greece	Kentriki Makedonia	Pella	0	
Greece	Kentriki Makedonia	Pieria	0	4
Greece	Kentriki Makedonia	Serrai	0	3
Greece	Kentriki Makedonia	Thessaloniki	0	
Greece	Peloponnisos	Arcadia	0	
Greece	Stereá Elláda	Euboea	0	
Greece	Stereá Elláda	Evritania	0	2
Greece	Stereá Elláda	Fthiotis	0	
Hungary	Borsod-Abaúj-Zemplén		443	
Hungary	Budapest		322	
Italy	Apulia	Foggia	0	8
Italy	Campania		802	
Italy	Campania	Avellino	0	
Italy	Campania	Caserta	0	
Italy	Emilia-Romagna	Forlè-Cesena	0	2
Italy	Emilia-Romagna	Ravenna	0	2
Italy	Lazio	Latina	0	
Italy	Lazio	Rieti	0	
Italy	Lazio	Roma	0	2
Italy	Marche	Macerata	0	
Italy	Molise	Campobasso	0	
Italy	Molise	Isernia	0	
Italy	Piemonte	Cuneo	0	
Italy	Sardegna	Nuoro	0	2
Italy	Sicily	Palermo	0	2
Italy	Sicily	Palermo	242	
Italy	Sicily	Palermo	366	
Italy	Sicily	Palermo	366	
Italy	Sicily	Palermo	366	
Italy	Sicily	Palermo	998	
Italy	Toscana		1331	
Italy	Toscana	Florence	0	
Italy	Umbria	Perugia	0	2
Macedonia	Skopje	Centar	0	5
Macedonia	Skopje	Zelenikovo	0	
Moldova	Anenii Noi		667	
Moldova	Balti		667	
Moldova	Basarabeasca		667	
Moldova	Bender		667	
Moldova	Briceni		667	
Moldova	Cahul		667	
Moldova	Calarasi		667	
Moldova	Cantemir		667	
Moldova	Causeni		667	
Moldova	Chisinau		667	
Moldova	Cimislia		667	

Continued

Table A1.22. Continued.

Country	Administrative region 1	Administrative region 2	Reference ID	No. entries[a]
Moldova	Criuleni		667	
Moldova	Donduseni		667	
Moldova	Drochia		667	
Moldova	Dubasari		667	
Moldova	Edinet		667	
Moldova	Falesti		667	
Moldova	Floresti		667	
Moldova	Gagauzia		667	
Moldova	Glodeni		667	
Moldova	Hîncesti		667	
Moldova	Ialoveni		667	
Moldova	Leova		667	
Moldova	Nisporeni		667	
Moldova	Ocnita		667	
Moldova	Orhei		667	
Moldova	Rezina		667	
Moldova	Rîscani		667	
Moldova	Sîngerei		667	
Moldova	Soldanesti		667	
Moldova	Soroca		667	
Moldova	Stefan Voda		667	
Moldova	Straseni		667	
Moldova	Taraclia		667	
Moldova	Telenesti		667	
Moldova	Transnistria		667	
Moldova	Ungheni		667	
Morocco	Meknès-Tafilalet	Ifrane	0	2
Morocco	Rabat-Salé-Zemmour-Zaer	Khémisset	0	4
Morocco	Rabat-Salé-Zemmour-Zaer	Skhirate-Témara	0	2
Morocco	Tanger-Tétouan	Larache	0	
Morocco	Tanger-Tétouan	Tanger-Assilah	0	
Portugal	Aveiro	Oliveira de Azeméis	0	
Portugal	Beja	Mértola	0	
Portugal	Bragança		1143	
Portugal	Castelo Branco	Castelo Branco	0	4
Portugal	Évora	Évora	0	
Portugal	Leiria		1143	
Portugal	Lisboa		1143	
Portugal	Portalegre		1143	
Portugal	Santarém		1143	
Portugal	Santarém	Benavente	0	6
Portugal	Santarém	Coruche	0	
Portugal	Santarém	Santarém	0	
Portugal	Setúbal		1143	
Portugal	Setúbal	Alcácer do Sal	0	10
Portugal	Setúbal	Grândola	0	
Portugal	Setúbal	Grândola	1143	
Portugal	Setúbal	Palmela	0	6
Romania	Suceava		460	

Continued

Table A1.22. Continued.

Country	Administrative region 1	Administrative region 2	Reference ID	No. entries[a]
Romania	Suceava		462	
Romania	Teleorman		460	
Romania	Teleorman		462	
Romania	Timis		177	
Romania	Tulcea		460	
Serbia	Macvanski	Ljubovija	0	
Serbia	Ni'avski	Aleksinac	0	
Serbia	Zlatiborski	Priboj	0	
Slovakia	Košický		1349	
Slovakia	Zilinský		1349	
Spain	Andalucía	Cádiz	0	7
Spain	Andalucía	Córdoba	0	12
Spain	Andalucía	Granada	0	11
Spain	Andalucía	Huelva	0	
Spain	Andalucía	Jaén	0	9
Spain	Andalucía	Jaén	611	
Spain	Andalucía	Jaén	611	
Spain	Andalucía	Málaga	0	3
Spain	Andalucía	Sevilla	0	
Spain	Aragón	Huesca	0	4
Spain	Aragón	Zaragoza	0	
Spain	Castilla y León	Soria	0	2
Spain	Castilla y León	Toledo	821	
Spain	Castilla-La Mancha	Albacete	0	
Spain	Castilla-La Mancha	Albacete	821	
Spain	Castilla-La Mancha	Cuenca	821	
Spain	Castilla-La Mancha	Guadalajara	821	
Spain	Castilla-La Mancha	Toledo	989	
Spain	Castilla-La Mancha	Toledo	989	
Spain	Cataluña	Girona	0	2
Spain	Cataluña	Lleida	0	4
Spain	Cataluña	Tarragona	0	
Spain	Comunidad de Madrid	Madrid	0	
Spain	Comunidad Foral de Navarra	Navarra	0	2
Spain	Extremadura	Badajoz	0	8
Spain	Extremadura	Cáceres	0	19
Spain	Islas Baleares	Baleares	1345	
Spain	La Rioja	La Rioja	0	3
Spain	País Vasco		77	
Spain	País Vasco	Álava	0	4
Spain	País Vasco	Guipúzcoa	0	5
Spain	País Vasco	Vizcaya	0	3
Switzerland	Graubünden		435	
Switzerland	Graubünden		435	
Switzerland	Graubünden		435	
Tunisia	Béja	Nefza	0	
Tunisia	Jendouba	Aïn Draham	0	2
Tunisia	Nabeul	Haouaria	0	
Tunisia	Nabeul	Takelsa	0	
Tunisia	Zaghouan	Zriba	0	2

Continued

Table A1.22. Continued.

Country	Administrative region 1	Administrative region 2	Reference ID	No. entries[a]
Turkey	Aksaray		1353	
Turkey	Ankara		0	20
Turkey	Ankara		1353	
Turkey	Ankara		1353	
Turkey	Bolu		0	3
Turkey	Burdur		0	9
Turkey	Burdur		1008	
Turkey	Çorum		0	
Turkey	Elâziğ		0	6
Turkey	Giresun		0	
Turkey	Izmir		0	3
Turkey	Kayseri		0	
Turkey	Konya		0	
Turkey	Nevşehir		0	3
Turkey	Samsun		0	3
Turkey	Sivas		0	
Turkey	Tokat		0	
Turkey	Van		0	8

[a]If more than one.

Table A1.23. Geographic distribution data for *Dermacentor reticulatus*. See Appendix 2 (hosted on http://www.cabi.org/vetmedresource/) for complete reference indicated by the Reference ID.

Country	Administrative region 1	Administrative region 2	Reference ID	No. entries[a]
Czech Republic	Jihomoravsky		427	3
France	Alsace	Bas-Rhin	1280	
France	Aquitaine	Dordogne	1280	
France	Aquitaine	Gironde	1280	
France	Aquitaine	Landes	1280	
France	Aquitaine	Lot-et-Garonne	1280	
France	Aquitaine	Pyrénées-Atlantiques	1280	
France	Auvergne	Allier	1280	
France	Auvergne	Haute-Loire	1280	
France	Bourgogne	Côte-d'Or	1280	
France	Bourgogne	Nièvre	1280	
France	Bourgogne	Saône-et-Loire	1280	
France	Bourgogne	Yonne	1280	
France	Bretagne	Côtes-d'Armor	1280	
France	Bretagne	Finistère	1280	
France	Bretagne	Ille-Et-Vilaine	1280	
France	Centre	Indre	1280	
France	Centre	Indre-et-Loire	1280	
France	Haute-Normandie	Eure	1280	
France	Île-de-France	Essonne	1280	
France	Île-de-France	Hauts-de-Seine	1280	
France	Île-de-France	Seine-et-Marne	1280	
France	Île-de-France	Seine-Saint-Denis	1280	

Continued

Table A1.23. Continued.

Country	Administrative region 1	Administrative region 2	Reference ID	No. entries[a]
France	Île-de-France	Val-d'Oise	1280	
France	Île-de-France	Val-de-Marne	1280	
France	Île-de-France	Ville de Paris	1280	
France	Île-de-France	Yvelines	1280	
France	Limousin	Corrèze	1280	
France	Limousin	Haute-Vienne	1280	
France	Lorraine	Moselle	1280	
France	Midi-Pyrénées	Ariège	1280	
France	Midi-Pyrénées	Aveyron	1280	
France	Midi-Pyrénées	Gers	1280	
France	Midi-Pyrénées	Haute-Garonne	1280	
France	Midi-Pyrénées	Hautes-Pyrénées	1280	
France	Midi-Pyrénées	Tarn-et-Garonne	1280	
France	Nord-Pas-de-Calais	Pas-de-Calais	1280	
France	Pays de la Loire	Loire-Atlantique	1280	
France	Pays de la Loire	Maine-et-Loire	1280	
France	Pays de la Loire	Sarthe	1280	
France	Poitou-Charentes	Charente	1280	
France	Poitou-Charentes	Charente-Maritime	1280	
France	Poitou-Charentes	Vienne	1280	
France	Rhône-Alpes	Ain	1280	
France	Rhône-Alpes	Drôme	1280	
France	Rhône-Alpes	Isère	1280	
France	Rhône-Alpes	Loire	1280	
France	Rhône-Alpes	Rhône	1280	
Germany	Baden-Württemberg		576	
Germany	Bayern		236	3
Germany	Bayern	Oberpfalz	1075	3
Germany	Berlin		236	4
Germany	Brandenburg		236	
Germany	Hessen		236	3
Germany	Sachsen		236	4
Germany	Sachsen-Anhalt		236	4
Germany	Thüringen		576	
Hungary	Bács-Kiskun		341	
Hungary	Baranya		322	2
Hungary	Borsod-Abaúj-Zemplén		341	2
Hungary	Budapest		322	3
Hungary	Csongrád		341	
Hungary	Fejér		322	2
Hungary	Fejér		341	
Hungary	Gyor-Moson-Sopron		322	
Hungary	Gyor-Moson-Sopron		341	
Hungary	Komárom-Esztergom		341	
Hungary	Somogy		341	
Hungary	Szabolcs-Szatmár-Bereg		341	
Hungary	Vas		341	
Hungary	Veszprém		341	
Moldova	Anenii Noi		667	

Continued

Table A1.23. Continued.

Country	Administrative region 1	Administrative region 2	Reference ID	No. entries[a]
Moldova	Calarasi		667	
Moldova	Chisinau		667	
Moldova	Criuleni		667	
Moldova	Hîncesti		667	
Moldova	Ialoveni		667	
Moldova	Nisporeni		667	
Moldova	Orhei		667	
Moldova	Straseni		667	
Moldova	Ungheni		667	
Netherlands	Zeeland		686	
Netherlands	Zuid-Holland		686	
Poland	Lublin		103	
Poland	Lublin		105	2
Poland	Lublin		1282	
Poland	Masovian	Warsaw	1099	
Poland	Podlachian		1282	
Poland	Podlachian		464	
Poland	Podlachian	Hajnówka	385	
Poland	Warmian-Masurian	Gizycko	118	
Poland	Warmian-Masurian	Gizycko	120	3
Portugal	Bragança		1143	
Romania	Buzau		769	
Romania	Caras-Severin		769	
Romania	Gorj		769	
Romania	Mures		769	
Romania	Tulcea		769	
Slovakia	Banskobystrický		140	12
Slovakia	Košický		140	29
Slovakia	Trenčianský		140	5
Slovakia	Trnavský		140	5
Slovakia	Žilinský		140	4
Spain	País Vasco		77	
Spain	Principado de Asturias		821	

[a]If more than one.

Table A1.24. Data from serological surveillance of Crimean-Congo haemorrhagic fever (CCHF) in domestic animals.

Location	Species	Serology	Year/s	Reference
Albania	Goats	Serological surveillance	2003–2005	Papa, A., Velo, E., Papadimitriou, E., Cahani, G., Kota, M. and Bino, S. (2009) Ecology of the Crimean-Congo hemorrhagic fever endemic area in Albania. *Vector-Borne and Zoonotic Diseases* 9, 713–716.
Central African Republic	Zebu cattle	Serological surveillance	1992–1993	Guilherme, J.M., Gonella-Legall, C., Legall, F., Nakoume, E. and Vincent, J. (1996) Seroprevalence of five arboviruses in Zebu cattle in the Central African Republic. *Transactions of the Royal Society of Tropical Medicine and Hygiene* 90, 31–33.
China	Sheep, camels	Serological surveillance	2004–2005	Sun, S., Dai, X., Ashian, M., Wang, X., Meng, W., Fend, C., Zhang, F., Hang, C., Hu, Z. and Zhang, Y. (2009) Epidemiology and phylogenetic analysis of the Crimean-Congo hemorrhagic fever viruses in Xinjiang, China. *Journal of Clinical Microbiology* 47, 2536–2543.
China	Sheep	Serological surveillance	2001	Qing, T., Saijo, M., Lei, H., Niikura, M., Maeda, A., Ikegami, T., Xinjung, W., Kurane, I. and Morikawa, S. (2003) Detection of immunoglobulin G to Crimean-Congo hemorrhagic fever virus in sheep sera by recombinant nucleoprotein-based enzyme linked immunosorbent and immunofluorescence assays. *Journal of Virological Methods* 108, 111–116.
Egypt	Domestic buffaloes, cattle, sheep, goats, donkeys, horses, mules, pigs, dogs	Serological surveillance	1977	Darwish, M.A., Imam, I.Z.E., Omar, F.M. and Hoogstraal, H. (1977) A seroepidemiological survey for Crimean-Congo hemorrhagic fever virus in humans and domestic animals in Egypt. *Journal of the Egyptian Public Health Association* 52, 156–163.
Greece	Sheep	Serological surveillance	1971–1976	Avšič-Županc, T. (2007) Epidemiology of Crimean-Congo hemorrhagic fever in the Balkans. In: Ergönül, Ö. and Whitehouse, C.A. (eds) *Crimean Congo Haemorrhagic Fever: A Global Perspective*. Springer, Dordrecht, pp. 75–88.
Iran	Dairy cattle	Serological surveillance	2006–2008	Lotfollahzadeh, S., Nikbakht Boroujeni, G.R., Mokhber Dezfouli, M.R. and Bokaei, S. (2009) A serosurvey of Crimean-Congo haemorrhagic fever virus in dairy cattle in Iran. *Zoonoses and Public Health* 58, 54–59.
Iran	Livestock	Serological surveillance	2000–2007	Chinikar, S., Goya, M.M., Shirzadi, M.R., Ghiasi, M.S., Mirahmadi, R., Haeri, A., Moradi, M., Afzali, N., Rahpeyma, M., Zeinali, M. and Meshkat, M. (2008) Surveillance and laboratory detection system of Crimean-Congo haemorrhagic fever in Iran. *Transboundary and Emerging Diseases* 55, 200–204.
Iran	Sheep, goats	Serological surveillance	2003–2005	Bokaie, S., Mostafavi, E., Haghdoost, A.A., Keyvanfar, H., Gooya, M.M., Meshkat, M., Davari, A. and Chinikar, S. (2008) Crimean Congo hemorrhagic fever in northeast of Iran. *Journal of Animal and Veterinary Advances* 7, 343–350.

Continued

Table A1.24. Conitnued.

Location	Species	Serology	Year/s	Reference
Iran	Sheep	Serological surveillance	2008	Telmadarraiy, Z., Moradi, A.R., Vatandoost, H., Mostafavi, E., Oshaghi, M.A., Zahirnia, A.H., Haeri, A. and Chinikar, S. (2008) Crimean-Congo hemorrhagic fever: a seroepidemiological and molecular survey in Bahar, Hamadan Province of Iran. *Asian Journal of Animal and Veterinary Advances* 3, 321–327.
Iran	Cattle, sheep, goats	Serological surveillance	1975	Saidi, S., Casals, J. and Faghih, M.A. (1975) Crimean hemorrhagic fever-Congo (CHF-C) virus antibodies in man, and in domestic and small mammals, in Iran. *American Journal of Tropical Medicine and Hygiene* 24, 353–357.
Iraq	Cattle, sheep, goats	Serological surveillance	1975	Tantawi, H.H., Shony, M.O. and Al-Tikriti, S.K. (1981) Antibodies to Crimean-Congo haemorrhagic fever virus in domestic animals in Iraq: a seroepidemiological survey. *International Journal of Zoonoses* 8, 115–120.
Kosovo	Sheep, cattle	Serological surveillance	1977	Avšič-Županc (2007); see Greece above for full reference.
Kosovo	Cattle, sheep	Serological surveillance	1995–2009	Humolli, I., Dedushaj, I., Zupanac, T.A. and Muçaj, S. (2010) Epidemiological, serological and herd immunity of Crimean-Congo haemorrhagic fever in Kosovo. *Medicinski Arhiv* 64, 91–93.
Oman	Cattle, sheep, goat, camels	Serological surveillance	1995–1996	Williams, R.J., Al-Busaidy, S., Mehta, F.R., Maupin, G.O., Wagoner, K.D., Al-Awaidy, S., Suleiman, A.J., Khan, A.S., Peters, C.J. and Ksiazek, T.G. (2000) Crimean-Congo haemorrhagic fever: a seroepidemiological and tick survey in the Sultanate of Oman. *Tropical Medicine and International Health* 5, 99–106.
Niger	Cattle, sheep, goats, camels	Serological surveillance	1984–1988	Mariner, J.C., Morrill, J. and Ksiazek, T.G. (1995) Antibodies to hemorrhagic fever viruses in domestic livestock in Niger: Rift Valley fever and Crimean-Congo hemorrhagic fever. *American Journal of Tropical Medicine and Hygiene* 53, 217–221.
Saudi Arabia	Sheep and goats	Serological surveillance	1993–1994	Agag, A.E. (1998) Rapid detection of Congo-Crimean haemorrhagic fever antibodies. *Assiut Veterinary Medical Journal* 39, 280–291.
Senegal	Birds	Serological surveillance	1991–1992	Zeller, H.G., Cornet, J.P. and Camicas, J.L. (1994) Crimean-Congo haemorrhagic fever virus infection in birds: field investigations in Senegal. *Research in Virology* 145, 105–109.

Senegal	Sheep	Serological surveillance	1986–1987	Wilson, M.L., LeGuenno, B., Guillaud, M., Desoutter, D., Gonzalez, J.P. and Camicas, J.L. (1990) Distribution of Crimean-Congo hemorrhagic fever viral antibody in Senegal: environmental and vectorial correlates. *American Journal of Tropical Medicine and Hygiene* 43, 557–566.
South Africa	Sheep and cattle, wild vertebrates	Serological surveillance	1993	Burt, F.J., Swanepoel, R. and Braack, L.E.O. (1993) Enzyme-linked immunosorbent assays for the detection of antibody to Crimean-Congo haemorrhagic fever virus in the sera of livestock and wild vertebrates. *Epidemiology and Infection* 111, 547–558.
South Africa	Cattle	Serological surveillance	1981–1984	Swanepoel, R., Shepherd, A.J., Leman, P.A., Shepherd, S.P., McGillivray, G.M., Erasmus, M.J., Searle, L.A. and Gill, D.E. (1987) Epidemiologic and clinical features of Crimean-Congo hemorrhagic fever in southern Africa. *American Journal of Tropical Medicine and Hygiene* 36, 120–132.
South Africa	Ostriches	Serological surveillance	1984	Shepherd, A.J., Swanepoel, R., Leman, P.A. and Shepherd, S.P. (1987) Field and laboratory investigation of Crimean-Congo haemorrhagic fever virus (*Nairovirus*, family *Bunyaviridae*) infection in birds. *Transactions of the Royal Society of Tropical Medicine and Hygiene* 81, 1004–1007.
Sudan	Ruminants	Serological surveillance	1997	Hassanein, K.M., El-Azazy, O.M., Yousef, H.M. (1997) Detection of Crimean-Congo haemorrhagic fever virus antibodies in humans and imported livestock in Saudia Arabia. *Transactions of the Royal Society of Tropical Medicine and Hygiene* 91, 356–537.
United Arab Emirates	Cattle, sheep, goats	Serological surveillance	1994–1995	Khan, A.S., Maupin, G.O., Rollin, P.E., Noor, A.M., Shurie, H.H., Shalabi, A.G., Wasef, S., Haddad, Y.M., Sadek, R., Ijaz, K., Peters, C.J. and Ksiazek, T.G. (1997). An outbreak of Crimean-Congo hemorrhagic fever in the United Arab Emirates, 1994–1995. *American Journal of Tropical Medicine and Hygiene* 57, 519–525.
Zimbabwe	Cattle	Serological surveillance	1981–1984	Swanepoel *et al.* (1987); see South Africa above for full reference.

Table A1.25. Geographic distribution data for *Rickettsia* spp. Data were extracted from the references that are listed by country at the end of the table.

Country	*Rickettsia* species	Region	Identified vector	Human disease (reported cases)
Algeria	*R. aeschlimannii*	Northern Algeria (Ain Ennoussour, Jijel, Tafaraoui)	*Hyalomma* (*Hy.*) *detritum*, *Hy. marginatum*	Yes
Algeria	*R. conorii* Malish	Oran, Northern Algeria (Algiers, El Tarf)	*Rhipicephalus* (*Rh.*) *sanguineus*	Yes
Algeria	*R. massiliae*	Northern Algeria (Tizi Ouzou)	*Rh. turanicus*	No
Algeria	*R. monacensis*	Northern Algeria (El Tarf)	*Ixodes ricinus*	No
Algeria	*R. sibirica mongolitimonae*	Southern Algeria	NA[a]	Yes
Armenia	*R. slovaca*	Armenia	*Dermacentor marginatus*	No
Austria	*R. helvetica*	All mainland	*I. ricinus*	No
Austria	*R. slovaca*	Alpine region	NA	Yes
Bulgaria	*R. conorii* Malish	Burgas, Haskovo, Pazardzhik, Plovdiv, Stara Zagora, Varna	*Rh. sanguineus*	Yes
Bulgaria	*R. helvetica*	Sofia region	*I. ricinus*	No
Bulgaria	*R. monacensis*	Sofia region	*I. ricinus*	No
Bulgaria	*R. slovaca*	NA	NA	Yes
Croatia	*R. aeschlimannii*	Southern Croatia	*Hy. marginatum*	No
Croatia	*R. conorii* Malish	Dalmatia region, Split	NA	Yes
Croatia	*R. helvetica*	North-western Croatia	*D. reticulatus*	No
Croatia	*R. hoogstraali*	North-western Croatia	*Haemaphysalis* (*Ha.*) *sulcata*	No
Croatia	*R. rhipicephali*	North-western Croatia	*Rh. turanicus*	No
Croatia	*R. slovaca*	Southern Croatia	*D. marginatus*	Yes
Cyprus	*R. conorii* Malish	NA	*Rh. sanguineus*	NA
Denmark	*R. helvetica*	Jutland county	*I. ricinus*	No
Egypt	*R. aeschlimannii*	NA	*Hy. marginatum*	No
Egypt	*R. conorii* Malish	NA	NA	Yes
Egypt	*R. sibirica mongolitimonae*	NA	NA	Yes
France	*R. aeschlimannii*	Corsica	*Hy. marginatum*, *Hy. rufipes* (on migratory birds)	No
France	*R. conorii* Malish	South of France, Corsica	*Rh. sanguineus*	Yes
France	*R. helvetica*	Central France, Puy-de-Dôme	*I. ricinus*	Yes
France	*R. massiliae*	Camargue, Marseille, Sisteron	*Rh. sanguineus*, *Rh. turanicus*	No
France	*R. raoulti*	Marne region	*D. reticulatus*	No
France	*R. sibirica mongolitimonae*	Southern France, Camargue, Marseille	NA	Yes
France	*R. slovaca*	Southern France	*D. marginatus*	Yes
Germany	*R. aeschlimannii*	Saxony-Anhalt, Zerbst	*D. reticulatus*	No
Germany	*R. helvetica*	Southern Germany, Baden-Württemberg and Bavaria	*I. ricinus*	No
Germany	*R. monacensis*	Southern Germany, Bavaria	*I. ricinus*	No

Continued

Table A1.25. Continued.

Country	*Rickettsia* species	Region	Identified vector	Human disease (reported cases)
Germany	*R. raoulti*	Southern Germany	*D. reticulatus*	No
Germany	*R. slovaca*	Southern Germany, Alpine region	*D. marginatus*	No
Greece	*R. aeschlimannii*	Cephalonia Island	*Hy. anatolicum*	No
Greece	*R. conorii* Malish	All mainland, Cephalonia Island, Crete Island	*Rh. sanguineus*	Yes
Greece	*R. massiliae*	Central Greece, Cephalonia Island	*Rh. turanicus*	No
Greece	*R. rhipicephali*	Cephalonia Island	*Rh. sanguineus*	No
Greece	*R. sibirica mongolitimonae*	Crete Island (Sitia)	*Hy. anatolicum*	Yes
Greece	*R. slovaca*	Northern Greece (Halkidiki)	NA	No
Hungary	*R. helvetica*	NA	*I. ricinus*	No
Hungary	*R. monacensis*	NA	*I. ricinus*	No
Hungary	*R. raoulti*	NA	*D. reticulatus*	No
Hungary	*R. slovaca*	NA	*D. marginatus*, *D. reticulatus*	Yes
Israel	*R. conorii* Israeli spotted fever (ISF) strain	Beersheva, Broil Hail, Hadera–Caesarea, Haifa Bay, Kfar Saba, Kibbutz, Reim–Tzeelim	*Rh. sanguineus*, *Rh. turanicus*	Yes
Israel	*R. massiliae*	NA	*Rh. turanicus*	No
Israel	*R. sibirica mongolitimonae*	South Israel (Or Haner)	*Hyalomma* sp.	No
Italy	*R. aeschlimannii*	Sardinia, Sicily	*Hy. marginatum*	No
Italy	*R. conorii* ISF	Sicily	*Rh. sanguineus*	Yes
Italy	*R. conorii* Malish	All mainland, Sardinia, Sicily	*Rh. sanguineus*, *Rh. turanicus*	Yes
Italy	*R. helvetica*	Central and north-eastern Italy	*I. ricinus*	No
Italy	*R. massiliae*	Sardinia, Sicily (Palermo)	*Rh. sanguineus*, *Rh. turanicus*	Yes
Italy	*R. monacensis*	Central and north-eastern Italy (Drena)	*I. ricinus*	No
Italy	*R. raoulti*	Tuscany	*D. marginatus*	Yes
Italy	*R. slovaca*	Sicily, Tuscany	*D. marginatus*, *Ha. punctata*	Yes
Italy	*Rickettsia* PoTi 169	Sardinia	*Rh. turanicus*	No
Kosovo	*R. conorii* Astrakhan	NA	*Rh. sanguineus*	NA
Libya	*R. conorii* ISF	NA	NA	Yes
Malta	*R. conorii* Malish	NA	NA	Yes
Moldova	*R. helvetica*	NA	*I. ricinus* (on migratory birds)	No
Moldova	*R. monacensis*	NA	*I. ricinus* (on migratory birds)	No
Morocco	*R. aeschlimannii*	Taza region (between Babboudir and Babezhare)	*Hy. marginatum*	Yes
Morocco	*R. conorii* Malish	NA	*Rh. sanguineus*	Yes

Continued

Table A1.25. Continued.

Country	*Rickettsia* species	Region	Identified vector	Human disease (reported cases)
Morocco	*R. helvetica*	Taza region (between Babboudir and Babezhare)	*I. ricinus*	No
Morocco	*R. massiliae*	NA	*Rh. sanguineus*	No
Morocco	*R. monacensis*	Taza region (between Babboudir and Babezhare)	*I. ricinus*	No
Morocco	*R. raoulti*	Taza region (between Babboudir and Babezhare)	*D. marginatus*	No
Morocco	*R. slovaca*	Taza region (between Babboudir and Babezhare)	*D. marginatus*	No
Netherlands	*R. helvetica*	NA	*I. hexagonus, I. ricinus*	No
Netherlands	*R. raoulti*	NA	*D. reticulatus*	No
Poland	*R. helvetica*	Central and southern Poland	*I. ricinus*	No
Poland	*R. raoulti*	Eastern, central and southern Poland (Białowieża, Radomsko, Warszawa)	*D. reticulatus, I. ricinus*	No
Poland	*R. slovaca*	Central and southern Poland	*I. ricinus*	No
Portugal	*R. aeschlimannii*	Alentejo	*H. marginatum*	No
Portugal	*R. conorii* ISF	All mainland	*Rh. sanguineus*	Yes
Portugal	*R. conorii* Malish	All mainland	*Rh. sanguineus*	Yes
Portugal	*R. helvetica*	Alentejo and Lisbon regions	*I. ricinus, I. ventalloi*	No
Portugal	*R. massiliae*	Alentejo and Trás-os-Montes regions	*Rh. sanguineus*	No
Portugal	*R. monacensis*	Alentejo and Lisboa regions, Madeira Island	*I. ricinus*	No
Portugal	*R. raoulti*	Alentejo and Trás-os-Montes regions	*H. marginatum*	No
Portugal	*R. sibirica mongolitimonae*	Alentejo region, Coimbra and Lisbon districts	*Rh. pusillus*	Yes
Portugal	*R. slovaca*	Alentejo and Trás-os-Montes regions, Coimbra district	*D. marginatus, D. reticulatus*	Yes
Portugal	*Rickettsia* PoTi 169	Alentejo region	*Rh. pusillus*	No
Romania	*R. conorii* Malish	Southern Romania	NA	Yes
Russia (Astrakhan region)	*R. conorii* Astrakhan strain	Astrakhan region	*Rh. pumilio, Rh. sanguineus*	Yes
Russia (Astrakhan region)	*R. raoulti*	Astrakhan region	*Rh. pumilio*	No
Russia (European part)	*R. aeschlimannii*	NA	*Ha. punctata*	No

Continued

Table A1.25. Continued.

Country	*Rickettsia* species	Region	Identified vector	Human disease (reported cases)
Russia (European part)	*R. raoulti*	Voronezh	*D. reticulatus*	No
Russia (European part)	*R. slovaca*	NA	*D. marginatus*	No
Serbia	*R. conorii*	NA	NA	NA
Serbia	*R. helvetica*	NA	*I. ricinus*	No
Serbia	*R. monacensis*	NA	*I. ricinus*	
Slovakia	*R. monacensis*	North-eastern (Becherov) and south-western (Pounajske) regions	*I. ricinus*	No
Slovakia	*R. raoulti*	Białowieża	*D. reticulatus*	No
Slovakia	*R. slovaca*	NA	*D. marginatus, D. reticulatus, Ha. inermis, Ha. punctata, I. ricinus*	Yes
Slovenia	*R. conorii* Malish	NA	*Rh. sanguineus*	Yes
Slovenia	*R. helvetica*	NA	*I. ricinus*	No
Slovenia	*R. monacensis*	NA	*I. ricinus*	No
Spain	*R. aeschlimannii*	Castilla y Léon, Madrid, Rioja	*Ha. inermis, Ha. punctata, Hy. marginatum, Rh. bursa, Rh sanguineus, Rh. turanicus*	No
Spain	*R. conorii* Malish	Andalucía, Canary Islands, Castilla-La Mancha, Castilla y Léon, Catalonia, Madrid, Mallorca Island	*Rh. sanguineus*	Yes
Spain	*R. helvetica*	Castilla y Léon, Jaen	*I. ricinus, I. ventalloi*	No
Spain	*R. massiliae*	Catalonia	*Rh. pusillus, Rh. sanguineus, Rh. turanicus*	No
Spain	*R. monacensis*	Basque country, Castilla y Léon, Jaen, Rioja	*Ha. punctata, I. ricinus*	Yes
Spain	*R. raoulti*	Basque country, Jaen, Rioja	*D. marginatus, D. reticulatus, Ha. punctata*	Yes
Spain	*R. rioja*	NA	NA	Yes
Spain	*R. sibirica mongolitimonae*	NA	*Rh. pusillus*	Yes
Spain	*R. slovaca*	Castilla-La Mancha, Castilla y Léon, Granada, Jaen, Madrid, Rioja	*D. marginatus, D. reticulatus*	Yes
Sweden	*R. helvetica*	NA	*I. ricinus*	Yes
Switzerland	*R. helvetica*	Alpine region	*I. ricinus*	No
Switzerland	*R. massiliae*	Canton Ticino	*Rh. sanguineus*	No

Continued

Table A1.25. Continued.

Country	*Rickettsia* species	Region	Identified vector	Human disease (reported cases)
Switzerland	*R. monacensis*	NA	*I. ricinus*	No
Switzerland	*R. slovaca*	Alpine region	*D. marginatus*	No
Tunisia	*R. conorii* ISF	Sfax	NA	Yes
Tunisia	*R. conorii* Malish	Monastir, Sousse, Tunis	*Rh. sanguineus*	Yes
Tunisia	*R. monacensis*	NA	*I. ricinus*	No
Turkey	*R. conorii* Malish	Trakya	NA	Yes
UK	*R. helvetica*	NA	*I. ricinus*	No
Ukraine, Crimea	*R. aeschlimannii*	NA	*Hy. marginatum*	No
Ukraine, Crimea	*R. conorii*	Basin of the Black and Caspian Seas	*Rh. sanguineus*	Yes
Ukraine, Crimea	*R. helvetica*	Eastern Ukraine	*I. ricinus*	No
Ukraine, Crimea	*R. monacensis*	Eastern Ukraine	*I. ricinus*	No

[a]NA, not available.

List of scientific papers from which data were extracted for the table above.

Algeria

Bitam, I., Parola, P., Matsumoto, K., Rolain, J.M., Baziz, B., Boubidi, S.C., Harrat, Z., Belkaid, M. and Raoult, D. (2006) First molecular detection of *R. conorii, R. aeschlimannii,* and *R. massiliae* in ticks from Algeria. *Annals of the New York Academy of Sciences* 1078, 368–372.

Bitam, I., Kernif, T., Harrat, Z., Parola, P. and Raoult, D. (2009) First detection of *Rickettsia aeschlimannii* in *Hyalomma aegyptium* from Algeria. *Clinical Microbiology and Infection* 15(Suppl. 2), 253–254.

Dib, L., Bitam, I., Bensouilah, M., Parola, P. and Raoult, D. (2009) First description of *Rickettsia monacensis* in *Ixodes ricinus* in Algeria. *Clinical Microbiology and Infection* 15(Suppl. 2), 261–262.

Mokrani, N., Parola, P., Tebbal, S., Dalichaouche, M., Aouati, A. and Raoult, D. (2008) *Rickettsia aeschlimannii* infection, Algeria. *Emerging Infectious Diseases* 14, 1814–1815.

Mouffok, N., Parola, P., Lepidi, H. and Raoult, D. (2009) Mediterranean spotted fever in Algeria – new trends. *International Journal of Infectious Diseases* 13, 227–235.

Armenia

Balayeva, N.M., Eremeeva, M.E. and Raoult, D. (1994) Genomic identification of *Rickettsia slovaca* among spotted fever group rickettsia isolates from *Dermacentor marginatus* in Armenia. *Acta Virologica* 38, 321–325.

Austria

Rehácek, J. (1993) Rickettsiae and their ecology in the Alpine region. *Acta Virologica* 37, 290–301.

Blaschitz, M., Narodoslavsky-Gföller, M., Kanzler, M., Walochnik, J. and Stanek, G. (2008) First detection of *Rickettsia helvetica* in *Ixodes ricinus* ticks in Austria. *Vector-Borne Zoonotic Diseases* 8, 561–563.

Bulgaria

Alexandrov, E., Alexandrova, D., Kamarinchev, B., Kantardjiev, T. and Dimitrova, Z. (2007) Marseilles fever (Mediterranean spotted fever) in Bulgaria – state and problems. In: Alexandrov, E., Kazar, J., Hechemy, K. and Kantardjiev, T. (eds) *Contemporary State of the Rickettsioses in the World and in Bulgaria*. Prof. Marin Drinov Academic Publishing House. Sofia, Bulgaria, pp. 80–93.

Christova, I., Van De Pol, J., Yazar, S., Velo, E. and Schouls, L. (2003) Identification of *Borrelia burgdorferi sensu lato, Anaplasma* and *Ehrlichia* species, and spotted fever group rickettsiae in ticks from southeastern Europe. *European Journal of Clinical Microbiology and Infection* 22, 535–542.

Komitova, R., Lakos, A., Aleksandrov, A., Christova, I. and Murdjeva, M. (2003) A case of tick-transmitted lymphadenopathy in Bulgaria associated with *Rickettsia slovaca* [Letter to the editor]. *Scandinavian Journal of Infectious Diseases* 35, 213.

Croatia

Dobec, M., Golubic, D., Punda-Polić, V., Kaeppeli, F. and Sievers, M. (2009) *Rickettsia helvetica* in *Dermacentor reticulatus* ticks. *Emerging Infectious Diseases* 15, 98–100.

Duh, D., Petrovec, M., Trilar, T., Punda-Polić, V., Bradarić, N., Klismanić, Z. and Avšič-Županc, T. (2003) A follow-up study on newly recognized spotted fever group rickettsiae in ticks collected in southern Croatia. *Annals of the New York Academy of Sciences* 990, 149–151.

Duh, D., Punda-Polić, V., Avšič-Županc, T., Bouyer, D., Walker, D.H., Popov, V.L., Jelovsek, M., Gracner, M., Trilar, T., Bradaric, N., Kurtti, T.J. and Strus, J. (2010) *Rickettsia hoogstraalii* sp. nov., isolated from hard- and soft-bodied ticks. *International Journal of Systematic and Evolutionary Microbiology* 60, 977–984.

Punda-Polić, V., Petrovec, M., Trilar, T., Duh, D., Bradaric, N., Klismanic, Z. and Avšič-Županc, T. (2002) Detection and identification of spotted fever group rickettsiae in ticks collected in southern Croatia. *Experimental and Applied Acarology* 28, 169–176.

Punda-Polić, V., Luksić, B. and Capkun, V. (2008) Epidemiological features of Mediterranean spotted fever, murine typhus, and Q fever in Split-Dalmatia County (Croatia), 1982–2002. *Epidemiology and Infection* 136, 972–979.

Sardelic, S., Fournier, P.E., Punda-Polić, V., Bradaric, N., Grgic, D., Ivic, I., Ledina, D., Luksić, B., Milas, I. and Raoult D. (2003) First isolation of *Rickettsia conorii* from human blood in Croatia. *Croatian Medical Journal* 44, 630–634.

Cyprus

Psaroulaki, A., Loukaidis, F., Hadjichristodoulou, C. and Tselentis, Y. (1999) Detection and identification of the aetiological agent of Mediterranean spotted fever (MSF) in two genera of ticks in Cyprus. *Transactions of the Royal Society of Tropical Medicine and Hygiene* 93, 597–598.

Denmark

Skarphédinsson, S., Lyholm, B.F., Ljungberg, M., Søgaard, P., Kolmos. H.J., Nielsen, L.P. (2007) Detection and identification of *Anaplasma phagocytophilum, Borrelia burgdorferi,* and *Rickettsia helvetica* in Danish *Ixodes ricinus* ticks. *APMIS* 115, 225–230.

Egypt

Botros, B.A., Soliman, A.K., Darwish, M., el Said, S., Morrill, J.C. and Ksiazek, T.G. (1989) Seroprevalence of murine typhus and fièvre boutonneuse in certain human populations in Egypt. *Journal of Tropical Medicine and Hygiene* 92, 373–378.

Loftis, A.D., Reeves, W.K., Szumlas, D.E., Abbassy, M.M., Helmy, I.M., Moriarity, J.R. and Dasch, G.A. (2006) Population survey of Egyptian arthropods for rickettsial agents. *Annals of the New York Academy of Sciences* 1078, 364–367.

Socolovschi, C., Barbarot, S., Lefebvre, M., Parola, P. and Raoult, D. (2010) *Rickettsia sibirica mongolitimonae* in traveler from Egypt [Letter to the editor]. *Emerging Infectious Disease* 16, 1495–1496.

France

Beati, L. and Raoult, D. (1993) *Rickettsia massiliae* sp. nov., a new spotted fever group *Rickettsia*. *International Journal of Systematic and Evolutionary Bacteriology* 43, 839–840.

Beati, L., Finidori, J.P. and Raoult, D. (1993) First isolation of *Rickettsia slovaca* from *Dermacentor marginatus* in France. *American Journal of Tropical Medicine and Hygiene* 48, 257–268.

Drancourt, M., Kelly, P.J., Regnery, R. and Raoult, D. (1992) Identification of spotted fever group rickettsiae using polymerase chain reaction and restriction-endonuclease length polymorphism analysis. *Acta Virologica* 36, 1–6.

Fournier, P.E., Allombert, C., Supputamongkol, Y., Caruso, G., Brouqui, P. and Raoult, D. (2004) An eruptive fever associated with antibodies to *Rickettsia helvetica* in Europe and Thailand. *Journal of Clinical Microbiology* 42, 816–818.

Fournier, P.E., Gouriet, F., Brouqui, P., Lucht, F. and Raoult, D. (2005) Lymphangitis-associated rickettsiosis, a new rickettsiosis caused by *Rickettsia sibirica mongolotimonae*: seven new cases and review of the literature. *Clinical Infectious Diseases* 40, 1435–1444.

Matsumoto, K., Parola, P., Brouqui, P. and Raoult D. (2004) *Rickettsia aeschlimannii* in *Hyalomma* ticks from Corsica. *European Journal of Clinical Microbiology and Infectious Diseases* 23, 732–734.

Mediannikov, O., Matsumoto, K., Samoylenko, I., Drancourt, M., Roux, V., Rydkina, E., Davoust, B., Tarasevich, I., Brouqui, P. and Fournier, P.E. (2008) *Rickettsia raoultii* sp. nov., a spotted fever group rickettsia associated with *Dermacentor* ticks in Europe and Russia. *International Journal of Systematic and Evolutionary Microbiology* 58, 1635–1639.

Parola, P., Beati, L., Cambon, M. and Raoult, D. (1998) First isolation of *Rickettsia helvetica* from *Ixodes ricinus* ticks in France. *European Journal of Clinical Microbiology and Infectious Diseases* 17, 95–100.

Raoult, D., Nicolas, D., De Micco, P., Gallais, H. and Casanova, P. (1985) Aspects epidémiologiques de la fièvre boutonneuse Méditerranéenne en Corse du Sud [Epidemiological aspects of Mediterranean boutonneuse fever in the south of Corsica]. *Bulletin de la Société de Pathologie Exotique et de ses Filiales* 78, 446–451.

Raoult, D., Brouqui, P. and Roux, V. (1996) A new spotted-fever-group rickettsiosis. *The Lancet* 348, 412.

Germany

Hildebrandt, A., Fritzsch, J., Franke, J., Sachse, S., Dorn, W. and Straube, E. (2010) Co-circulation of emerging tick-borne pathogens in Middle Germany. *Vector-Borne Zoonotic Diseases* 11, 533–537.

Pluta, S., Tewald, F., Hartelt, K., Oehme, R., Kimmig, P. and Mackenstedt, U. (2009) *Rickettsia slovaca* in *Dermacentor marginatus* ticks, Germany. *Emerging Infectious Diseases* 15, 2077–2078.

Pluta, S., Hartelt, K., Oehme, R., Mackenstedt, U. and Kimmig, P. (2010) Prevalence of *Coxiella burnetii* and *Rickettsia* spp. in ticks and rodents in southern Germany. *Ticks and Tick-borne Diseases* 1, 145–147.

Rumer, L., Graser, E., Hillebrand, T., Talaska, T., Dautel, H., Mediannikov, O., Roy-Chowdhury, P., Sheshukova, O., Mantke, O.D. and Niedrig, M. (2011) *Rickettsia aeschlimannii* in *Hyalomma marginatum* ticks, Germany. *Emerging Infectious Diseases* 17, 325–326.

Silaghi, C., Hamel, D., Thiel, C., Pfister, K. and Pfeffer, M. (2011) Spotted fever group rickettsiae in ticks, Germany. *Emerging Infectious Diseases* 17, 890–892.

Simser, J.A., Palmer, A.T., Fingerle, V., Wilske, B., Kurtti, T.J., Munderloh, U.G. (2002) *Rickettsia monacensis* sp. nov., a spotted fever group *Rickettsia*, from ticks (*Ixodes ricinus*) collected in a European city park. *Applied and Environmental Microbiology* 68, 4559–4566.

Greece

Babalis, T., Tselentis, Y., Roux, V., Psaroulaki, A. and Raoult, D. (1994) Isolation and identification of a rickettsial strain related to *Rickettsia massiliae* in Greek ticks. *American Journal of Tropical Medicine and Hygiene* 50, 365–372.

Germanakis, A., Psaroulaki, A., Gikas, A. and Tselentis, Y. (2006) Mediterranean spotted fever in Crete, Greece: clinical and therapeutic data of 15 consecutive patients. *Annals of the New York Academy of Sciences* 1078, 263–269.

Kachrimanidou, M., Souliou, E., Pavlidou, V., Antoniadis, A. and Papa, A. (2010) First detection of *Rickettsia slovaca* in Greece. *Experimental and Applied Acarology* 50, 93–96.

Psaroulaki, A., Spyridaki, I., Ioannidis, A., Babalis, T., Gikas, A. and Tselentis Y. (2003) First isolation and identification of *Rickettsia conorii* from ticks collected in the region of Fokida in Central Greece. *Journal of Clinical Microbiology* 41, 3317–3319.

Psaroulaki, A., Germanakis, A., Gikas, A., Scoulica, E. and Tselentis, Y. (2005a) First isolation and genotypic identification of *Rickettsia conorii* Malish 7 from a patient in Greece. *European Journal of Clinical Microbiology and Infectious Diseases* 24, 297–298.

Psaroulaki, A., Germanakis, A., Gikas, A., Scoulica, E. and Tselentis, Y. (2005b) Simultaneous detection of *Rickettsia mongolotimonae* in a patient and a tick in Greece. *Journal of Clinical Microbiology* 43, 3558–3559.

Psaroulaki, A., Ragiadakou, D., Kouris, G., Papadopoulos, B., Chaniotis, B. and Tselentis, Y. (2006) Ticks, tick-borne rickettsiae, and *Coxiella burnetii* in the Greek Island of Cephalonia. *Annals of the New York Academy of Sciences* 1078, 389–399.

Hungary

Lakos, A. (1997) Tick-borne lymphadenopathy – a new rickettsial disease? *The Lancet* 350(9083), 1006.

Raoult, D., Lakos, A., Fenollar, F., Beytout, J., Brouqui, P. and Fournier, P.E. (2002) Spotless rickettsiosis caused by *Rickettsia slovaca* and associated with *Dermacentor* ticks. *Clinical and Infectious Diseases* 34, 1331–1336.

Rehácek, J., Nosek, J., Urvölgyi, J. and Sztankay, M. (1979) Rickettsiae of the spotted fever group in Hungary. *Folia Parasitologica* (Praha) 26, 367–371.

Sréter-Lancz, Z., Sréter, T., Széll, Z. and Egyed, L. (2005) Molecular evidence of *Rickettsia helvetica* and *R. monacensis* infections in *Ixodes ricinus* from Hungary. *Annals of Tropical Medicine and Parasitology* 99, 325–330.

Sréter-Lancz, Z., Széll, Z., Kovács, G., Egyed, L., Márialigeti, K. and Sréter T. (2006) Rickettsiae of the spotted-fever group in ixodid ticks from Hungary: identification of a new genotype ('Candidatus *Rickettsia kotlanii*'). *Annals of Tropical Medicine and Parasitology* 100, 229–236.

Israel

Goldwasser, R.A., Steiman, Y., Klingberg, W., Swartz, T.A. and Klingberg, M.A. (1974) The isolation of strains of rickettsiae of the spotted fever group in Israel and their differentiation from other members of the group by immunofluorescence methods. *Scandinavian Journal of Infectious Diseases* 6, 53–62.

Gross, E.M. and Yagupsky, P. (1987) Israeli rickettsial spotted fever in children. A review of 54 cases. *Acta Tropica* 44, 91–96.

Harrus, S., Perlman-Avrahami, A., Mumcuoglu, K.Y., Morick, D. and Baneth, G. (2011) Molecular detection of *Rickettsia massiliae, Rickettsia sibirica mongolitimonae* and *Rickettsia conorii israelensis* in ticks from Israel. *Clinical Microbiology and Infection* 17, 176–180.

Italy

Beninati, T., Lo, N., Noda, H., Esposito, F., Rizzoli, A., Favia, G. and Genchi, C. (2002) First detection of spotted fever group rickettsiae in *Ixodes ricinus* from Italy. *Emerging Infectious Diseases* 8, 983–986.

Beninati, T., Genchi, C., Torina, A., Caracappa, S., Bandi, C. and Lo, N. (2005) Rickettsiae in ixodid ticks, Sicily. *Emerging Infectious Diseases* 11, 509–511.

Cascio, A., Dones, P., Romano, A. and Titone, L. (1998) Clinical and laboratory findings of boutonneuse fever in 645 Sicilian children. *European Journal of Pediatrics* 157, 482–486.

Ciceroni, L., Pinto, A., Ciarrocchi, S. and Ciervo, A. (2006) Current knowledge of rickettsial diseases in Italy. *Annals of the New York Academy of Sciences* 1078, 143–149.

Floris, R., Yurtman, A.N., Margoni, E.F., Mignozzi, K., Boemo, B., Altobelli, A. and Cinco, M. (2008) Detection and identification of *Rickettsia* species in the northeast of Italy. *Vector-Borne and Zoonotic Diseases* 8, 777–782.

Giammanco, G.M., Mansueto, S., Ammatuna, P. and Vitale, G. (2003) Israeli spotted fever *Rickettsia* in Sicilian *Rhipicephalus sanguineus* ticks. *Emerging Infectious Diseases* 9, 892–893.

Giammanco, G.M., Vitale, G., Mansueto, S., Capra, G., Caleca, M.P. and Ammatuna, P. (2005) Presence of *Rickettsia conorii* subsp. *israelensis*, the causative agent of Israeli spotted fever, in Sicily, Italy, ascertained in a retrospective study. *Journal of Clinical Microbiology* 43, 6027–6031.

Mura, A., Masala, G., Tola, S., Satta, G., Fois, F., Piras, P., Rolain, J.M., Raoult, D. and Parola, P. (2008) First direct detection of rickettsial pathogens and a new rickettsia, 'Candidatus Rickettsia barbariae', in ticks from Sardinia, Italy. *Clinical Microbiology and Infection* 14, 1028–1033.

Selmi, M., Bertolotti, L., Tomassone, L. and Mannelli, A. (2008) *Rickettsia slovaca* in *Dermacentor marginatus* and tick-borne lymphadenopathy, Tuscany, Italy. *Emerging Infectious Diseases* 14, 817–820.

Selmi, M., Martello, E., Bertolotti, L., Bisanzio, D. and Tomassone, L. (2009) *Rickettsia slovaca* and *Rickettsia raoultii* in *Dermacentor marginatus* ticks collected on wild boars in Tuscany, Italy. *Journal of Medical Entomology* 46, 1490–1493.

Vitale, G., Mansuelo, S., Rolain, J.M. and Raoult, D. (2006) *Rickettsia massiliae* human isolation. *Emerging Infectious Diseases* 12, 174–175.

Kosovo

Fournier, P.E., Durand, J.P., Rolain, J.M., Camicas, J.L., Tolou, H., Raoult, D. (2003) Detection of Astrakhan fever rickettsia from ticks in Kosovo. *Annals of the New York Academy of Sciences* 990, 158–61.

Libya

Boillat, N., Genton, B., D'Acremont, V., Raoult, D. and Greub, G. (2008) Fatal case of Israeli spotted fever after Mediterranean cruise. *Emerging Infectious Diseases* 14, 1944–1946.

Malta

Tonna, I., Mallia Azzopardi, C., Piscopo, T., Cuschieri, P., Fenollar, F. and Raoult, D. (2006) Characterisation of rickettsial diseases in a hospital-based population in Malta. *Journal of Infection* 53, 394–402.

Moldova

Movila, A., Reye, A.L., Dubinina, H.V., Tolstenkov, O.O., Toderas, I., Hübschen, J.M., Muller, C.P. and Alekseev, A. (2011) Detection of *Babesia* sp. EU1 and members of spotted fever group rickettsiae in ticks collected from migratory birds at Curonian Spit, North-Western Russia. *Vector-Borne and Zoonotic Diseases* 11, 89–91.

Morocco

Beati, L., Meskini, M., Thiers, B. and Raoult, D. (1997) *Rickettsia aeschlimannii* sp. nov., a new spotted fever group rickettsia associated with *Hyalomma marginatum* ticks. *International Journal of Systematic Bacteriology* 47, 548–554.

Demeester, R., Claus, M., Hildebrand, M., Vlieghe, E. and Bottieau, E. (2010) Diversity of life-threatening complications due to Mediterranean spotted fever in returning travelers. *Journal of Travel Medicine* 17, 100–104.

Raoult, D., Fournier, P.E., Abboud, P. and Caron, F. (2002) First documented human *Rickettsia aeschlimannii* infection. *Emerging Infectious Diseases* 8, 748–749.

Sarih, M., Socolovschi, C., Boudebouch, N., Hassar, M., Raoult, D. and Parola, P. (2008) Spotted fever group rickettsiae in ticks, Morocco. *Emerging Infectious Diseases* 14, 1067–1073.

Netherlands

Nijhof, A.M., Bodaan, C., Postigo, M., Nieuwenhuijs, H., Opsteegh, M., Franssen, L., Jebbink, F. and Jongejan, F. (2007) Ticks and associated pathogens collected from domestic animals in the Netherlands. *Vector-Borne and Zoonotic Diseases* 7, 585–595.

Poland

Chmielewski, T., Podsiadly, E., Karbowiak, G. and Tylewska-Wierzbanowska, S. (2009) *Rickettsia* spp. in ticks, Poland. *Emerging Infectious Diseases* 15, 486–488.

Portugal

Bacellar, F., Núncio, M.S., Alves, M.J. and Filipe, A.R. (1995) [*Rickettsia slovaca*: an agent of the group of exanthematous fevers, in Portugal]. *Enfermedades Infecciosas y Microbiologia Clinica* 13, 218–223.

Bacellar, F., Regnery, R.L., Núncio, M.S. and Filipe, A.R. (1995) Genotypic evaluation of rickettsial isolates recovered from various species of ticks in Portugal. *Epidemiology and Infection* 114, 169–178.

Bacellar, F., Beati, L., França, A., Poças, J., Regnery, R. and Filipe, A. (1999) Israeli spotted fever *Rickettsia* (*Rickettsia conorii* complex) associated with human diseases in Portugal. *Emerging Infectious Diseases* 5, 835–836.

Bacellar, F., Sousa, R., Santos, A., Santos-Silva, M. and Parola, P. (2003) Boutonneuse fever in Portugal: 1995–2000. Data of a state laboratory. *European Journal of Epidemiology* 18, 275–277.

de Sousa, R., Barata, C., Vitorino, L., Santos-Silva, M., Carrapato, C., Torgal, J., Walker, D. and Bacellar, F. (2006) *Rickettsia sibirica* isolation from a patient and detection in ticks, Portugal. *Emerging Infectious Diseases* 12, 1103–1108.

de Sousa, R., Santos-Silva, M., Santos, A.S., Barros, S., Torgal, J., Walker, D.H. and Bacellar, F. (2007) *Rickettsia conorii* Israeli tick typhus strain isolated from *Rhipicephalus sanguineus* ticks in Portugal. *Vector-Borne and Zoonotic Diseases* 7, 444–447.

de Sousa, R., Duque, L., Anes, M., Poças, J., Torgal, J., Bacellar, F., Olano, J.P. and Walker, D.H. (2008a) Lymphangitis in a Portuguese patient infected with *Rickettsia sibirica*. *Emerging Infectious Diseases* 14, 529–530.

de Sousa, R., França, A., Nóbrega-Dória, S., Belo, A., Amaro, M., Abreu, T., Proença, P., Poças, J., Vaz, J., Torgal, J., Bacellar, F., Ismail, N. and Walker, D.H. (2008b) Host- and microbe-related risk factors for and pathophysiology of fatal *Rickettsia conorii* infection in Portuguese patients. *Journal of Infectious Diseases* 198, 576–585.

Lopes de Carvalho, I., Milhano, N., Santos, A.S., Almeida, V., de Sousa, R., Núncio, M.S. (2008) Detection of *Borrelia lusitaniae*, *Rickettsia* sp. IRS3, *Rickettsia monacensis* and *Anaplasma phagocytophilum* in *Ixodes ricinus* collected in Madeira Island, Portugal. *Vector Borne and Zoonotic Diseases* 8, 575–579.

Milhano, N., Lopes de Carvalho, I., Alves, A.S., Arroube, S., Soares, J., Rodriguez, P., Carolino, M., Núncio, M.S., Piesman, J. and de Sousa, R. (2010) Coinfections of *Rickettsia slovaca* and *Rickettsia helvetica* with *Borrelia lusitaniae* in ticks collected in a Safari Park, Portugal. *Ticks and Tickborne Diseases* 1, 172–177.

Santos-Silva, M., Sousa, R., Santos, A.S., Lopes, D., Queijo, E., Doreta, A., Vitorino, L. and Bacellar, F. (2006a) Ticks and tick-borne rickettsiae surveillance in Montesinho Natural Park – Portugal. *Annals of the New York Academy of Sciences* 1078, 137–142.

Santos-Silva, M., Sousa, R., Santos, A., Melo, P., Encarnação, V. and Bacellar, F. (2006b) Ticks parasitizing wild birds in Portugal: detection of *Rickettsia aeschlimannii*, *R. helvetica* and *R. massiliae*. *Experimental and Applied Acarology* 39, 331–338.

Romania

Serban, R., Pistol, A., Neguţ, M. and Cucuiu, R. (2009) [Rickettsia conorii infection in Romania, 2000–2008]. *Bacteriologia, Virusologia, Parazitologia, Epidemiologia* (Bucarest, Romania) 54, 177–183.

Russia, Ukraine and Astrakhan region

Eremeeva, M.E., Beati, L., Makarova, V., Fetisova, N.F., Tarasevich, I.V., Balayeva, N.M. and Raoult, D. (1994) Astrakhan fever rickettsiae: antigenic and genotypic analysis of isolates obtained from human and *Rhipicephalus pumilio* ticks. *American Journal Tropical Medicine and Hygiene* 51, 697–706.

Mediannikov, O., Matsumoto, K., Samoylenko, I., Drancourt, M., Roux, V., Rydkina, E., Davoust, B., Tarasevich, I., Brouqui, P. and Fournier, P.E. (2008) *Rickettsia raoultii* sp. nov., a spotted fever group rickettsia associated with *Dermacentor* ticks in Europe and Russia. *International Journal of Systematic and Evolutionary Microbiology* 58, 1635–1639.

Movila, A., Rolain, J.M., Podavalenko, A., Toderas, I., Tkachenco. L., Naglov, V. and Raoult, D. (2009) Detection of spotted fever group rickettsiae and family Anaplasmataceae in *Ixodes ricinus* ticks from Republic of Moldova and Eastern Ukraine. *Clinical Microbiology and Infection* 15(Suppl. 2), 32–33.

Rydkina, E., Roux, V., Rudakov, N., Gafarova, M., Tarasevich, I. and Raoult, D. (1999) New rickettsiae in ticks collected in territories of the former Soviet Union. *Emerging Infectious Diseases* 5, 811–814.

Shpynov, S., Rudakov, N., Tohkov, Y., Matushchenko, A., Tarasevich, I. and Raoult, D. (2009) Detection of *Rickettsia aeschlimannii* in *Hyalomma marginatum* ticks in western Russia. *Clinical Microbiology and Infection* 15(Suppl. 2), S315–S316.

Tarasevich, I.V. and Mediannikov, O.Y. (2006) Rickettsial diseases in Russia. *Annals of the New York Academy of Sciences* 1078, 48–59.

Tarasevich, I.V., Makarova, V.A., Fetisova, N.F., Stepanov, A.V., Miskarova, E.D. and Raoult, D. (1991) Studies of a "new" rickettsiosis "Astrakhan" spotted fever. *European Journal of Epidemiology* 7, 294–298.

Serbia

Radulović, Z., Chochlakis, D., Tomanović, S., Milutinović, M., Tselentis, Y. and Psaroulaki, A. (2011) First detection of spotted fever group rickettsiae in ticks in Serbia. *Vector-Borne and Zoonotic Diseases* 11, 111–115.

Samardzic, S., Marinkovic, T., Marinkovic, D., Djuricic, B., Ristanovic, E., Simovic, T., Lako, B., Vukov, B., Bozovic, B. and Gligic, A. (2008) Prevalence of antibodies to rickettsiae in different regions of Serbia. *Vector-Borne and Zoonotic Diseases* 8, 219–224.

Slovakia

Matsumoto, K., Grzeszczuk, A., Brouqui, P. and Raoult, D. (2009) *Rickettsia raoultii* and *Anaplasma phagocytophilum* in *Dermacentor reticulatus* ticks collected from Białowieża Primeval Forest European bison (*Bison bonasus bonasus*), Poland. *Clinical Microbiology and Infection* 15(Suppl. 2), 286–287.

Mittermayer, T., Brezina, R. and Urvölgyi, J. (1980) First report of an infection with *Rickettsia slovaca*. *Folia Parasitologica* (Praha) 27, 373–376.

Rehácek, J. (1984) *Rickettsia slovaca*, the organism and its ecology. *Acta Scientifica National Academy Scientifica Bohemoslovacae Brno* 18, 1–50.

Rehácek, J., Brezina, R., Ac, P., Zupancicova, M. and Kovacova, E. (1972) Contribution to the natural focality of rickettsiae belonging to the Rocky Mountain spotted fever (RMSF) group in Slovakia. *Folia Parasitologica* 19, 41–52.

Sekeyová, Z., Roux, V., Xu, W., Rehácek, J. and Raoult, D. (1998) *Rickettsia slovaca* sp. nov., a member of the spotted fever group rickettsiae. *International Journal of Systematic Bacteriology* 48, 1455–1462.

Sekeyová, Z., Fournier, P.E., Rehácek, J. and Raoult, D. (2000) Characterization of a new spotted fever group rickettsia detected in *Ixodes ricinus* (Acari: Ixodidae) collected in Slovakia. *Journal of Medical Entomology* 37, 707–713.

Urvolgyi, J. and Brezina, R. (1978) *Rickettsia slovaca*: a new member of spotted fever group rickettsiae. In: Kazar, J., Ormsbee, R.A. and Tarasevich, I.V. (eds) *Rickettsiae and Rickettsial Diseases. Proceedings of the Second International Symposium, Slovak Academy of Sciences*. Bratislava, Veda, Slovakia, pp. 299–305.

Slovenia

Prosenc, K., Petrovec, M., Trilar, T., Duh, D. and Avšič-Županc, T. (2003) Detection of rickettsiae in *Ixodes ricinus* ticks in Slovenia. *Annals of the New York Academy of Sciences* 990, 201–204.

Spain

Aguirrebengoa, K., Portillo, A., Santibáñez, S., Marín, J.J., Montejo, M. and Oteo, J.A. (2008) Human *Rickettsia sibirica mongolitimonae* infection, Spain. *Emerging Infectious Diseases* 14, 528–529.

Barandika, J.F., Hurtado, A., García-Sanmartín, J., Juste, R.A., Anda, P. and García-Pérez, A.L. (2008) Prevalence of tick-borne zoonotic bacteria in questing adult ticks from northern Spain. *Vector-Borne and Zoonotic Diseases* 8, 829–835.

Bartolomé, J., Lorente, S., Hernández-Pérez, N., Martínez-Alfaro, E., Marín-Ors, A. and Crespo, M.D. (2005) Clinical and epidemiological study of spotted fever group rickettsiosis in Albacete, Spain. *Enfermades Infecciosas y Microbiologia Clinica* 23, 194–196.

Bassa Malondra, A. and Mulet, V.T. (1987) [Mediterranean boutonneuse fever in Mallorca]. *Revista Clinica Espanola* 181, 230.

Beati, L., Roux, V., Ortuño, A., Castella, J., Porta, F.S. and Raoult, D. (1996) Phenotypic and genotypic characterization of spotted fever group rickettsiae isolated from Catalan *Rhipicephalus sanguineus* ticks. *Journal of Clinical Microbiology* 34, 2688–2694.

Cardeñosa, N., Roux, V., Font, V., San Feliu, I., Raoult, D. and Segura, F. (2000) Isolation and identification of two spotted fever rickettsial strains from patients in Catalonia, Spain. *American Journal of Tropical Medicine and Hygiene* 62, 142–144.

Espejo-Arenas, E. and Raoult, D. (1989) First isolates of *Rickettsia conorii* in Spain using a centrifugation-shell vial assay. *Journal of Infectious Diseases* 159, 1158–1159.

Fernández-Soto, P., Encinas-Grandes, A. and Pérez-Sánchez, R. (2003) *Rickettsia aeschlimannii* in Spain: molecular evidence in *Hyalomma marginatum* and five other tick species that feed on humans. *Emerging Infectious Diseases* 9, 889–890.

Fernández-Soto, P., Pérez-Sánchez, R., Encinas-Grandes, A. and Sanz, R.A. (2004) Detection and identification of *Rickettsia helvetica* and *Rickettsia* sp. IRS3/IRS4 in *Ixodes ricinus* ticks found on humans in Spain. *European Journal of Clinical Microbiology and Infectious Diseases* 23, 648–649.

Fernández-Soto, P., Pérez-Sánchez, R., Díaz Martín, V., Encinas-Grandes, A. and Alamo Sanz, R. (2006) *Rickettsia massiliae* in ticks removed from humans in Castilla y León, Spain. *European Journal of Clinical Microbiology and Infectious Diseases* 25, 811–813.

Jado, I., Oteo, J.A., Aldámiz, M., Gil, H., Escudero, R., Ibarra, V., Portu, J., Portillo, A., Lezaun, M.J., García-Amil, C., Rodríguez-Moreno, I. and Anda, P. (2007) *Rickettsia monacensis* and human disease, Spain. *Emerging Infectious Diseases* 13, 1405–1407.

Laynez Cerdeña, P., Pascual Velasco, F., Guemes Artiles, M. and Cervera Pérez, A. (1986) [Mediterranean boutonneuse fever in the Canary Islands] [Case reports, letter]. *Medicina Clinica* (Barcelona) 87, 564.

Márquez, F.J. (2008) Spotted fever group *Rickettsia* in ticks from southeastern Spain natural parks. *Experimental and Applied Acarology* 45, 185–194.

Oteo, J.A., Ibarra, V., Blanco, J.R., Martínez de Artola, V., Márquez, F.J., Portillo, A., Raoult, D. and Anda, P. (2004) *Dermacentor*-borne necrosis erythema and lymphadenopathy: clinical and epidemiological features of a new tick-borne disease. *Clinical Microbiology and Infection* 10, 327–331.

Oteo, J.A., Portillo, A., Santibáñez, S., Pérez-Martínez, L., Blanco, J.R., Jiménez, S., Ibarra, V., Pérez-Palacios, A. and Sanz, M. (2006) Prevalence of spotted fever group *Rickettsia* species detected in ticks in La Rioja, Spain. *Annals of the New York Academy of Sciences* 1078, 320–323.

Portillo, A., Ibarra, V., Santibáñez, S., Pérez-Martínez, L., Blanco, J.R. and Oteo, J.A. (2009) Genetic characterisation of *ompA*, *ompB* and *gltA* genes from Candidatus *Rickettsia rioja*. *Clinical Microbiology and Infection* 15(Suppl. 2), 307–308.

Toledo, A., Olmeda, A.S., Escudero, R., Jado, I., Valcárcel, F., Casado-Nistal, M.A., Rodríguez-Vargas, M., Gil, H. and Anda, P. (2009) Tick-borne zoonotic bacteria in ticks collected from central Spain. *American Journal of Tropical Medicine and Hygiene* 81, 67–74.

Sweden

Nilsson, K. (2009) Septicaemia with *Rickettsia helvetica* in a patient with acute febrile illness, rash and myasthenia. *Journal of Infection* 58, 79–82.

Nilsson, K., Lindquist, O., Liu, A.J., Jaenson, T.G., Friman, G. and Påhlson, C. (1999) Association of *R. helvetica* with chronic periomyocarditis in sudden cardiac death. *The Lancet* 354, 1169–1173.

Nilsson, K., Elfving, K. and Pahlson, C. (2010) *Rickettsia helvetica* in patient with meningitis, Sweden, 2006. *Emerging Infectious Diseases* 16, 490–492.

Severinsson, K., Jaenson, T.G., Pettersson, J., Falk, K. and Nilsson, K. (2010) Detection and prevalence of *Anaplasma phagocytophilum* and *Rickettsia helvetica* in *Ixodes ricinus* ticks in seven study areas in Sweden. *Parasites and Vectors* 3, 66, doi:10.1186/1756-3305-3-66.

Skarphédinsson, S., Lyholm, B.F., Ljungberg, M., Søgaard, P., Kolmos, H.J. and Nielsen, L.P. (2007) Detection and identification of *Anaplasma phagocytophilum*, *Borrelia burgdorferi*, and *Rickettsia helvetica* in Danish *Ixodes ricinus* ticks. *Acta Pathologica, Microbiologica et Immunologica Scandinavica* 115, 225–230.

Switzerland

Beati, L., Péter, O., Burgdorfer, W., Aeschlimann, A. and Raoult, D. (1993) Confirmation that *Rickettsia helvetica* sp. nov. is a distinct species of the spotted fever group of rickettsiae. *International Journal of Systematic Bacteriology* 43, 521–526.

Beati, L., Humair, P.F., Aeschlimann, A. and Raoult, D. (1994) Identification of spotted fever group rickettsiae isolated from *Dermacentor marginatus* and *Ixodes ricinus* ticks collected in Switzerland. *American Journal of Tropical Medicine and Hygiene* 51, 138–148.

Bernasconi, M.V., Casati, S., Péter, O. and Piffaretti, J.C. (2002) *Rhipicephalus* ticks infected with *Rickettsia* and *Coxiella* in southern Switzerland (Canton Ticino). *Infection, Genetics and Evolution* 2, 111–120.

Rehácek, J. (1993) Rickettsiae and their ecology in the Alpine region. *Acta Virologica* 37, 290–301.

Tunisia

Romdhane, F.B., Loussaief, C., Toumi, A., Yahia, S.B., Khaiyrallah, M., Bouzouaïa, N. and Chakroun, M. (2009) Mediterranean spotted fever: a report of 200 cases in Tunisia. *Clinical Microbiology and Infection* 15(Suppl. 2), 209–210.

Sfar, N., M'ghirbi, Y., Letaïef, A., Parola, P., Bouattour, A. and Raoult, D. (2008) First report of *Rickettsia monacensis* and *Rickettsia helvetica* from Tunisia. *Annals of Tropical Medicine and Parasitology* 102, 561–564.

Sfar, N., Kaabia, N., Letaïef, A., Rolain, J.M., Parola, P., Bouattour, A. and Raoult, D. (2009) First molecular detection of *R. conorii* subsp. *conorii* 99 years after the Conor description of Mediterranean spotted fever, in Tunisia. *Clinical Microbiology and Infection* 15(Suppl.2), 309–310.

Znazen, A., Hammami, B., Lahiani, D., Ben-Jemaa, M. and Hammami, A. (2011) Molecular confirmation of human infection caused by Israeli spotted fever rickettsia in Tunisia. *Emerging Infectious Diseases* 17, 1328–1330.

Turkey

Kuloglu, F., Rolain, J.M., Fournier, P.E., Akata, F., Tugrul, M. and Raoult, D. (2004) First isolation of *Rickettsia conorii* from humans in the Trakya (European) region of Turkey. *European Journal of Clinical Microbiology and Infectious Diseases* 23, 609–614.

Kuloglu, F., Rolain, J.M., Celik, A.D., Akata, F., Tugrul, M. and Raoult, D. (2009) Prospective evaluation of rickettsioses in the Trakya (European) Region of Turkey in 2005. *Clinical Microbiology and Infection* 15(Suppl. 2), 220–221.

UK

Tijsse-Klasen, E., Jameson, L.J., Fonville, M., Leach, S., Sprong, H. and Medlock, J.M. (2011) First detection of spotted fever group rickettsiae in *Ixodes ricinus* and *Dermacentor reticulatus* ticks in the UK. *Epidemiology and Infection* 139, 524–529.

Appendices 2 and 3

Appendix 2

List (By ID Number) of References from which Geographic Data Were Extracted (hosted on http://www.cabi.org/resources/tatbd).

Appendix 3

List of Scientific Papers That Could Not Be Retrieved (hosted on http://www.cabi.org/resources/tatbd).

Appendix 4 Keys for Genera and Species of Ticks in the Mediterranean Basin

Agustín Estrada-Peña,[1] Alan R. Walker[2] and Maxime Madder[3]
[1]University of Zaragoza, Spain; [2]University of Edinburgh, Scotland, UK; and [3]Institute of Tropical Medicine, Antwerp, Belgium and University of Pretoria, South Africa

The reliable determination of the genera and species of ticks that are involved is the first step towards any further knowledge on the ecology, distribution or epidemiology of tick-borne pathogens. Current DNA-based methods are not yet accurate enough to be systematically used for this purpose. The 'old-fashioned' methods using a stereomicroscope and systematic keys are still the best approach to determining tick taxonomy. We want to provide in this work an overview of the most important features of interest that can be used to identify the ticks present in the Mediterranean basin and most of Europe. It must be understood, however, that our work provides information only about the species of ticks that have an impact on domestic animals, as well as those which are similar and could be easily confused. It must also be stressed that the investigation of immature stages might provide more information than that of adults. Immature stages are more abundant, and their importance may be overlooked by focusing only on adult stages.

Keys for genera: ***Amblyomma*** (A) and ***Dermacentor*** (B)

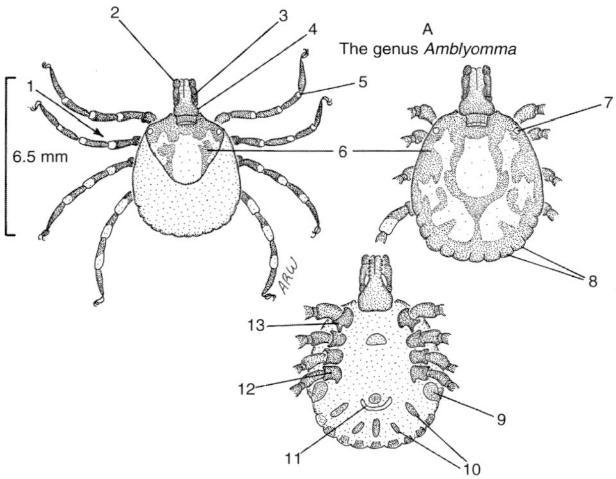

The only ornate ticks that are in Europe or regions in its proximity are the adults of the genera *Amblyomma* and *Dermacentor*. The former is absent in Europe, although isolated specimens have been collected in northern Africa.

Two species of the genus *Dermacentor* are very common in Europe, namely *D. marginatus* and *D. reticulatus*. The first is a common parasite present in Mediterranean-type areas, while the second is a serious concern because of repeated reports of its spread into areas that were formerly free of the species.

The size of *Amblyomma* ticks (Plate A) is large (see 1), the mouthparts are anterior (2) and palpal articles II are longer than articles I and III (3).

The basis capituli has straight lateral margins (4). Legs usually have pale rings (5) with enamel ornamentation in the scutum (6). Eyes and festoons are present (7). Spiracular plates are large (9), and ventral plates in males are indistinct (10) as flat plates posterior to the anus. An anal groove is posterior to the anus (11), coxae IV are of normal size (12) and coxae I have unequal paired spurs (13).

Dermacentor ticks (Plate B) are medium sized (see 1) with anterior mouthparts (2) and broad palpal articles (3) with straight lateral margins (4). Legs have no pale rings (5) and the conscutum has enamel (6). Eyes are present (7) and festoons are present in males (8). Spiracular plates are large (9) and ventral plates are absent in males (10). The anal groove (11) is posterior to the anus, coxae IV are very large (12) and coxae I have large and equal paired spurs.

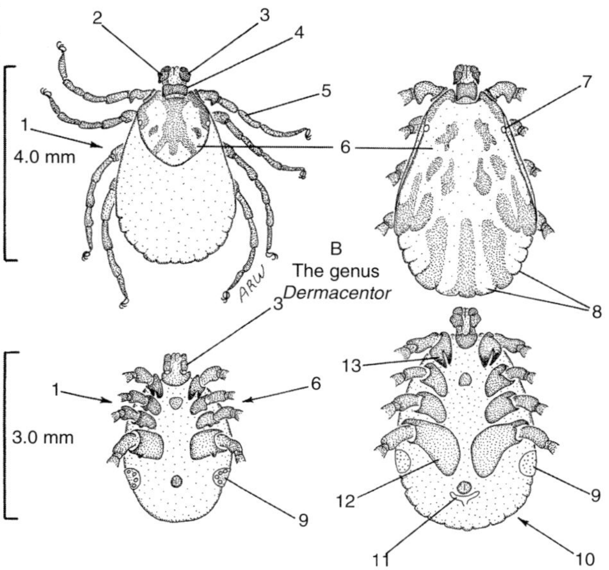

Keys for genera: *Rhipicephalus* (A) and *Haemaphysalis* (B)

Adults of the genus *Rhipicephalus* are some of the most common ticks in the Mediterranean region (Plate A). They are medium sized (see 1) and mouthparts are anterior (2). The palp articles are all small (3) and the basis capituli has distinctly angular lateral margins, making a hexagonal shape of the entire basis capituli (4), which is a diagnostic character. Legs have no pale rings (5) and the scutum has no enamel (6). Eyes are present (7) and festoons are present in males (8) and in females but unclear if fed. Spiracular plates are large (9) and ventral plates are present in the males (10), usually as two pairs of plates. Anal groove is posterior to anus (11), coxae IV are of normal size (12) and coxae I have large and equal paired spurs (13).

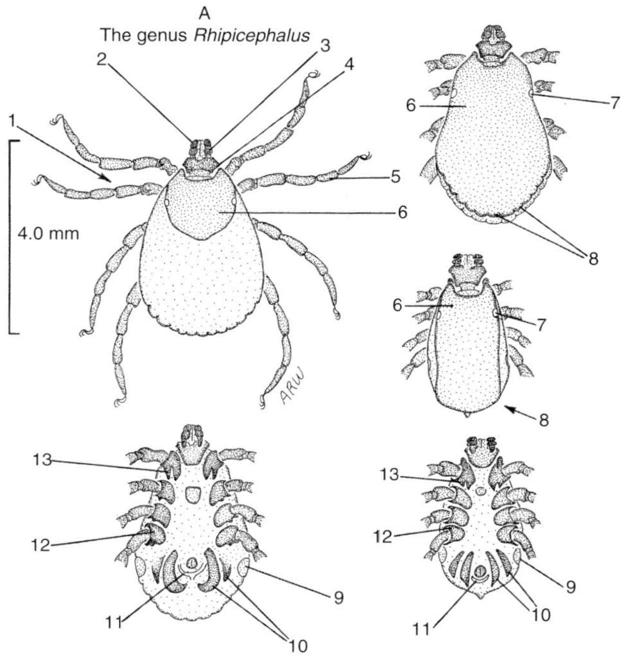

Adults of the genus *Haemaphysalis* (Plate B) are small (see 1), with anterior mouthparts (2) and broad palpal articles (3). The basis capituli has straight lateral margins (4). Legs have no pale rings (5) and the scutum lacks ornamentation (6). Eyes are always absent (7) and festoons are always present (8) but may be difficult to see in engorged females. The spiracular plates are large (9) and ventral plates are absent from the males (10). The anal groove is posterior to the anus (11), coxae IV are of normal size (12) and coxae I have no paired spurs – there is only one single spur present (13).

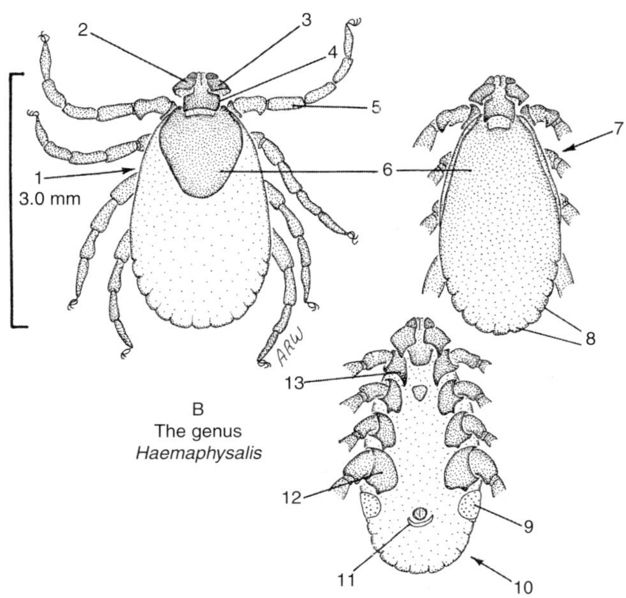

Keys for genera: *Hyalomma* (A) and *Ixodes* (B)

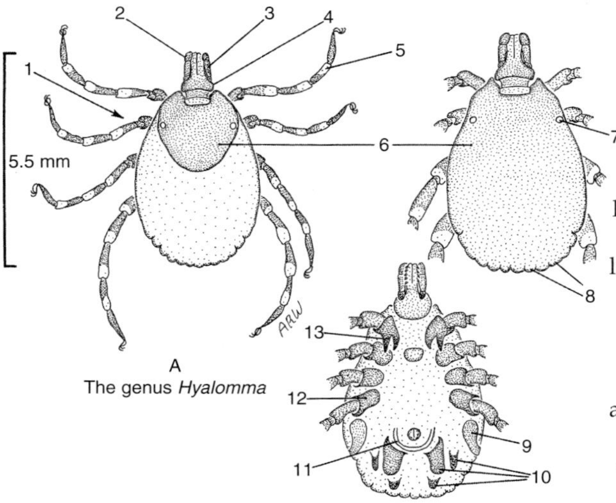

A
The genus *Hyalomma*

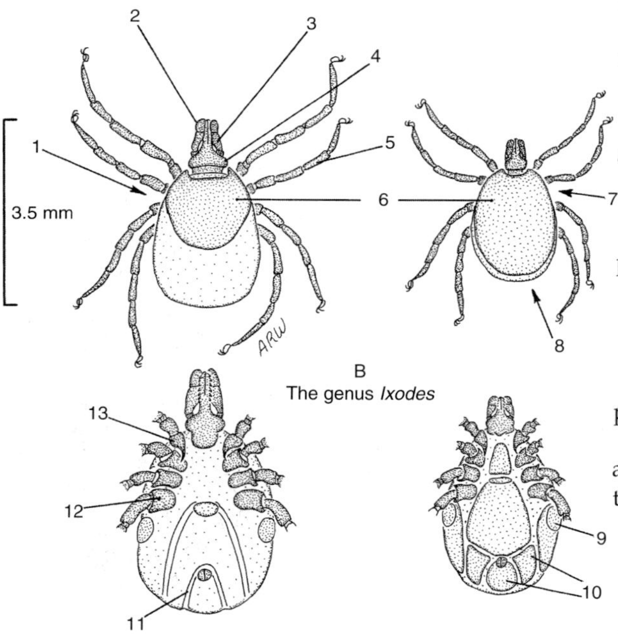

B
The genus *Ixodes*

Both *Hyalomma* and *Ixodes* are genera of ticks with long mouthparts and which lack enamel in the scutum.

Ticks in the genus *Hyalomma* (Plate A) are large (see 1) with anterior mouthparts (2) and palp articles II longer than articles I and III (3). The basis capituli has medium angular lateral margins (4) and the legs have pale rings (5). The scutum has no ornamentation (6) but *Hy. lusitanicum* may be an exception to this in some specimens. The eyes are always very convex (7) and festoons are present in both sexes (8), but difficult to see in engorged females. Spiracular plates are large (9) and ventral plates are present only in males, usually as three distinct pairs (10). The anal groove is posterior to the anus (11), coxae IV are of normal size (12) and coxae I have large and equal paired spurs (13).

Adults of the genus *Ixodes* (Plate B) are medium sized (see 1) with anterior mouthparts (2) and palp articles II longer than articles I and III (3). The basis capituli has straight lateral margins (4). Legs never have pale rings (5) and the scutum lacks enamel (6). Eyes and festoons are always absent (7 and 8). Spiracular plates are large (9) and ventral plates are present in males only (10) in the form of large flat plates. The anal groove forms a loop anterior to the anus, and in some species forms a circle around the anus.

Keys for species

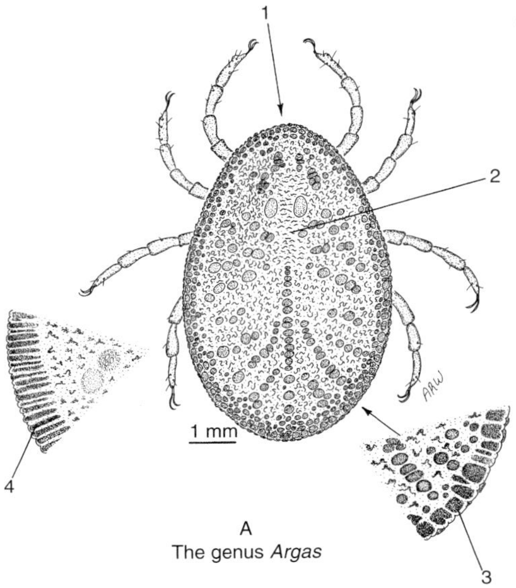

A
The genus *Argas*

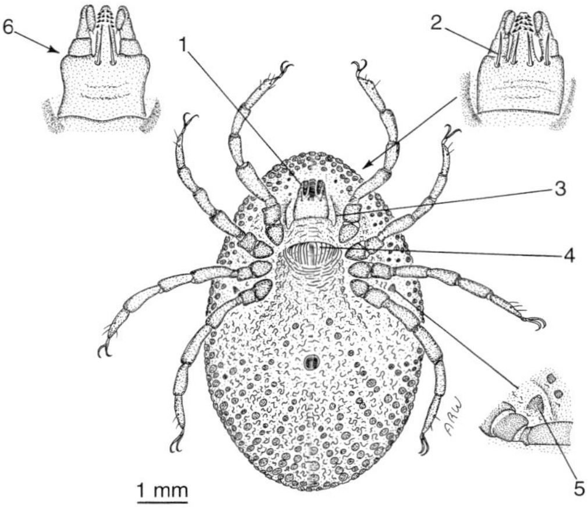

B
The genus *Ornithodoros*

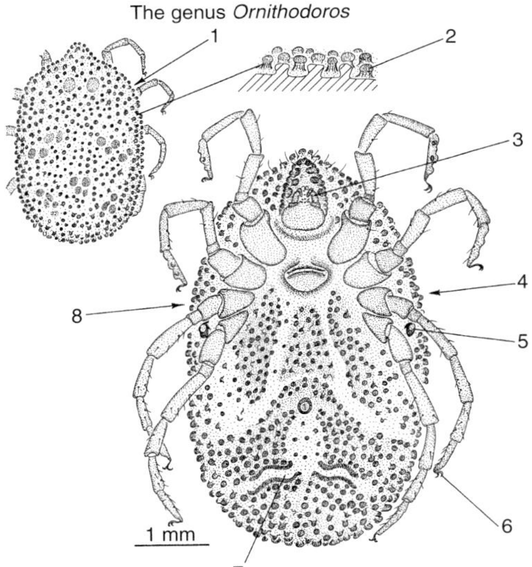

Three main species of soft ticks (family Argasidae) are commonly reported in the western Palaearctic. The genera *Argas* and *Ornithodoros* are characterized by their ventral mouthparts (Plate A, right, 1; Plate B, right 3) and the lack of a scutum or conscutum (Plate A, left, 2) in the adults.

Two of these species are part of the genus *Argas*, a tick commonly found on birds, which is illustrated in plate A. The genus *Argas* can be separated from the genus *Ornithodoros* by the presence of a lateral margin, which is marked by rectangular plates in *A. persicus* and by lateral fine ridges in *A. reflexus* (Plate A, left, 3 and 4). Such a lateral suture is absent in the genus *Ornithodoros* (Plate B, left, 1). Both genera have a mammillate pattern on the dorsum (Plate B, 2), which in *Argas* is convoluted, with raised discs present, smoothly flat and well defined.

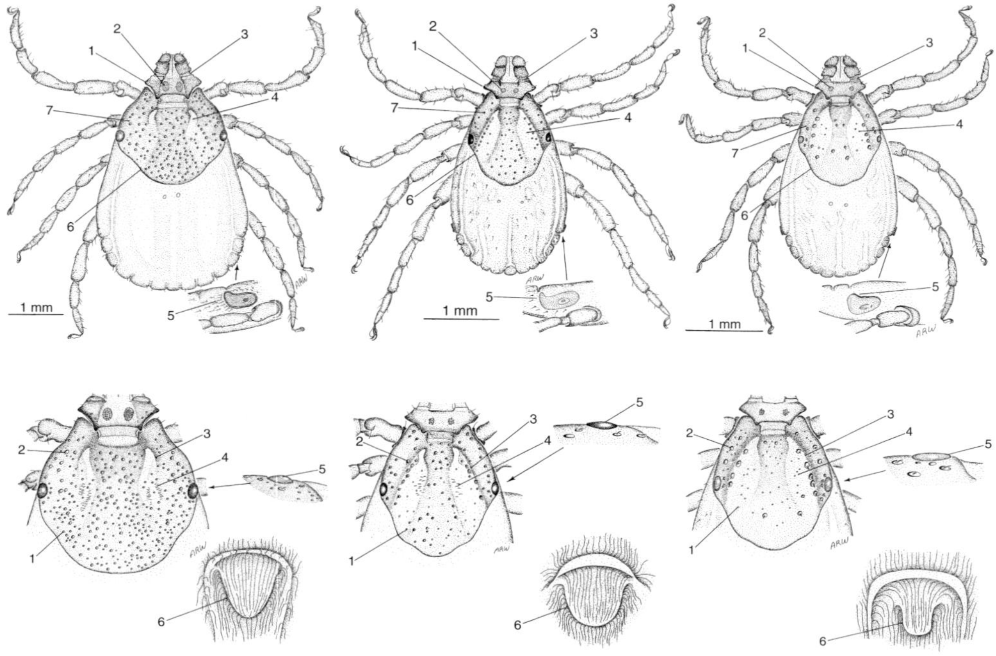

The females of the genus *Rhipicephalus*, including *Rh. bursa* (left), *Rh. sanguineus* (centre) and *Rh. turanicus* (right). The female of *Rh. annulatus* is illustrated below this legend.

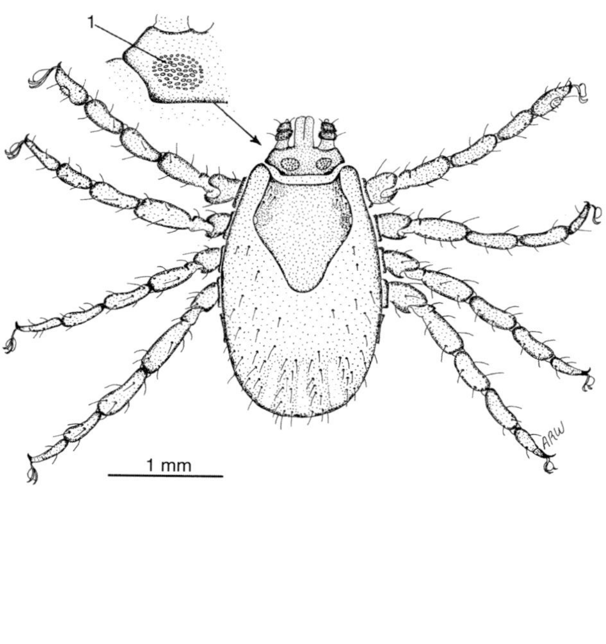

It is difficult to distinguish between females of different species of the genus *Rhipicephalus*. A first separation can be made by the use of the relative size of the spurs in coxae I (not illustrated). They are very short and indistinct in *Rh. annulatus* and very large and well visible (even with the naked eye) for the rest of the species. The separation of the females of the other species can be achieved by investigating the details of the genital opening, the spiracular plate, the size and position of the eyes, and the dorsal punctations in the scutum, as illustrated herein. Females of *Rh. bursa* are commonly easily separated because the large size of their dorsal punctations. More difficult is an adequate and reliable separation of *Rh. sanguineus* and *Rh. turanicus*. If difficulties are found, it is preferably to report these species as '*Rh. sanguineus* complex' to avoid confusion.

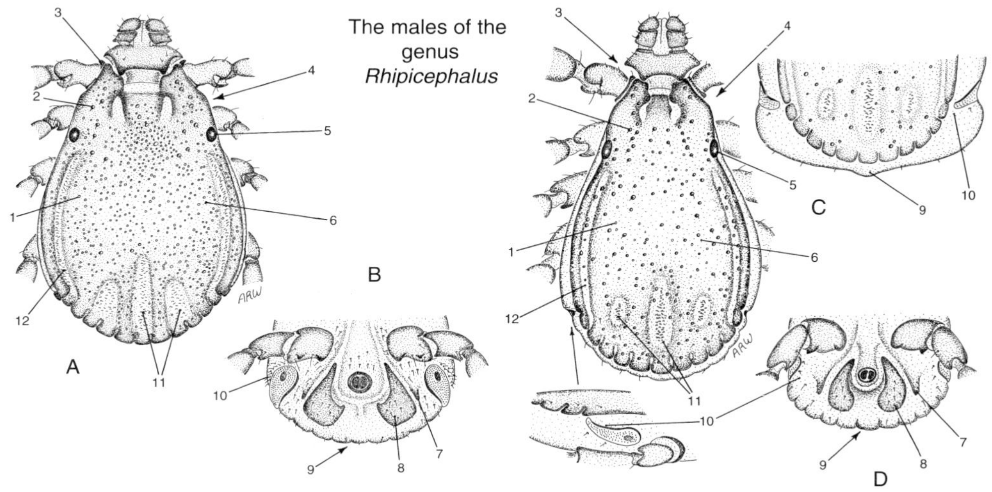

As with the females, differentiating between males of species of the genus *Rhipicephalus* in the western Palaearctic may represent a considerable challenge. Special attention must to be paid details such as the size and distribution of the conscutum punctations (Plates A, C, E, 1), the observation of coxae I anterior spurs from a top position (Plates A, C, F, 3) the size and shape of the adanal plates (Plates B, D, 7, 8), the presence or absence of a caudal appendage (Plates B, D, E, F, 9) and the aspect of the spiracular plates (Plates B, D, E, F, 10). All these features are illustrated herein. The final determination of species depends not only on the presence or absence of a given character, but also on the combination of the different features of each species. Illustrated are the dorsal and ventral views of the males of *Rh. bursa* (Plates A, B), *Rh. sanguineus* (Plates C, D), *Rh. turanicus* (Plates E, F) and *Rh. annulatus* (Plates G, H).

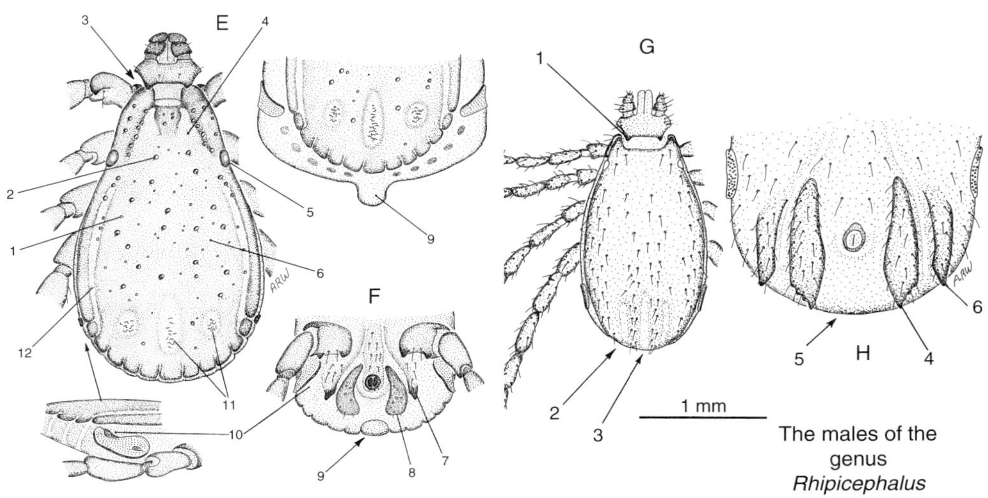

The females of the genus *Hyalomma*

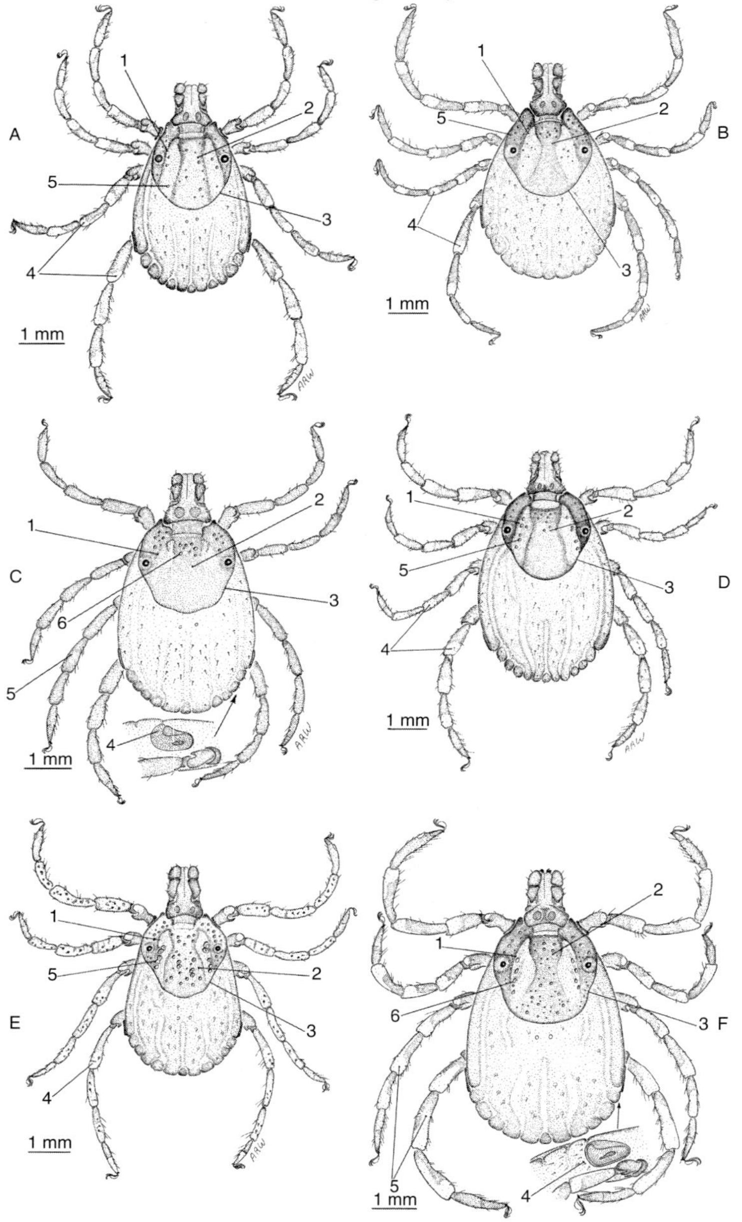

The previous page illustrates the females of the genus *Hyalomma*. As for the genus *Rhipicephalus*, species in the genus *Hyalomma* are very difficult to determine to a high degree of reliability. We have included herein the drawings of six species of the genus, which may be not present in the whole European Mediterranean, but are common in the African coast. It is important to obtain adequate material for differentiation of species, i.e. avoid engorged specimens. If engorged immature stages are available, the most useful are to left to moult and the newly moulted adults are examined.

The drawings on the previous page refer to the females of *Hy. anatolicum* (Plate A), *Hy. excavatum* (Plate B), *Hy. scupense* (Plate C), *Hy. dromedarii* (Plate D), *Hy. lusitanicum* (Plate E) and *Hy. marginatum* (Plate F). Similar species, such as *Hy. rufipes* and *Hy. turanicum* are not illustrated, although they may be found in places like northern Africa (i.e. Egypt) or Turkey. The precise distribution of all these species is not yet well known. *Hy. dromedarii* is an introduced species, and a parasite of the one-humped camel, with morphological similarities to other species in the genus.

In *Hy. anatolicum* (Plate A on the previous page), the scapular grooves profile is shallow (1) and the grooves reach the posterior margin of scutum, which is pale coloured (2). The scutum posterior margin is smooth (3) and the leg colouration has pale rings (4), but the legs are also pale, so the rings may be difficult to observe. Punctation size is small and distribution of punctation is sparse (5).

In *Hy. excavatum* (Plate B on the previous page) the scapular grooves profile is steep (1) and the grooves reach close the posterior margin of the scutum, which is dark coloured (2). The scutum posterior margin is slightly sinuous (3), leg colouration is with pale rings (4) and the legs are mainly dark. The distribution of punctation is localized, with medium numbers mainly on the anterior of the scutum (5).

In *Hy. scupense* (Plate C on the previous page), the scapular grooves profile is shallow (1) and the grooves reach the posterior margin of the scutum, which is dark coloured (2). The scutum posterior margin is slightly sinuous (3). The leg colouration is without pale rings (5) although there may be pale patches on the dorsal surface of leg segments. The size of punctations is size is small, and their distribution is localized (6) with small numbers on the anterior of the scutum, giving it a smooth shiny appearance.

In *Hy. dromedarii* (Plate D on the previous page), the scapular grooves profile is steep (1) and the grooves reach the posterior margin of the scutum, which is dark coloured (2). The scutum posterior margin is slightly sinuous (3). The leg colouration is with pale rings (4). The size of punctations is large (5) and their distribution is sparse.

The female of *Hy. lusitanicum* (Plate E on the previous page) has a steep scapular grooves profile (1) and the grooves are obscured by lines or large punctations, but they reach the posterior margin of the scutum. The scutum is dark coloured but has faint white enamel ornamentation on the scapulae (2). The scutum posterior margin is distinctly sinuous, with distinctly concave outlines posterior to the eyes (3). The leg colouration is with white enamel (4) as irregular patterns on the dorsal and lateral leg surfaces. Punctation size is large (5) but there are some small ones and some large ones. These large punctations are shallow, often joined together and with a rough internal surface. The distribution of punctation is localized, with the large ones in the cervical fields and central areas.

The female of *H. marginatum* (Plate F on the previous page) has a steep scapular grooves profile (1) and the grooves do not reach the posterior margin of the scutum. The scutum is dark coloured (2). The scutum posterior margin is distinctly sinuous (3). The leg colouration is with pale rings (5) and there are also patches of pale colour along the dorsal surfaces of the central segments of all legs. The size of punctations is small (6) and their distribution is sparse.

The illustrations in the following two pages include males of the genus *Hyalomma*.

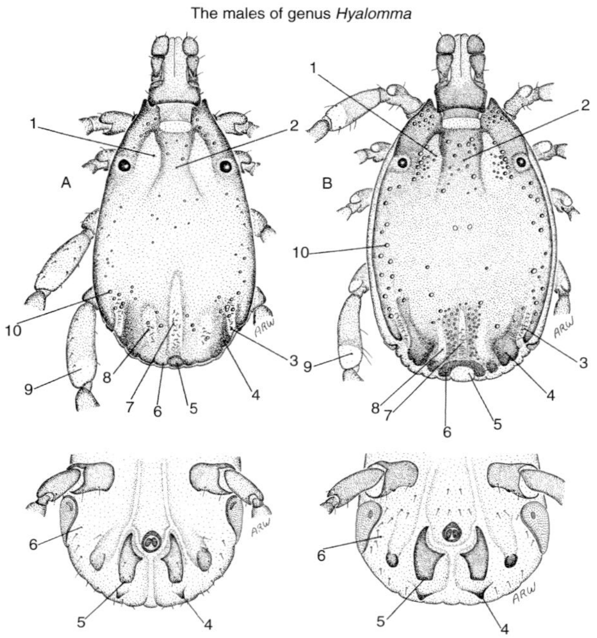

The male of *Hy. anatolicum* (Plate A, this page) has cervical fields with evident depression (1), a pale coloured conscutum (2) and short lateral grooves (3). There are two posterior ridges (4) and a caudal depression is present (5). The paracentral festoons are separated anteriorly (6) with a posteromedian groove that is long and narrow (7). The paramedian grooves are small (8). The legs are pale in a patchy or marbled pattern (9) and punctations are small and their distribution, sparse (10).

The male of *Hy. excavatum* (Plate B) has an apparent cervical depression (1) with the conscutum dark and heavily sclerotized (2). Lateral grooves are short (3) but distinct with a rough surface. There are two posterior ridges (4) and a caudal depression is present. The central festoon is pale (5), paracentral festoons are joined anteriorly (6) and a posteromedian groove is present (7). The paramedian grooves are small or indistinct (8). Legs are mainly dark and the pale rings are well visible (9). Punctations are large (10).

The male of *Hy. scupense* is pictured in Plate C. Its cervical fields depression is apparent (1), with a dark coloured conscutum (2) and long lateral grooves (3). Posterior ridges number four (4) and a caudal depression is present. The central festoon is pale coloured (but may be dark) and paracentral festoons are separated anteriorly. There is a posteromedian groove (6). Paramedian grooves are small (7). Leg colouration is without pale rings (8). Punctations are small (9).

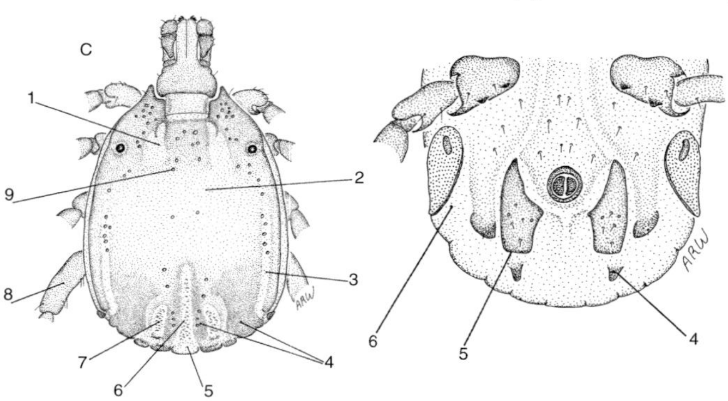

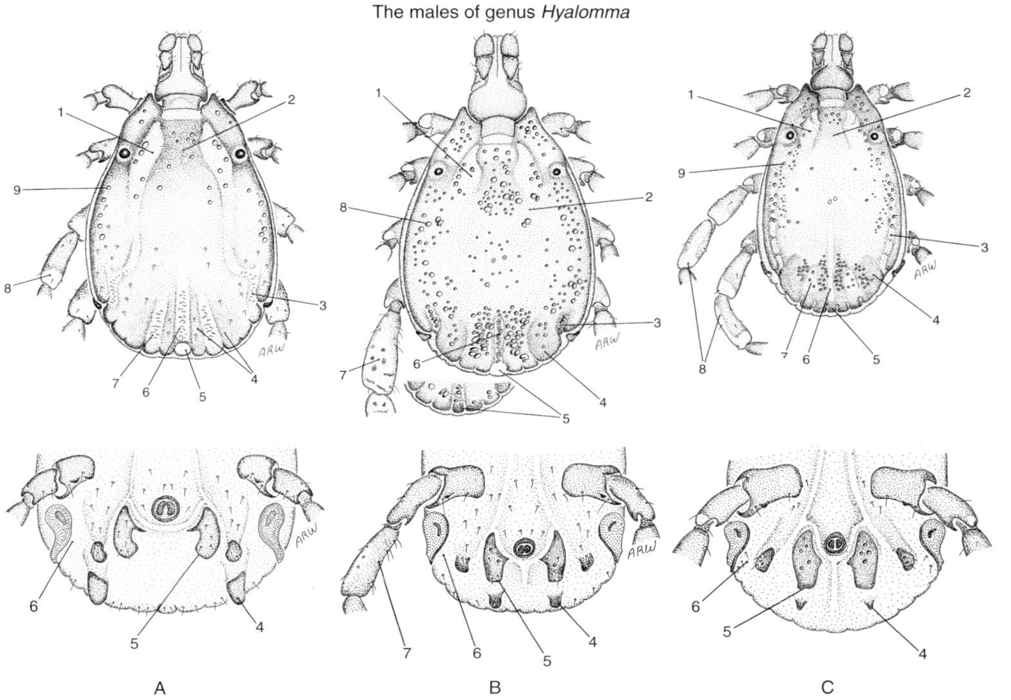

In the male of *Hy. dromedarii* (Plate A, this page) the cervical field depression is apparent (1) and the conscutum is dark coloured (2). Lateral grooves are short (3) and the posterior ridges number four (4). A caudal depression is present but partially obscured by posterior ridges. The central festoon is pale coloured (5). The paracentral festoons are separate anteriorly. The posteromedian groove is present (6), the paramedian grooves are large (7). The leg colouration is with pale rings (8). Punctations are large and sparse (9).

The male of *Hy. lusitanicum* is pictured in Plate B. There is an apparent depression in the cervical fields (1) and a dark coloured conscutum (2) which may have faint white enamel ornamentation on the scapulae. Lateral grooves are short (3). Posterior ridges number two and a caudal depression is present (4). The central festoon is pale coloured (but may be dark coloured). Paracentral festoons are separate anteriorly (5). A posteromedian groove is present. Paramedian grooves are absent (6). Leg colouration is with enamel ornamentation, as irregular patterns of marbling on the dorsal and lateral leg surfaces (7). Punctations are large (although there are also small and very large punctations) and their distribution is localized (8).

The male of *Hy. marginatum* is shown in Plate C. There is an apparent cervical depression (1) and the conscutum is dark coloured (2). Lateral grooves are long and they continue towards the eyes as lines of punctations (3). Posterior ridges number two and a caudal depression is present but shallow (4). The central festoon is dark coloured (5). Paracentral festoons are separate anteriorly (5) and a posteromedian groove is present (6). Paramedian grooves are small (7), but all the posterior grooves are shallow and may be indistinct. Leg colouration is with pale rings; there are also patches of pale colour along the dorsal surfaces of the legs (8). Punctations are small and their distribution is sparse (9).

The males of the genus *Dermacentor*

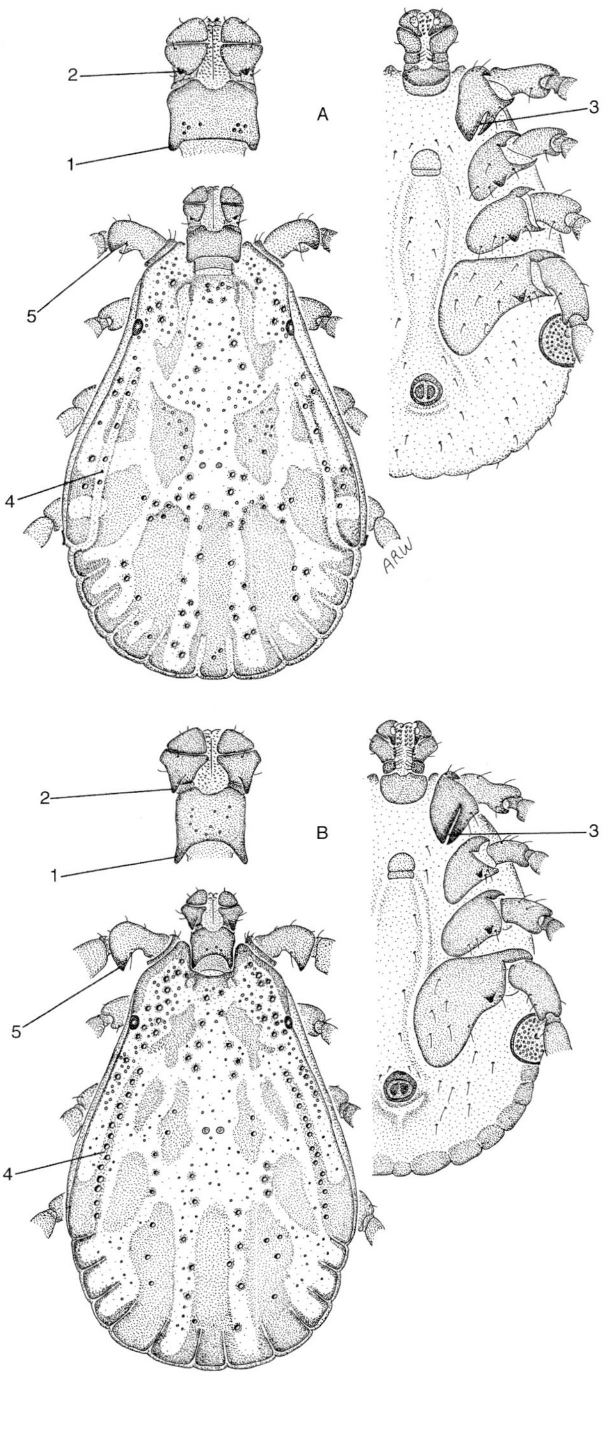

There are only two species of ticks in the genus *Dermacentor* in the western Palaearctic, namely *D. marginatus* (Plate A) and *D. reticulatus* (Plate B). Both are large, very ornate ticks whose identities are easily determined.

The males of *D. marginatus* (Plate A) have short cornua (1). The posterior spur on palpal articles II is short on the dorsal surface (2). In coxae I, the gap between internal and external spurs is medium, and the external spur is slightly shorter than the internal spur (3). The lateral groove is indistinct, only more conspicuous than the punctations (4). The posterior spur on trochanter I is short on the dorsal surface (5).

The male of *D. reticulatus* (Plate B) has long cornua (1). The posterior spur on palpal articles II is long on the dorsal surface (2). In coxae I, the gap between external and internal spurs is narrow, and the external spur is as long as the internal spur (3). The lateral groove is in the form of punctations only (there is no groove visible) (4). The posterior spur on trochanter I is long on the dorsal surface (5).

As with the males, differentiating between females of the two species of the genus *Dermacentor* in the western Palaearctic is easy because they are the only ornate ticks found. Both species are included in this page: *D. marginatus* (Plate A) and *D. reticulatus* (Plate B).

In the females of *D. marginatus*, the shape of the porose areas is a narrow oval, and they are slanting (1). The posterior spur of palp articles II is absent from the dorsal surface (2). In coxae I, the gap between the external and internal spurs is medium, and the external spur is slightly shorter than the internal spur (3). The posterior lips of the genital aperture have a narrow V shape (4) and are not truncated at the posterior margin.

The females of *D. reticulatus* have porose areas shaped in a broad oval, nearly circular, pattern (1). The posterior spur of palp articles II is present on the dorsal surface (2). The gap between coxae I external and internal spurs is narrow, and the external spur is as long as the internal spur, as displayed in (3). The posterior lips of the genital aperture have a broad U shape, which is truncated posteriorly (4).

The females of the genus *Dermacentor*

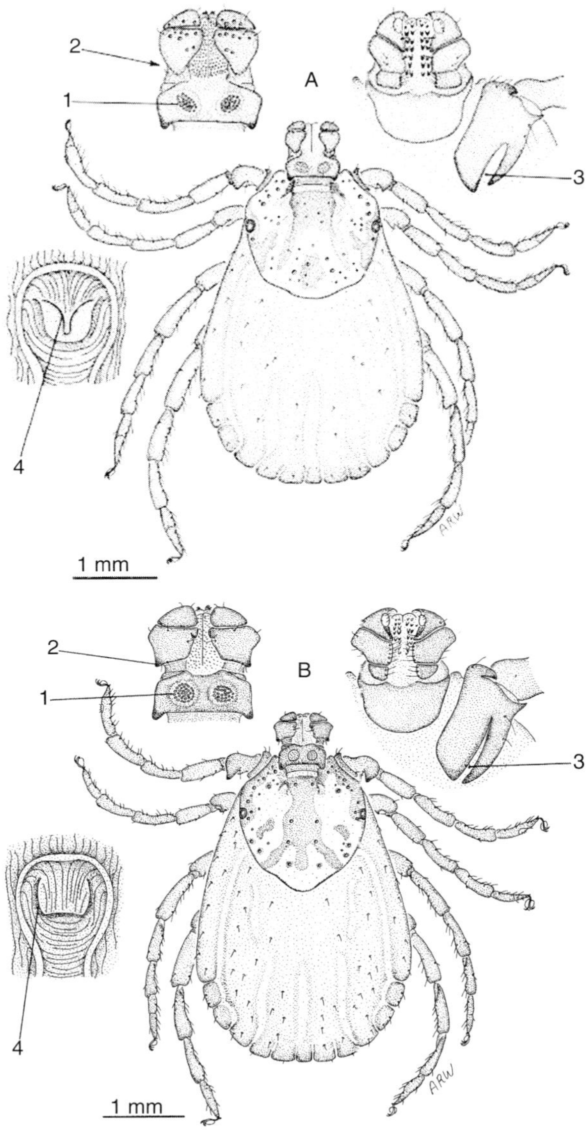

The male of the genus *Haemaphysalis*

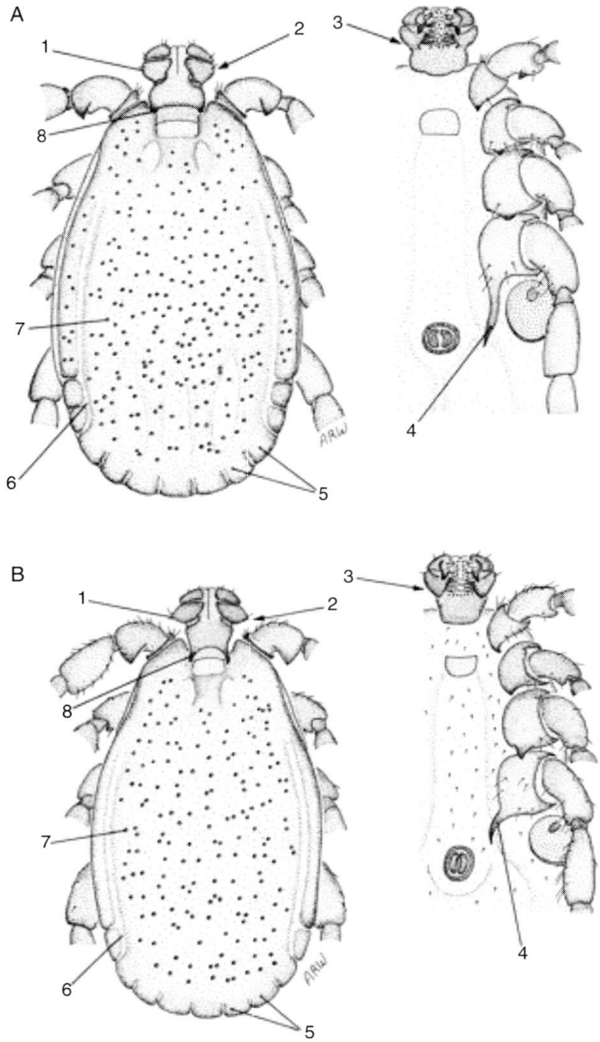

There are two prominent species of the genus *Haemaphysalis* in the western Palaearctic, namely *Ha. punctata* (Plate A) and *Ha. sulcata* (Plate B). Other species, such as *Ha. numidiana*, *Ha. erinacei*, *Ha. concinna*, etc. are not so common or of importance to the health of domestic animals or humans, as far as we know currently. New studies and additional collections of these species will most likely provide interesting data about the importance and distribution of this tick genus.

The males of *Ha. punctata* have a short lateral extension (1) on palp articles II, and the dorsal spur is absent from these articles (2). The ventral spur is also absent from palp segments II (3). The spurs of coxae IV are long, extending to the level of the posterior margin of the anus and curving inwards (4). There are 11 festoons (5), and two of these are enclosed by a lateral groove (6). Distribution of the punctuations is dense (7). The cornua are short (8).

In the males of *Ha. sulcata*, the lateral extension of palp segments II is longer than in *Ha. punctata* (1) and the dorsal spur is absent from these articles (2). Palp segments 2 The ventral spur is also absent from palp segments II (3). The spurs of coxae IV are long, extending to the level of the posterior margin of the anus, and curving outwards (4). There are nine festoons (5), and only one is enclosed by a lateral groove (6). Distribution of the punctuations is sparse (7). The cornua are long (8).

In the females of *Ha. punctata* (Plate A, this page), the lateral extension of palp articles II is small (1), and the dorsal spur is absent (2). The ventral spur of palp segments III is also absent (3). The spurs of coxae I to II are of medium length, and those of coxae IV are distinct (4). There are 11 festoons (5) and three of these are enclosed by each lateral groove (6). The distribution of punctuation is dense (7).

In the females of *Ha. sulcata* (Plate B, this page), the lateral extension to palp articles II is small (1) and the dorsal spur is absent from these articles (2). The ventral spur is present on palp articles III (3). The spurs of coxae I to III are short and those of coxae IV are indistinct (4). There are nine festoons (5), and only one festoon is enclosed by each lateral groove (6). The distribution of punctation is sparse (7).

There are several species of the genus *Ixodes* in the western Palaearctic. They are rare in northern Africa, and much more common in Europe. Details are given here for female ticks only, and for the species *I. ricinus*, *I. gibbosus* and *I. hexagonus*. Illustrations of the females of the genus *Ixodes* are included on the next page (Plates A to F) in both dorsal (A, C, E) and ventral views (B, D, F).

The females of the genus *Haemaphysalis*

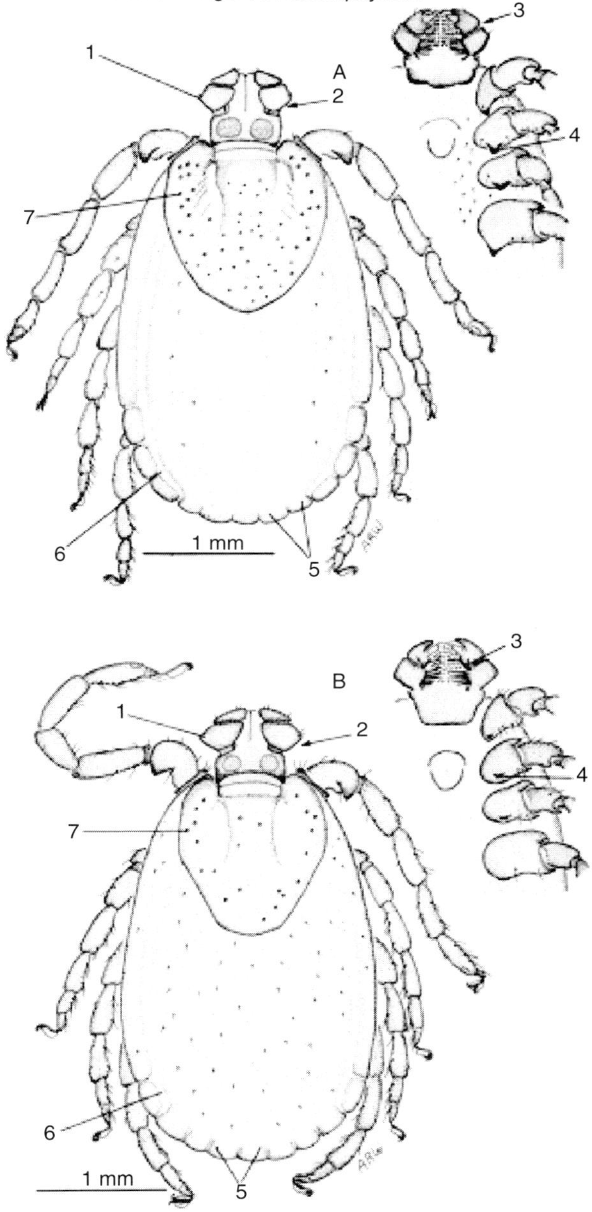

The females of the genus *Ixodes*

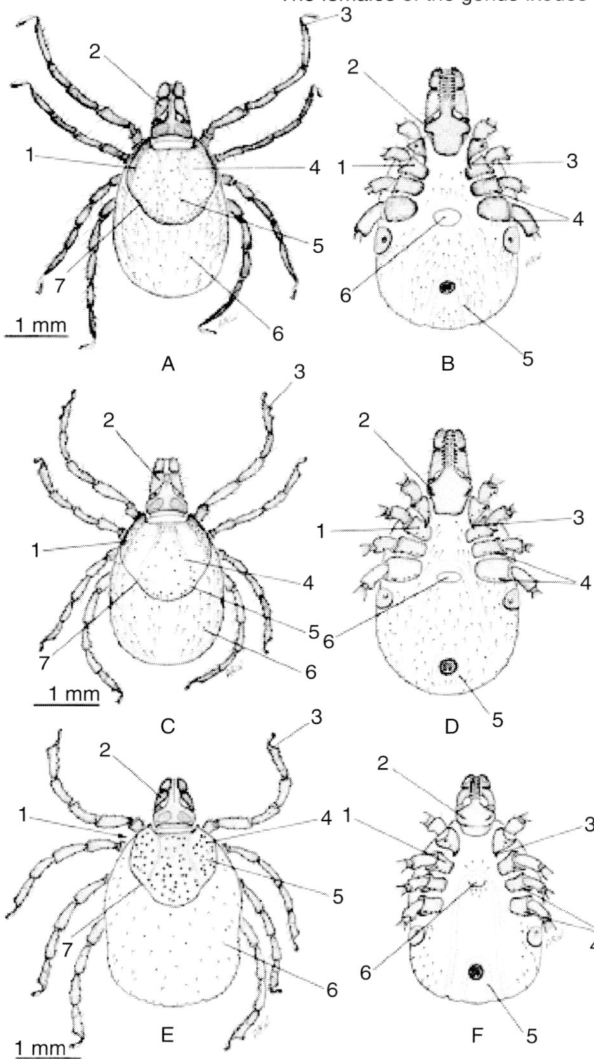

In female *I. ricinus* (Plate A), scapular grooves are present (1), palp alignment slopes inward (2), the tarsi are slightly tapered towards the claws (3), punctations are distinct (4), setae are present on the scutum (5), setae on the alloscutum are thin and colourless (6), and the scutum posterior margin is slightly sinuous (7). On its ventral surface (Plate B), *I. ricinus* has coxae of normal type (1), auriculae are distinct (2), coxae I internal spurs are very long (3), coxae II to IV external spurs are distinct (4), alignment of the posterior anal groove is long and diverging (5), and the genital aperture is between coxae IV (6).

The female of *I. gibbosus* (Plates C, D) is very similar to that of *I. ricinus*, but there are important differences in the tarsi, which are very tapered Plate C, 3), the auriculae are short (Plate D, 2) and the internal spurs of coxae I are shorter. The posterior alignment of the anal groove is long and converging (D, 5).

I. hexagonus is a very different species (Plates E, F). The dorsal surface (Plate E) scapular grooves are absent (1), palp alignment slopes inward (2), the tarsi are blunt stepped towards the taris (3), punctations are distinct (4), setae are present on the scutum but small and sparse (5), setae on the alloscutum are thin and colourless (6). On the ventral surface (Plate F), coxae are of normal type (1), auriculae are indistinct (2), internal spurs of coxae I are short (3), external spurs of coxae II to IV are indistinct (4), posterior alignment of the anal groove is long and diverging (5), and the genital aperture is between coxae III (6).

Index